T0378926

Agricultural Education for Development in the Arab Countries

Mohamed M. Samy
Independent Researcher, USA

R. Kirby Barrick
University of Florida, USA (retired)

A volume in the Advances in Educational Technologies and Instructional Design (AETID) Book Series

Published in the United States of America by
IGI Global
Information Science Reference (an imprint of IGI Global)
701 E. Chocolate Avenue
Hershey PA, USA 17033
Tel: 717-533-8845
Fax: 717-533-8661
E-mail: cust@igi-global.com
Web site: http://www.igi-global.com

Library of Congress Cataloging-in-Publication Data

Names: Samy, Mohamed, 1953- editor. | Barrick, R. Kirby, editor.
Title: Agricultural education for development in the Arab countries / Mohamed Samy and R. Kirby Barrick, editors.
Description: Hershey, PA : Information Science Reference, 2023. | Includes bibliographical references and index. | Summary: "This book, with an emphasis on the Arab world, provides the reader with all the essential knowledge, know-how and ways and means to enhance all elements of agricultural secondary, post-secondary, and university education programs to effectively prepare students for successful careers in global agriculture, multi-national food supply chains and natural resource management"-- Provided by publisher.
Identifiers: LCCN 2022027223 (print) | LCCN 2022027224 (ebook) | ISBN 9781668440506 (hardcover) | ISBN 9781668440513 (paperback) | ISBN 9781668440520 (ebook)
Subjects: LCSH: Agricultural education--Arab countries. | Technical education--Arab countries.
Classification: LCC S535.A28 A37 2023 (print) | LCC S535.A28 (ebook) | DDC 630.71/1--dc23/eng/20220722
LC record available at https://lccn.loc.gov/2022027223
LC ebook record available at https://lccn.loc.gov/2022027224

This book is published in the IGI Global book series Advances in Educational Technologies and Instructional Design (AETID) (ISSN: 2326-8905; eISSN: 2326-8913)

British Cataloguing in Publication Data
A Cataloguing in Publication record for this book is available from the British Library.
All work contributed to this book is new, previously-unpublished material.
The views expressed in this book are those of the authors, but not necessarily of the publisher.
For electronic access to this publication, please contact: eresources@igi-global.com.

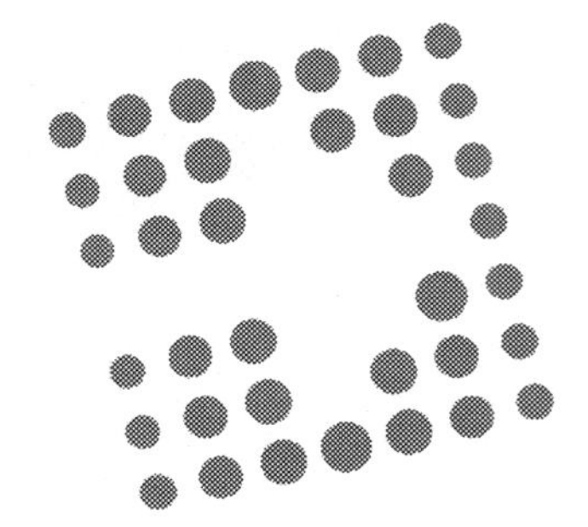

Advances in Educational Technologies and Instructional Design (AETID) Book Series

ISSN:2326-8905
EISSN:2326-8913

Editor-in-Chief: Lawrence A. Tomei, Robert Morris University, USA

MISSION

Education has undergone, and continues to undergo, immense changes in the way it is enacted and distributed to both child and adult learners. In modern education, the traditional classroom learning experience has evolved to include technological resources and to provide online classroom opportunities to students of all ages regardless of their geographical locations. From distance education, Massive-Open-Online-Courses (MOOCs), and electronic tablets in the classroom, technology is now an integral part of learning and is also affecting the way educators communicate information to students.

The **Advances in Educational Technologies & Instructional Design (AETID) Book Series** explores new research and theories for facilitating learning and improving educational performance utilizing technological processes and resources. The series examines technologies that can be integrated into K-12 classrooms to improve skills and learning abilities in all subjects including STEM education and language learning. Additionally, it studies the emergence of fully online classrooms for young and adult learners alike, and the communication and accountability challenges that can arise. Trending topics that are covered include adaptive learning, game-based learning, virtual school environments, and social media effects. School administrators, educators, academicians, researchers, and students will find this series to be an excellent resource for the effective design and implementation of learning technologies in their classes.

COVERAGE

- Bring-Your-Own-Device
- Adaptive Learning
- Web 2.0 and Education
- K-12 Educational Technologies
- Game-Based Learning
- Instructional Design Models
- Higher Education Technologies
- Educational Telecommunications
- Online Media in Classrooms
- Instructional Design

IGI Global is currently accepting manuscripts for publication within this series. To submit a proposal for a volume in this series, please contact our Acquisition Editors at Acquisitions@igi-global.com or visit: http://www.igi-global.com/publish/.

The Advances in Educational Technologies and Instructional Design (AETID) Book Series (ISSN 2326-8905) is published by IGI Global, 701 E. Chocolate Avenue, Hershey, PA 17033-1240, USA, www.igi-global.com. This series is composed of titles available for purchase individually; each title is edited to be contextually exclusive from any other title within the series. For pricing and ordering information please visit http://www.igi-global.com/book-series/advances-educational-technologies-instructional-design/73678. Postmaster: Send all address changes to above address.

Titles in this Series

For a list of additional titles in this series, please visit:
http://www.igi-global.com/book-series/advances-educational-technologies-instructional-design/73678

Handbook of Research on Perspectives in Foreign Language Assessment
Dinçay Köksal (Çanakkale 18 Mart University, Turkey) Nurdan Kavaklı Ulutaş (Izmir Demokrasi University, Turkey) and Sezen Arslan (Bandırma 17 Eylül University, Turkey)
Information Science Reference • © 2023 • 418pp • H/C (ISBN: 9781668456606) • US $270.00

Cases on Effective Universal Design for Learning Implementation Across Schools
Frederic Fovet (Royal Roads University, Canada)
Information Science Reference • © 2023 • 300pp • H/C (ISBN: 9781668447505) • US $215.00

Engaging Students With Disabilities in Remote Learning Environments
Manina Urgolo Huckvale (William Paterson University, USA) and Kelly McNeal (School of Education, Georgian Court University, Lakewood, USA)
Information Science Reference • © 2023 • 258pp • H/C (ISBN: 9781668455036) • US $215.00

Supporting Self-Regulated Learning and Student Success in Online Courses
Danny Glick (University of California, Irvine, USA) Jeff Bergin (General Assembly, USA) and Chi Chang (Michigan State University, USA)
Information Science Reference • © 2023 • 325pp • H/C (ISBN: 9781668465004) • US $215.00

Handbook of Research on Scripting, Media Coverage, and Implementation of E-Learning Training in LMS Platforms
Mohamed Khaldi (Abdelmalek Essaadi University, Morocco)
Information Science Reference • © 2023 • 506pp • H/C (ISBN: 9781668476345) • US $285.00

For an entire list of titles in this series, please visit:
http://www.igi-global.com/book-series/advances-educational-technologies-instructional-design/73678

701 East Chocolate Avenue, Hershey, PA 17033, USA
Tel: 717-533-8845 x100 • Fax: 717-533-8661
E-Mail: cust@igi-global.com • www.igi-global.com

Table of Contents

Detailed Table of Contents

Agriculture is an important sector of the economy in Arab countries. While a system of agricultural technical schools (ATS) is in place in some countries, there has not been an effort to help ensure that students in those programs are adequately prepared to enter the workforce and be productive. The development and implementation of a model that identifies the agricultural business and industry and how to prepare graduates to enter the workforce will transcend the Arab countries. This chapter will identify ways to conduct a skill gap analysis, provide ways to transform the curricula, maintain updated technical course content, identify ways to have improved teaching methods, model a train-the-trainer approach, and design ways to promote student involvement and growth.

Providing quality and innovative agricultural education programs (AEPs) to the rural communities should be a vital goal for the Arab countries to be able to accelerate rural development and achieve acceptable level of food security. A successful and effective AEP includes four overlapping, interactive components: 1) positive and encouraging learning environments, 2) effective supervised hands-on practical experiences, 3) supportive career development and placement services, and 4) enabling partnerships with agribusiness and community leaders. Effective program leaders are essential to the success of planning and implementing AEPs. Successful implementation of these programs will result in significant improvement of students' academic achievement and skill acquisitions, and increase the overall employability skills of agricultural graduates to allow them to actively participate in serving their communities; and assist in achieving acceptable levels of food security, increasing agricultural exports, and wisely managing the environment and natural resources, especially water.

Developed countries attach great importance to agricultural education of all kinds and stages. Agricultural education plays its role in creating skilled human forces capable of giving excellence, which supports the agricultural sector and enhances its economic and competitive standing in those countries. However, the philosophy and characteristics of agricultural technical education vary from state to state; as well as the role of the state and the private sector in shaping its policies and defining its objectives. Despite these differences, which are attributable to the nature of each state, practical practices have proven to be successful. However, there are some problems that impede progress and development in agricultural education in particular, and this chapter tries to highlight the features of these problems.

A community college, sometimes referred to as a junior college or technical college, is a taxpayer-supported two-year institution of higher education. The term "community" is at the heart of the mission of the community college. These schools offer a level of accessibility—in terms of time, money, and geography—that cannot be found at most liberal arts colleges and private universities. Community colleges are common in many different Arab countries, and just recently established within other countries. It is mostly responsible for graduating effective technicians in different disciplines. In addition, it plays an important role for students' manipulation for university enrollment in different Arab countries. Through this chapter there are many points and views relevant to concept, initiation, academic programs, employability, job destination, and benchmarking versus global community colleges.

Agricultural and economic development in Arab countries is dependent upon having a well-prepared agricultural workforce. Among the components that are central to that mission are the utilization of an active and supportive external advisory committee and the incorporation of real-life, hands-on experiences for students through internships. This chapter provides background information to assist school and college administrators in establishing and maintaining an advisory committee as well as guidance for school and college personnel to incorporate student internships into the curricula.

The agriculture sector is extremely important for Egypt's economy. For achieving food security objectives, within the SDGs 2030, the Agricultural Human Resources (AHRs) need systematic and regular training and lifelong learning. Faculties of agriculture (FAs) prepare and qualify researchers, technicians, and specialists in all areas of agriculture. Career advancement centers (CACs) need to be established in all FAs to provide lifelong opportunities to all AHRs' categories that need to cope with the ever advancing agricultural developments. To secure effective functioning of CACs, the different units needed, within their structures, are suggested. As a conclusion, establishing a CAC in each FA, is necessary and urgent for activating and maintaining its extension education, out-reach function of development, and serving the surrounding communities, especially rural communities. The currently working CACs need all types of support from the government, agri-business, and civil society organizations.

Teachers are at the heart of education systems because they directly interact with and influence learners in communities around the world. Teachers should be supported through dynamic professional development to engage them as critical partners in sustainable development. Professional development should exhibit the research-based characteristics of a strong content focus, active learning, collaboration, and sufficient duration. Connecting to teacher professional identities and supporting the development of teacher efficacy can help inspire teachers to become change agents in their communities. Communities of practice are one example of professional development practice that bring these elements together. Teacher professional development is an important component of the transformation, but teacher professional development needs to be supported by the educational system to fully realize the potential for impact.

This chapter sheds light on the main features of education in the universities of Arab countries. Limitations and obstacles were highlighted, with presentation of non-controversial keys for their resolution. The language used in education is largely discussed to reveal the great significance of using the Arabic language in the Arabic world. The discussion reveals clearly how this issue can guarantee easy and perfect education, keeping in mind that this is the situation in all advanced countries who are sticking to their national languages in education. The chapter demonstrates the nature of ornamental and medicinal plants as an application for the intended aim, together with highlighting their significance for human being. Expansion of essential knowledge in the field of ornamental and medicinal plants was also demonstrated. The economic significance of these plants was also mentioned.

Preface

Agricultural development, including production, processing, marketing, distribution, and natural resource management, is an important economic engine for development in the Arab countries in North Africa and the Middle East. An essential factor to accelerate agricultural development in these countries is a well-educated and trained agricultural workforce. However, the areas of secondary school, community college, and university agricultural education as well as non-formal agricultural education have lacked attention and resources for many years. Curriculum development, instructional enhancement, practical training, career advancement, and agricultural extension and advisory services are central to agricultural education development, technology utilization, and sustainable management of natural resources. Engagement of educators, extension agents, administrators, business and community leaders, and policy makers ensures that the graduates of all levels of agricultural education and small-scale producers are well prepared for the jobs of today and tomorrow.

This book provides the reader with all the essential knowledge, know-how, and ways and means to enhance all elements of agricultural secondary, post-secondary, and university education programs to effectively prepare students for successful careers in global agriculture, multi-national food supply chains, and natural resource management. In addition, this book provides information on improving the role of non-formal agricultural education in the forms of agricultural extension and advisory services (AEASs) to increase productivity and achieve acceptable levels of food security. This text, with emphasis on the Arab world, is of great value to higher education and agricultural technical school systems, ministries of education and agriculture, agricultural extension and advisory organizations, agribusiness and community leaders, policy makers, environmental agencies, and international research and development entities.

The first section of the book addresses the need for creating a well-developed and well-educated workforce to sustain and grow agricultural development in the Arab countries. Beginning with a conceptual model, the ensuing chapters provide additional guidance for the various segments of the model. The focus is on successful programs of agricultural education that should be present in universities, community colleges, agriculture technical schools, and extension and advisory service organizations throughout the region.

The agriculture sector can only improve when the necessary skills are taught to and developed by students and producers and when instructors and extension agents at all levels are experts in the content and in educational processes. An important step includes identifying the abilities of program graduates and producers, and comparing their skills and skill levels with the needs of the agriculture sector businesses, industries, and farms. Agricultural instructors often have received very little preparation and on-going support in becoming effective teachers and extension agents; the chapters on learning strategies and on professional development are designed to address that concern.

The second section of the book includes concepts and practices in non-formal education in agriculture. Examples of challenges and successful extension training programs are included. Of particular importance is a discussion of food security within the realm of agricultural development. Developing and expanding public-private partnerships to provide extension and advisory services can be a key component of successful non-school, community-based programs.

Agriculture is an important sector of the economy in Arab countries. While a system of Agricultural Technical Schools (ATS) is in place in some countries, there has not been put in place an effort to help ensure that students in those programs are adequately prepared to enter the workforce and be productive. The development and implementation of a model that identifies the agricultural business and industry and how to prepare graduates to enter the workforce will transcend the Arab countries. Chapter 1 will identify ways to conduct a skill gap analysis, provide ways to transform the curricula, maintain updated technical course content, identify ways to have improved teaching methods, model a train-the-trainer approach, and design ways to promote student involvement and growth.

Chapter 2 talks about how providing quality and innovative agricultural education programs (AEPs) to the rural communities should be a vital goal for the Arab countries to be able to accelerate rural development and achieve acceptable level of food security. A successful and effective AEP includes four overlapping, interactive components: 1) positive and encouraging learning environments, 2) effective supervised hands-on practical experiences 3) supportive career development and placement services and 4) enabling partnerships with agribusiness and community leaders. Effective program leaders are essential to the success of planning and implementing AEPs. Successful implementation of these programs will result in significant improvement of students' academic achievement and skill acquisitions, and increase the overall employability skills of agricultural graduates to allow them to actively participate in serving their communities; and assist in achieving acceptable levels of food security, increasing agricultural exports, and wisely managing the environment and natural resources, especially water.

Developed countries attach great importance to agricultural education of all kinds and stages. Agricultural education plays its role in creating skilled human forces capable of giving excellence, which supports the agricultural sector and enhances its economic and competitive standing in those countries. However, the philosophy and characteristics of agricultural technical education vary from State to State.as well as the role of the State and the private sector in shaping its policies and defining its objectives. Despite these differences, which are attributable to the nature of each State, practical practices have proven to be successful. However, there are some problems that impede progress and development in agricultural education in particular, and Chapter 3 to highlight the features of these problems.

A community college, sometimes referred to as a junior college or technical college, is a taxpayer-supported two-year institution of higher education. The term "community" is at the heart of the mission of the Community College. These schools offer a level of accessibility—in terms of time, money, and geography—that cannot be found at most liberal arts colleges and private universities. Community colleges are common in many different Arab countries, and just recently established within other countries. It is mostly responsible about graduating effective technicians in different disciplines. In addition, it plays an important role for students' manipulation for university enrolment in different Arab countries. Through Chapter 4, there are many points and views relevant to concept, initiation, academic programs, employability, job destination and benchmarking versus global community colleges.

The disconnect between the perceptions of the agriculture sector in terms of skills and competencies needed by workers and the abilities of agriculture graduates and others entering the workforce is existent throughout much of the world and especially in the Arab world. Economic development cannot be sustaining until this disconnect, a skills-gap, is resolved. Chapter 5 provides an overview of the skills and competencies needed in agriculture and the perceptions of the extent graduates possess those skills and competencies. A system for identifying the workforce development priorities, namely high importance and low worker preparedness, is described. The last section of the chapter provides a mechanism, the, to identify what skills and competencies agriculture teachers and faculty need to develop and enhance in order to teach students the skills they need to enter and be successful in the agriculture sector.

Through Chapter 6, there is reflection on using active learning strategies for agricultural education in the Arab world. There are difficulties with utilizing specific active learning strategies that encourage student learning and achieve important graduate attributes. Changing teachers' and faculty perceptions about learning strategies and focusing on students higher thinking order requires a lot of training, orientation, and mind shifting. Learning through specific active learning techniques, such as vocational, simulation, case studies, and using technology, supports student learning. It is particularly important to use instructional ways that encourages deep learning and analysis with agricultural problems and deficits. Professional development and hands-on learning changes perceptions of teachers through agricultural education in Arab world. Many aspects relevant for using active learning in agricultural education perfectly in Arab world countries are clarified.

Agricultural and economic development in Arab countries is dependent upon having a well-prepared agricultural workforce. Among the components that are central to that mission are the utilization of an active and supportive external advisory committee and the incorporation of real-life, hands-on experiences for students through internships. Chapter 7 provides background information to assist school and college administrators in establishing and maintaining an advisory committee as well as guidance for school and college personnel to incorporate student internships into the curricula.

Chapter 8 is focused on how agriculture sector is extremely important for Egypt's economy. For achieving food security objectives, within the SDGs 2030, the Agricultural Human Resources (AHRs) need systematic and regular training and lifelong learning. Faculties of Agriculture (FAs) prepare and qualify researchers, technicians and specialists in all areas of agriculture. Career Advancement Centers (CACs), need to be established in all FAs to provide lifelong opportunities to all AHRs' categories that need to cope with the ever-advancing agricultural developments. To secure effective functioning of CACs, the different units needed, within their structures, are suggested. As a conclusion, establishing a CAC in each FA, is necessary and urgent for activating and maintaining its extension education, out-reach function of development and serving the surrounding communities, especially rural communities. The currently working CACs, need all types of support from the government, agri-business, and civil society organizations.

Chapter 9 focuses on teachers being at the heart of education systems because they directly interact with and influence learners in communities around the world. Teachers should be supported through dynamic professional development to engage them as critical partners in sustainable development. Professional development should exhibit the research-based characteristics of a strong content focus, active learning, collaboration, and sufficient duration. Connecting to teacher professional identities and supporting the development of teacher efficacy can help inspire teachers to become change agents in their communities. Communities of Practice are one example of professional development practice that bring these elements together. Teacher professional development is an important component of the transformation, but teacher professional development needs to be supported by the educational system to fully realize the potential for impact.

Chapter 10 throws the light on the main features of education in the universities of Arab countries. Limitations and obstacles were highlighted, with presentation of non-controversial keys for their resolution. The language used in education is largely discussed to reveal the great significance of using the Arabic language in the Arabic world. The discussion reveals clearly how this issue can guarantee easy and perfect education, keeping in mind that this is the situation in all advanced countries who are sticking to their national languages in education. The chapter demonstrates the nature of ornamental and medicinal plants, as an application for the intended aim, together with highlighting their significance for human being. Expansion of essential knowledge in the field of ornamental and medicinal plants was also demonstrated. The economic significance of these plants was also mentioned.

Most Arab countries have been struggling to deal with the global food crisis and the shortage of food commodities. In spite of their efforts, they face a critical food insecurity crisis of increasing the production of food commodities less rapidly than their consumption. This chapter presents efforts of three Arab countries, Egypt, Saudi Arabia, and Morocco, in tackling the food insecurity crisis and the role of public and private agricultural extension in building agricultural human capital, and in turn increase agricultural productivity. These countries varied on their reliance on public agricultural extension to educate and train farmers, especially small-scale farmers, who are the majority of food crop producers. Chapter 11 identifies challenges facing public agricultural extension, and provides guidelines to reform and improve extension services to be able to assist in building agricultural human capital, increase productivity, and achieve acceptable levels of food security.

Chapter 12 offers a perspective on how agricultural education, research, and extension services address educational challenges and foster food security and sustainability in the Gulf Cooperation Council (GCC) countries. In the last two decades, significant progress has been made in Saudi Arabia, Oman, and the United Arab Emirates regarding agricultural education infrastructure and human capital development. However, agricultural education in the GCC countries face challenges, including stagnant enrollment, the need to offer more diverse programs, and enhanced education quality to satisfy the job market demand for qualified professionals. There is a need to formulate policies and initiatives to address the challenges facing agricultural education programs, research, and agricultural extension services. Graduates of agricultural higher education institutions must be trained and learn to be capable leaders to satisfy the needs for agricultural development, achieve food security goals and sustainability, and be industry entrepreneurs.

Agricultural education, both formal school- and college-based or non-formal training and advisory extension programs, is the key to enhanced agricultural development in the Arab region. Successful programs in countries like Egypt,

Morocco, and Saudi Arabia can serve as models for further development throughout the region.

Mohamed M. Samy
Independent Researcher, USA

R. Kirby Barrick
University of Florida, USA (retired)

Chapter 1
A Model for Developing a Well-Prepared Agricultural Workforce in Arab Countries

Andrew C. Thoron
https://orcid.org/0000-0002-9905-3692
Abraham Baldwin Agricultural College, USA

ABSTRACT

Agriculture is an important sector of the economy in Arab countries. While a system of agricultural technical schools (ATS) is in place in some countries, there has not been an effort to help ensure that students in those programs are adequately prepared to enter the workforce and be productive. The development and implementation of a model that identifies the agricultural business and industry and how to prepare graduates to enter the workforce will transcend the Arab countries. This chapter will identify ways to conduct a skill gap analysis, provide ways to transform the curricula, maintain updated technical course content, identify ways to have improved teaching methods, model a train-the-trainer approach, and design ways to promote student involvement and growth.

INTRODUCTION

Agriculture is an important sector of the economy in Arab countries. While a system of Agricultural Technical Schools (ATS) is in place in some countries, an effort has not been put in place to help ensure that students in those programs and in agricultural colleges are adequately prepared to enter the workforce and be productive. The development and implementation of a model that identifies the agricultural business

DOI: 10.4018/978-1-6684-4050-6.ch001

and industry and how to prepare graduates to enter the workforce will transcend the Arab countries. This chapter will identify ways to conduct a skill gap analysis, provide ways to transform the curricula, maintain updated technical course content, identify ways to have improved teaching methods, model a train-the-trainer approach, and design ways to promote student involvement and growth. The objectives for this chapter are 1) Describe a Model of Agricultural Education for Development in Arab Countries. 2) Describe components for assessment of student competence and agricultural industry needs. 3) Identify content focus areas for curriculum development. 4) Propose a process to enhance teaching and learning.

MODEL OF AGRICULTURAL EDUCATION FOR DEVELOPMENT IN ARAB COUNTRIES

Thinking holistically about an approach to agricultural education for development to meet industry needs is an important approach in Arab counties. Strong communication among industry, education at the secondary school level and post-secondary school level, effective teaching, appropriate agriculture content, and teacher development leads to a well-prepared agricultural workforce across Arab counties. The model outlines three main components: assessment, content, and process. Assessment of current student competence and assessing agricultural industry development needs leads to a skill gap analysis. Through the skill gap analysis, the focus then becomes the content delivery in the educational setting for development. Content areas of focus are internships, student decision-making skills, agriculture technical skills, leadership skills, and the development of an advisory committee. Additional consideration is upon the process. The process outlines a path for curricular focus and program development delivery that includes an internship developmental aspect, technical agriculture focus, teaching and learning through active learning and student engagement, and leadership development and competitive events for students to engage in their learning and put concepts into practice. Once followed, this model will promote a well-prepared agricultural workforce in the Arab countries.

Figure 1.

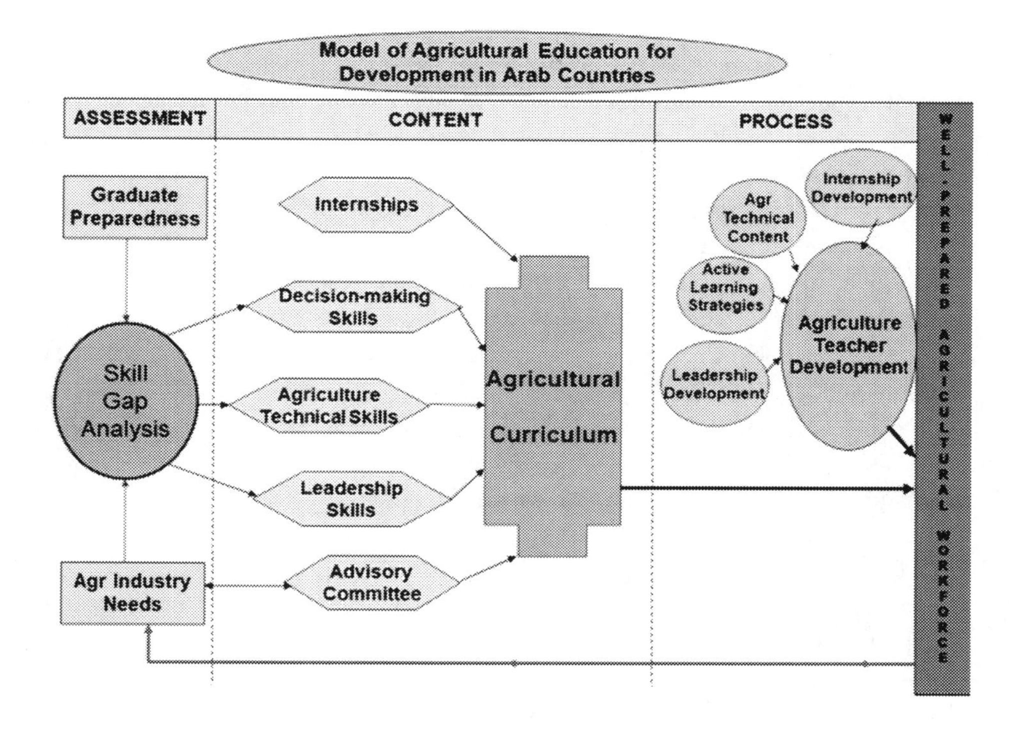

ASSESSMENT

Agricultural development in the Arab countries relies on the economic strength in the agriculture sector. Building human capital for sustainable value-chain development is essential throughout the region (Rasmussen et al., 2017). The first step in developing curricula that are responsive to human capital development and industry needs is to link the workforce needs as identified by industry leaders and the skills that students and perspective workers have developed for entering the workforce. The development of a survey and lists of potential skill competencies must be outlined and centered on the major agricultural enterprises in Arab countries and the programmatic areas of schools. An external advisory committee can be utilized for creating and providing leadership for the process. As the agriculture development needs of industry are determined, graduates are surveyed to determine their self-perceived preparedness to be able to perform tasks. The result is a skill-gap analysis that will assist in identifying changes needed in the curricula as well as the preparation of instructors to offer the new and revised instructional programs.

Agricultural Workforce Needs and Graduate Preparedness

Workforce development issues are common throughout the Arab region. High rates of unemployment and underemployment exist, especially among the youth and the relatively well-educated, in many areas of industry including agriculture. In many areas, unemployment rates for youth and university graduates approaches 30 percent. A low number of women in the workforce is also prevalent, creating additional unemployment concerns. In many countries, a large share of the workforce is employed in the public sector, often with wages and benefits greater than in much of the private sector. The World Economic Forum (2017) estimated that the Middle East and North Africa region captures only 62 percent of its full human capital potential.

Puckett et al. (2020) addressed what they termed as the "skills mismatch" between the goals of education systems and the needs of business. To fix the global skills mismatch, the researchers identified seven challenges.

- Insufficient focus on training for jobs that do not yet appear
- Members of the workforce not involved in lifelong learning and continuous retraining
- Lack of motivation and accountability for personal development
- Limited access to labor market opportunities
- Uneven human capital redistribution
- Lack of potential in certain labor categories
- Shifting labor force needs and values

In summary, focusing on industry needs and society-ready graduates is paramount to improving the economy in many sectors and locales, including agriculture development in the Arab countries. Human capital development leads to a functioning and dynamic economy.

If the agriculture economy in the Arab countries is to continue to develop and expand, the needs of the private sector must be clearly identified. A study report from Egypt (Swanson, Barrick, & Samy, 2007), expanding upon the work by Vreyens and Shaker (2005), discussed skill areas of graduates perceived as deficient by the agriculture private sector.

- Ability to analyze information
- Time management skills
- Able to develop a basic budget
- Able to access the Internet for resources and information
- Calculate rates of return on investments
- Write a (farm) business plan

· Conduct a cost-benefit analysis of an agricultural project

Agriculture and agribusiness employers additionally identified technical skills (horticulture, animal production, crop production, mechanization and others) they perceived to be deficient among agriculture graduates. A few examples of agriculture technical skills perceived by employers to be deficient included (Vreyens & Shaker, 2005) the following.

· Analyze the chemical composition of a feedstuff
· Design appropriate packaging for processed foods
· Regulate fertilizer application, and make recommendations for soil amendment
· Develop an irrigation management plan for multiple crops
· Select and breed high quality fruits and vegetables.

Equally important in the study, skills were identified that were not considered important for graduates as perceived by employers.

Ebner et al. (2020) conducted a related study that focused on the employability of agriculture university graduates in Egypt. University professors and employers in the agriculture sector identified areas of potential growth in job opportunities for agriculture graduates. While that study may not be applicable in other Arab countries, the results of their inquiry provided guidance for identifying 24 agriculture employment areas that should be addressed in agriculture programs and curricula in higher education. For example, the top-rated employment opportunities included the following.

· Animal production – poultry
· Food and beverage processing
· Energy, biofuels, alternative energy
· Chemicals, pesticides, fertilizers
· Horticulture – protected cultivation
· Water resource management
· Agriculture sales and marketing
· Horticulture – vegetable crops
· Post-harvest processing

Agriculture sector experts, then, should address these job opportunities when assessing educational programs and curricula and graduates' skills and competence.

El Ashmawi (2015) reported that "stakeholders in most of the Arab countries stress the need to strengthen the link between education and training institutes and planners representing the supply side and their partners from the demand side represented by social partners especially employers from both the public and private sectors" (p. 2). While the connect between both sides is weak and not institutionalized, some technical and vocational education and training (TVET) institutes and some employers may take the initiative to have joint committees to discuss the needs of employers. Work-based or apprenticeship programs at the employers' premises may be instituted, but the training programs may have already been designed and developed in isolation from employers. Large employers may approach TVET institutes to develop joint programs if the company recognizes a major skill shortage, but this is not common, considering that the majority of businesses (75 to 90 percent) in the region are small or informal enterprises that lack the means and awareness to take such initiatives (El Ashmawi, 2015).

A development project funded by the U.S. Agency for International Development in Egypt included three components for increasing and improving agriculture development, including capacity building, public-private partnership development, and biotechnology (Barrick et al., 2009). The capacity building component included developing competency-based curricula to better match workforce needs, developing new and updated courses in three agriculture sector programs, and developing internship programs to provide real-world experiences for students. The initial programmatic effort resulted in the following changes for faculties of agriculture and secondary school agriculture technical programs.

- Conducting a skill-gap analysis
- Transforming curricula
- Updating technical course content
- Improving teaching methods
- Developing supplemental instructional materials
- Providing train-the-trainer approaches
- Improving experiential learning
- Conducting overseas study tours for school administrators and teachers
- Promoting student involvement and growth

Skill-Gap Analysis

Othayman et al. (2022) conducted a study in Saudi Arabia that focused on the training needs assessment system. The researchers identified four components of Armstrong's Deficiency Model (2002) related to the potential gap between organizational needs and employee performance.

- Organizational analysis – short- and long-term goals, and factors affecting those goals
- Requirement analysis – requirements of the job

· Task and knowledge and ability analysis – specifying tasks required on the job; skills, knowledge, and attitudes needed to perform
· Person analysis – how well the employee performs

Information gained from such an analysis should lead to identifying the gap in skills and competencies among employees, leading to changes in training and development programs.

Skill-gap analysis is an outcomes assessment tool that is designed to measure skills and competencies that are more important to employers and in need of development by potential workforce participants (Vreyens & Shaker, 2007). The results of the skill-gap analysis provide guidance for curriculum development, utilizing the "backward design" approach of Wiggins and McTighe (1998).

To conduct a skill-gap analysis, assessors establish a comprehensive list of skills and competencies that are a part of a specific sector of agriculture development, such as field crop production. Industry and employer experts rate each skill and competency on the relative importance of each item on a 5-point Likert-type scale ranging from 5-Very Important to 1-Not Applicable for the Job. Additionally, the industry leaders and employers rate each item on the perceived level of competence of the graduates of the respective program, using the scale 5-Highly Competent to 1-Not Competent.

The second part of the skill-gap analysis process involves identifying recent graduates of related programs (following the example from above, field crop production). Program graduates are surveyed using the same list of skills and competencies on two scales. The first part of the instrument asks the graduates their perceptions of the importance of the skills and competencies on the scale 5-Very Important to 1-Not Applicable for the Job. The second part of the instrument identifies the perceptions of graduates regarding the level of preparedness to conduct or perform the skills and competencies on a scale of 5-Prepared to 1-Not Prepared. Table 1 provides a summary of the scales.

Table 1. Skill and competence importance, preparedness, and competence (adapted from Vreyens & Shaker, 2005)

Graduate Scale		Employer Scale	
Importance	**Level of Preparedness**	**Importance**	**Level of Competence**
1-Not applicable for the job	1-Not prepared	1-Not applicable for the job	1-Not competent
2-Not Important	2-Somewhat prepared	2-Not important	2-Somewhat competent
3-Somewhat important	3-Prepared	3-Somewhat important	3-Competent
4-Important	4-Very prepared	4-Important	4-Very competent
5-Very important	5-Highly prepared	5-Very important	5-Highly competent

Generally, the results of the two surveys can be summarized into several categories. When the industry and employer leaders rate a skill as Very Important and the graduates as Highly Competent, it can be concluded that there is no pressing need to adjust the curriculum of programs that are preparing graduates. Conversely, when the experts rate the skill as Very Important and graduates as Not Competent, major curriculum and program enhancements are warranted. In essence, if the industry experts rate the skill as Not Applicable for the Job, the level of competence of graduates is relatively unimportant. Perhaps the curriculum and program should consider removing that particular skill of competence from the preparation of graduates, providing time for including instruction for skills and competencies that need to be added.

Further guidance for program and curriculum development can result from the perceptions of graduates regarding importance of the skills and their perceived level of preparedness. A major disconnect occurs when graduates perceive the skill to be Very Important, but they perceive that they are Not Prepared to perform the skill. Again, major considerations for program and curriculum enhancement are warranted.

A culminating step in applying what was learned through the skill-gap analysis includes assessing the in-service needs of faculty and instructors who are responsible for curricular decisions and offering instruction in the skills and competencies needed for students to successfully enter and progress in the agriculture workforce. DiBenedetto, Willis, & Barrick (2018). Consistent and timely in-service needs assessment can determine the topics, skills, and competencies to be included in professional development programs for faculty and instructors, part of the Process section of the *Model of Agricultural Education for Development in Arab Countries.* Desimone (2009) proposed a conceptual framework for studying professional development, including five core features.

- Content focus
- Active learning
- Coherence
- Duration
- Collective participation

These core features are expanded upon in later sections of this chapter.

Figure 2.

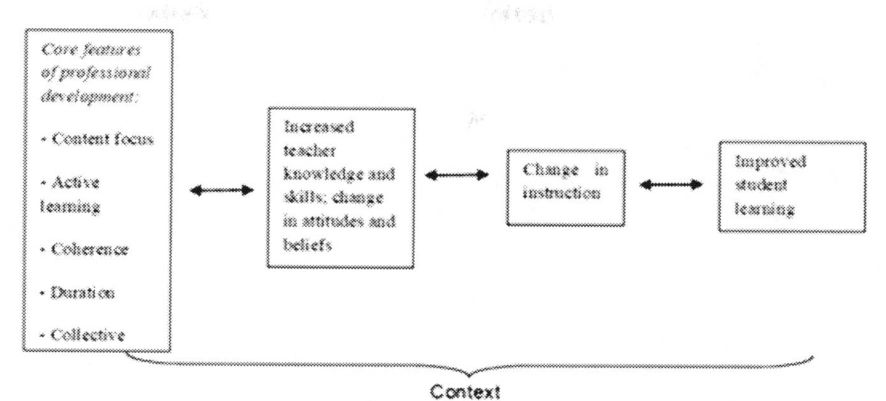

Desimone's framework indicates that the core features of successful professional development lead to increased teacher knowledge and skills and a positive change in attitudes and beliefs. Those improvements result in changes in instructional behaviors that lead to improved student learning. When students learn more and development new and enhanced skills and competencies in the agriculture sector, a well-educated workforce becomes available for the agriculture sector. Periodic changes in the core features and their adaptation to specific settings in Arab agriculture development ensures that new cadres of graduates will be relevant in the future.

CONTENT

During the content phase of the model, consideration for the implementation of internships, development of decision-making skills for students, technical skill acquisition, and providing leadership training and development for students to apply their knowledge in a work-based setting are all factors of their secondary-school and post-secondary school preparation. Continued engagement of the agricultural development needs should be consulted through the use of an advisory committee. The curricula for programs should be monitored and supported by the Ministry of Education. Therefore, significant change must be made within the parameters of the state curricula. Three components of the curricula are identified for enhancement, namely agricultural technical skills, decision-making skills, and leadership skills. The concept of establishing internships for students, beyond the typical summer experience at the school site, should be introduced, and industry involvement included through the establishment of advisory councils.

Internships

Internships are a concept that allow students to put skills, knowledge, and abilities into practice through a supportive learning environment that not only focuses on the development of skills, but also student learning and development. Internships provide an outlet for a student or trainee to work for an organization to gain work experience or skills for qualifications. Internships should be experiential in nature and involve a teacher or teaching supervisory role. Internships can be conducted on-site at the school or educational institution as well as through partnerships within the agricultural industry.

Students or trainees should seek an internship based on their area of interest and focus. The goal of a younger individual may be for career exploration, while the goal of an advanced individual should be for career development and experiential-based skills acquisition. This should be decided by both the trainee and the teacher. Once decided, the student can either be placed into an internship or seek a collaborator with whom they conduct their internship. A discussion between the agricultural employer, the trainee, and the teacher should be conducted to ensure learning goals, skills practice, and expectations while on-the-job. The focus is upon learning. Younger individuals who may be completing their first internship may experience general aspects of the agricultural business and assist with job performance tasks. For more advanced internships, trainees should be utilizing skills, developing new skills, and begin developing management decision-making competencies.

During a meeting, the three parties (student/trainee, industry host, and teacher) should outline learning goals and develop ways to assess the performance of the intern/trainee. Consideration of the skill gap analysis that identifies areas of focus for the student and needs of the agricultural industry should be utilized. During the initial meeting items that should be determined include number of hours expected on a typical day, start and end dates, expected travel, intern/trainee dress, safety considerations, to whom the intern/trainee will report to as their supervisor while on-the-job, and duties the intern/trainee will be expected to perform. The trainees should be part of the discussion and help identify areas in which they wish to learn more during their internship. However, it should be noted that the intern/trainee may not know all aspects or be able to identify all potential learning goals that the internship could provide. Thus, the teacher supervisor should work with the agricultural organization internship supervisor to outline additional learning goals based on the curricular focus and areas where the agricultural organization can benefit by hosting an intern.

The intern will benefit from the experience and the host organization should benefit from the intern's work creating a mutually beneficial experience. Internships will often lead to future employment opportunities for the intern and the agricultural organization could benefit by having a new employee that already knows the agricultural business/organization.

In the case of an internship being conducted on a school farm, there should be similar expectations and learning outcomes. Internships on the school farm should mirror real-life agribusinesses or production farming operations. Students should not only be associated with a labor focus and should gain and be included in production decisions. Internship focus should complement the student's curricula focus (Barrick et al., 2009).

Experiential learning, including activities such as supervised agricultural internships, focuses initially on the learner and follows the widely–accepted problem–solving approach to teaching and learning found in agricultural education. A synthesis of research on supervised agricultural experience in the United States, concluded that the teacher is central to the success of experience programs, and that employers can effectively help with programs such as internships. Preparing teachers to supervise programs and to work with potential employers to develop and enhance supervised experience programs in Arab countries is vital to the development of a well-prepared agricultural workforce.

Decision-making Skills

More advanced internships that contain learning goals for the students/trainees should allow for interns to have some decision-making responsibilities. Those decisions may be management decisions that are related to their job tasks. An intern may seek supervisory input before making decisions and as they grow during their internship experience, they may be allowed to make management decisions on their own. It is important that open communication, through weekly meetings, between the on-site supervisor and the intern take place where decisions are discussed.

Decision-making skills first should be developed in the teaching curricula through case studies and real-world applications. Identified areas of needs based on the students and agricultural industry will be identified in the skill gap analysis. Decision-making is a non-technical competency that should be promoted across curricular areas (Barrick et al., 2009). Decision-making ability includes problem-solving. Having a curricular focus on decision-making can be conducted through a five-step process: 1) Identify what decision needs to be made, 2) Gather information and data on potential solutions for the decision to be made, 3) Consider positives and negatives about each of the potential solutions; this may also include gathering others' input and perceptions about potential solutions, 4) Weigh evidence based upon the data and others' input, and 5) Make the decision and communicate that decision to others backed by the rationale. The purpose of teaching decision-making is to prepare future employees to have the ability to think through a process that they may not have encountered first-hand when on the job.

Agriculture Technical Skills

Agriculture technical skills are to be developed and enhanced in the major agricultural program areas identified through the skill-gap analysis. Instructors prepared or revised courses to reflect the changes needed by industry while also satisfying state requirements. The central focus and purpose of the curriculum is to teach agriculture content and technical skills. Teachers should outline their curriculum based upon several factors. First, consideration of the subject area is vital and then the teacher should outline their teaching to build on student prior knowledge and follow a logical sequence. Consideration should be given due to weather and when commodities may be grown or considered in-season so that laboratories and farms can be utilized during instruction.

Next, the consideration of skills to be developed and enhanced in the major program areas will be identified through the skill-gap analysis (Barrick et al. 2009) and should be addressed. Instructors should prepare and revise curricular focus based on the needs of the industry and students identified as needs of the skill gap. Areas noted as being met by the skill gap should remain and consideration of less repetition considered if those skills are repeated in the curriculum.

Connecting skills to concepts is important for classroom learning. This can be achieved through connecting lectures and laboratory experiences. It is well accepted that student motivation is positively affected by skill acquisition (Dale, 1969). Furthermore, abstract concepts become more difficult for students to grasp until they have had a concrete experience. For example, if cotton planting is being taught, considerations for planting depth would be one focus of the lesson. Taking students to a field laboratory to identify the factors of the soil that determine planting depth will be easier conceptualized through a field-based experiment or experience than a simple lecture where students memorize the factors that determine planting depth.

Dale's Cone of Experience (1969) should be consulted when teaching content and skills to students in the agricultural development areas. As the focus of Dale's cone moves down, expectations of the learner move from passive to active. Further, sustained learning based on their learning remains with them longer because learning is more concrete.

Figure 3.

Levels of abstraction increase as student complexity of tasks increase. In a more abstract approach students may only be asked to be verbal or visual receivers. As teaching involves learners to be active in their learning through skills, they have a more concrete approach and this requires them to watch demonstrations, participate in the demonstration, and even conduct the skills themselves.

Another consideration for the instructor is the student learning objective. During skill development, there should be a focus on teaching the process or the product. Teaching a process skill is for the focus to be on the steps that the learner goes through in the creation of an answer/solution. Teaching for the product is focused on the outcome or finished product and how it compares to a standard. When teaching for a product outcome, the process is embedded within, but learning outcomes need to be determined prior to assessing student learning. For example, when teaching the technical skill of trouble shooting why an engine does not work, is it more important to teach students the steps to trouble shooting or to be solely focused on the product – which would be to have an engine that will run? Noted, that a running engine is optimal, but in order to know why the engine may not have been running and how to prevent other engines from having similar issues would be discovered by identifying the process. On the other hand, if students have been taught technical skills of how to operate hand tools and the learning goal is to develop wooden feed bunks for dairy cattle to sell to farmers, instructors might focus on the final product and noting that students meet all the parameters outlined in the original design or model that was provided to them.

The development of agricultural technical skills will require first-hand experience. This can be achieved by utilizing field-based experiences in learning laboratory situations, through internships, within a traditional classroom that provides learners the opportunity to practice, through simulations, and by allowing students to develop models and supplemental ways to acquire skills. Input from the agricultural industry and skill gap analysis will supply at what level students need to be able to achieve. Ways to showcase technical skill development is through internship experiences and through competitive events where students compete against peers in completing tasks that require technical skills to complete. This will increase student motivation and create a since of pride and belonging among peers as they learn new skills. Rewards should be supplied to students that achieve at a predetermined level or successfully reach educational goals.

Leadership Skills

Leadership skills are identified within the skill gap analysis and transcend across all curricular areas. This is a developmental focus for students to become valued employees and enable them to provide leadership based upon experience. Leadership skill areas are personal skills and working as part of a group. McKinley et al. (1993) outlined four factor areas of leadership skills for agricultural students. The authors outlined interpersonal, administration, management, and communications as the four factor areas of focus.

Interpersonal Factors

Interpersonal focuses on the student's self-development as a future leader. Examples include flexibility in the workplace, willingness to listen to others, consideration of alternative viewpoints, ability to see both sides to an argument, having a good sense of humor, trustworthiness, conflict resolution, tact, respectfulness, cordial, empathy, working relationships.

Administration Factors

Administration includes the ability to motivate others, inspire others, acceptability as a leader, having others seek you for advice, others seek you for guidance when they encounter difficulties, decision making ability, willingness to take charge, ability to persuade others, acceptability of your ideas, and how you are viewed as a professional.

Management Factors

Management includes concepts associated with the ability to complete projects, enjoyment of achieving success, use of intelligence, follow through, professional goal setting, enthusiasm, persistence, and proficiency in conducting work related to the job description.

Communications Factors

Communication areas include involvement in the company or organization, concern for others and the employee workplace, sharing of information and keeping others informed, encouragement to keep others involved, confidence building, having a good relationship with others, sharing of ideas, group dynamics, sense of belonging, and willingness to meet new people.

Creating leadership opportunities for students can be showcased through case studies, role play, and additional active learning strategies outlined in the active learning strategies section within this book. Leadership can be developed through real-world experience through the creation of student/youth leadership organizations. Organizing a youth leadership organization at the local or regional level will help address factors outlined by McKinley et al. (1993). Examples include practicums where students prepare for and compete against others in job interviewing skills, problem-solving, working as a team, and communicating effectively. Highlighting leadership skills required to be successful in an internship provides first-hand experience in working with others and making decisions where input is sought from others. Creation of factors and focusing on components over time and embedding them into the technical skill context is the most effective way to include leadership skills.

Asking students to reflect upon experiences and thinking about how the group achieved or did not achieve in a working group situation or problem-solving activity is a great way to embed leadership skills into current teaching. During student organizational development, leadership skills can be addressed upfront so that students know the skills they are focusing on. Allowing students to have more control of their learning can happen during student organizational development for leadership skills. For example, having students make presentations to peers and seeking feedback creates communication skills, electing student organization officers helps develop strategies for students to build working relationships. Once elected, officers will have to work together as a team, yet have tact to include all students that are part of the organization. Students will address administrative factors by taking charge and creating buy-in among peers when making decisions. Interpersonal factors can be addressed through the student organization by being a trustworthy individual that respects other's opinions when resolving a conflict. Management will be addressed in the organization accomplishing tasks they wish to do and if it is done effectively the students will have communicated well.

There are many ways to address leadership skills. They can be addressed throughout the curriculum and completed within the context of agricultural technical skills, through the creation of a student leadership organization where officers are elected, by focused time during competitive team events that showcase agricultural content skills but contain a reflective aspect that requires student reflection on the learning process, and through internships that focus developing the individual trainee self-perception and ability to work with others. Leadership skills can also be thought of as employability skills and qualities that make a person that others think highly of and find to be a good individual to collaborate. Rewards can play a significant role in motivating students to achieve and remain motivated to participate and learn in leadership development. The creation of a certificate or recognition in front of peers should be developed. Therefore, skills such as oral expression, team building, and goal setting are taught within the context of agriculture.

Advisory Committee

Agricultural industry leaders play an important role in guiding the agricultural development model. The model provides the opportunity for local agricultural leaders to assist in making needed changes in the educational programs and can also add value to a continued effort of identifying industry needs for the agricultural workforce. The agricultural industry identifies needs that form the skill gap analysis and address the decision-making skills, technical skills, and leadership skills that are desired in the workplace. Further, the agriculture industry will partner with teachers in offering internship opportunities for students to develop real-life workplace experiences. Therefore, agriculture industry partners should make up the advisory committee.

The purpose of an advisory committee is to bring unique knowledge, skills, and perspectives to aid in the development of students and a well-prepared agricultural workforce. An advisory committee does not have formal authority to govern or make decisions, but they should provide directions and assist in needs of the educational focus. The advisory committee should be made up of agricultural industry professionals and leaders that have first-hand knowledge and experience in working with entry level and management employees.

When forming an advisory committee, teachers and administration should seek individuals that have an interest and participated in the skill gap analysis. A typical advisory committee would be made up of eight to fifteen individuals that represent different focus areas within the agriculture industry. For example, an advisory committee should not be made up of only individuals from the horticulture industry if the school teaches animal science, horticulture, food science, and mechanics. The advisory committee should be made up of individuals from each content focus area. Once individuals have been asked to and agreed to serve on the advisory committee, it is the teachers' responsibility to make them aware of their role on the committee. Teachers should develop a rotation of who serves on the advisory committee so that new ideas and new people are rotating off and coming onto the advisory committee. For example, have individuals serve a two- or three-year term and be sure they are aware of this when they begin serving on the committee.

Advisory committees should meet at least twice a year but not more than four times a year. There should be a meeting agenda developed and supplied to the advisory committee ahead of the meeting. The meetings should not last longer than two hours, as members are working professionals and have their job and own working responsibilities. Teachers should present updates and data based on previous work that addresses the skill gap analysis. Teachers may showcase successes they have had with students, internships, student organizations, competitive events, and agriculture technical skill development. Data should be shared, and recommendations sought from the committee. The environment in which the advisory committee operates is important so that suggestions are taken for improvement and not evaluation.

Advisory committees can also serve as a great way to develop partnerships with new agriculture-based companies that lead to internship opportunities for students in the future. Adding new members as partnerships develop should be sought.

The USDE (2008) outlines five steps to develop an advisory committee. 1) Establish a purpose for the group. The purpose for an advisory committee would be to supply knowledge and feedback based on the skill gap analysis and advocate for the internships and impact of the educational program. 2) Recruit members that fit the purpose; committees should include industry professionals in agriculture who have an interest in the development of a stronger workforce, and parents could also serve on the committee as well as other educational professionals such as university faculty. 3) Develop structure, by implementing tasks, roles, guidelines, and election of length of terms. 4) Prepare members for their role; during this step members would be provided information that they serve in an advisory role and serve to strengthen programs, not evaluate the program. 5) Empower the group to have a clear scope; committee members need to feel empowered to share their thoughts and visions and be supplied with clear directions through effective communication of meetings and agendas

Finally, effective committees are allowed to provide thoughts and suggestions in which following meetings should address outcomes based on previous meetings and advice provided. Committee members are more supportive when there is evidence their ideas are considered and/or utilized.

PROCESS

The process phase of the model addresses actionable items for developing a process to carry out the results of the skill gap analysis, preparation for student skill development, active learning strategies, and suggestions for the development of leadership and technical skills.

Internship Development

During the creation of internships teachers will need to address opportunities with potential industry partners and with students. Teachers will have the responsibility of securing agricultural partnerships. To best accomplish this, teachers need to develop clear objectives for the internship and note benefits to host organizations. In-person visits to agricultural companies and sharing curricular goals are vital in communicating the focus of internship work. Asking industry partners their desires and goals as well as addressing the skill gap analysis will assist. The teacher will need to develop responsibilities of the employer and selection criteria for placement of the student that will serve in that internship.

The teacher will then address the following points with the student prior to placement in an internship.

- Objectives, eligibility, credit, grading
- Responsibility of the student
- Responsibility of the faculty supervisor
- Responsibility of the employer
- Application, selection, and placement

An Example Outline for Responsibilities for Interns/Trainees

Carefully study and consider the suggestions and instructions offered by the teacher and on-site supervisor. The undertaking which you are beginning is without doubt the most important phase of your preparation as an agricultural professional.

Candidates engaged in an internship represent their school and future profession. Interns are expected to abide by professional guidelines and maintain a good work and learning ethic. First impressions are important. Be genuinely courteous, cooperative, and sincere in working with others. The ability to work well with other people and to maintain desirable relationships is important. Appearance and conduct should be acceptable for professionals in the Agricultural profession.

The internship is an opportunity for you to learn. Students acquire more and better skills and competences when they follow these concepts.

- Carefully observe not only what is done and reflect on how things are done. This should be the basis for much of your reflection journal entries.
- Attendance and preparation during the internship showcase your interest and commitment to learning and becoming a future professional in this industry.
- You should dress, talk, and act as a professional.
- You should exemplify professionalism in dealing with confidential information.
- Demonstrate a professional attitude in all contacts within the business/organization and community.
- You should assume responsibility for the quality of your experience. Seek out opportunities to be involved and ask for new tasks as soon as you feel able to master them.
- Do not hesitate to ask for assistance from your on-site supervisor or teacher. Learn all you can during your internship.
- Get all the experience possible in all phases of the job – program organization and management, skills for agriculture performance, and field work.
- Take constructive criticism in the spirit in which it is offered. Suggestions will be offered for your professional improvement.
- Invite suggestions and professional grow as a young professional.
- Be prompt in meeting all your classes and appointments. It is better to arrive ahead of time than to rush into a room just ahead of the bell.
- Be consistent in your methods and relationships.
- Develop patience and self-control.
- Set aside time at the end of each day for a conference with your on-site supervisor, at least for the first three weeks and at least on a weekly basis thereafter.

An Example Outline for Responsibilities On-site Internship Supervisors

The on-site internship supervisor is the individual who supervises the intern/trainee. The role of the on-site supervisor is to help the intern have a professionally rewarding experience while helping to prepare them for a career in agriculture. The internship is regarded as one of the most important phases in the development of a young professional. There is no doubt that this person will greatly influence the professional attitude of the intern as well as provide them with the opportunity to increase their professional knowledge and skill. It is with this challenge in mind that the following suggestions are made.

- Be sure that the intern is introduced to managers and employees.
- Establish daily working hours, responsibilities, appropriate attire, expected expenses, personal conduct, and absences.
- Acquaint the intern with facilities and procedures. Plan the intern at ease and make them feel useful and important to the business.
- Plan for time to meet with the intern on a daily basis until they are comfortable in their working environment. Then plan for weekly meetings.
- In cooperation with the intern, complete a list of tasks and experiences the intern should experience during their time in the internship.
- Maintain contact with the internship teacher.
- Be a salesperson for the profession. Portray a positive image of the profession and help the intern see the positive rewards of entering the profession.
- Establish an atmosphere where the intern is not afraid to try something new or to occasionally fail. Be a support person for the intern.
- Monitor progress of the intern and provide feedback for improvement as well as feedback on where the intern is doing well.
- Conduct evaluations as needed for the teacher.

Example Outline for Responsibilities of the Supervising Teacher

Supervising teachers are a liaison between the school and the agriculture business. In addition, they are charged with the following responsibilities:

- Working with the on-site supervisor and the intern in planning, executing, and evaluating the experience.
- The supervising teacher will make observation visits on at least two separate occasions. Each observation will be followed by a conference to evaluate progress, make constructive suggestions, and provide help as requested by the on-site supervisor and/or intern.
- The teaching supervisor will calculate the intern's final grade based on the on-site supervisor recommendations and their own observations and interactions during site visits.
- The teaching supervisor will act as a resource person for the intern.
- The teaching supervisor should be contacted immediately when a problem or concern arises.
- Assist on-site supervisor in providing feedback and internship development.

Items for development also include rubrics for evaluating interns during their experience, a way for interns to communicate back to the teacher supervisor by keeping a journal of work and logging working hours. These items will help document the experiences and technical skills developed that the intern could use to showcase their experiences for future employment.

Agricultural Technical Content

The technical content taught in schools should reflect the agricultural opportunities of the community or region. This will lead to interest from local industry partners and thus greater internship opportunities for students and jobs for graduates. General areas of consideration in agricultural development include, but not limited to:

- Animal science
- Agronomy
- Horticulture
- Agricultural Mechanics
- Agricultural Education and Communication
- Agricultural Business

Each of the above could be broken down further to develop grade-level focus. For example, animal science could be broken down by species to include poultry, beef, and equine or be broken down by systems including nutrition, husbandry, reproduction, and animal health.

The focus should be based on the needs of the community and the job opportunities in the region. Ways to engage in further learning are by holding competitive events where students can showcase their skills. The teacher can hold competitions at the local school or regionally where students can compete against peers on skills identified in the skill gap analysis.

Active Learning Strategies

Teacher development and use of active learning strategies are important for the engagement of students. Engaged learning that is experiential in nature leads to students retaining greater information and transfer of learning. Teachers should seek ways to engage students through active learning, keeping in mind that active learning needs to be purposeful and planned. Active learning strategies are not simply activities in the classroom or laboratory for the sake of having an activity. A chapter of this text is devoted to utilizing active learning. Teachers should develop and practice a combination of strategies to enhance their teaching and student learning.

Leadership Development

Ways to develop leaders in the classroom should be a focus of the teacher. This can be achieved through daily classroom teaching that incorporates leadership development earlier described, stand-alone lessons that focus on leadership components, or highlighted through the engagement of student leadership organizations. Recognition that leadership is the teacher's and the program's responsibility to instill into the students is important in the development of a workforce that creates problem-solvers and work-ready graduates.

Teacher Development

As Chickering and Gamson (1987, p. 3) stated, "Learning is not a spectator sport. Students do not learn much just by sitting in class listening to teachers, memorizing prepackaged assignments, and spitting out answers. They must talk about what they are learning, write about it, relate it to past experiences, and apply it to their daily lives. They must make what they learn part of themselves." This is an overall educational philosophy that can guide preparing for, conducting, and assessing agriculture teacher development programming.

Many teachers assert that all learning is inherently active and that students are, therefore, actively involved while listening to formal presentations in the classroom (Swanson et al., 2007). Chickering and Gamson (1987) suggest that students must do more than just listen. Students must read, write, discuss, or be engaged in solving problems. Within this context, it is proposed that strategies promoting active learning be defined as instructional activities involving students in doing things and thinking about what they are doing.

Bloom's Taxonomy of Educational Objectives (1956) is a foundational and widely accepted tool for educators to use in promoting student learning. The creation of educational objectives then should help address how students should learn the new content and be assessed. Lower levels of Bloom are knowledge, recall and comprehension and more advanced cognitive skills (higher levels) include analysis, synthesis, and evaluation of information and knowledge. This aids in students solving problems and making informed decisions. To clarify educational objectives, Bloom divided educational objectives into three domains: affective, cognitive, and psychomotor. The affective domain focuses on feeling and emotion; the psychomotor domain addresses motor skills; and the cognitive domain applies to thinking skills. Within the context of agriculture content, educational objectives typically focus on the cognitive and psychomotor domain.

Educational objectives organize what should be taught, the level at which the concepts should be taught and lead to ways to assess learning. Following, the teacher should review content with the consideration of utilizing active learning. Active learning in the lesson begins with an interest approach. An interest approach prepares and connects learners to what they are about to learn. Creating interest increases student motivation and draws learners into the lesson by having them consider previous experiences or potential impacts on their learning.

Active learning should be utilized throughout the lesson. Examples of active learning activities include: carousel brainstorming, case studies, clarification pauses, cooperative groups, concept mapping, daily journal, frequent short quizzes and feedback, jigsaw procedure, learning cycle, muddiest point, one-minute paper, moveable magnetic diagrams, field exercise, and think/pair/share, among others. In the end, students should be engaged in those learning activities as they move toward accomplishing the learning objectives.

Rosenshine and Furst (1971) conducted a meta-analysis of educational studies and identified 11 teaching behaviors for effective teaching. Of those 11 behaviors, Garton, Miller, and Torres (1992) identified 5 that could readily improve teacher performance in the agriculture classroom. These 5 teaching behaviors were: 1) being business like, 2) being enthusiastic, 3) being clear, 4) providing students with opportunities to learn current material, and 5) varying teaching methods to maintain student interest.

The creation of lesson plans acts as a guide for teachers to follow throughout teaching and active engagement of students during a daily lesson. Lesson plans can incorporate four of the desirable teaching procedures identified by Garton et al. (1992). The enthusiasm demonstrated by the teacher in the classroom is a behavior that comes from within the teacher. The other four behaviors can be incorporated into each lesson plan. For example, by following the lesson plan, staying on course, and being prepared, the teacher will have business like behavior. When teachers provide clear directions and are prepared with active learning strategies the teacher's communication with students will be clear. Clarity will enable teachers to assess student understanding before transitioning to the next learning activity. Lesson plans that include active learning provide students the opportunity to practice and learn the current concepts/material. Activities that are closely aligned with desired objectives will directly assist the teacher in being more effective. Finally, lesson plans that outline different teaching methods and learning strategies will help keep students engaged and interested.

When developing lesson plans, the desired outcome must be the guiding force. The guiding force must be presented as student learning objectives, and then the objectives should dictate the way the content is taught, the active learning strategies utilized, and how learning is assessed. This outcome is determined by what the students will be expected to do as a result of the lesson, unit, course or internship. This will lead to the improvement of production practices on the student's home farm or in gaining employment in the agriculture industry. In short, lesson plans are a guide for the teacher to lead learners through the learning process, with a central emphasis on active learning methods and techniques, so that students will achieve the desired learning outcomes.

Class time must be utilized to be as relevant and authentic to real-life as possible. While lecturing can be important and efficient, it is often not the best way to engage students in the learning process. Teaching at higher-level cognitive skills will engage students in experiential learning that leads to higher cognition and application of concepts that will lead to retention of the concepts. Lecturing induces passivity of thought, even in the best of students. Students hurriedly take notes but have little time to reflect on or question the content being recorded. If lecture is utilized, active learning can still be used through think/pair/share and through the use of questions to allow students to synthesize concepts. The key concept to remember is that the more students become engaged as a part of their learning, the more likely they will develop the desired cognitive and psychomotor skills they will need and the greater opportunity that leads to a better prepared workforce in agriculture.

The key point of the process portion of the model is teacher and faculty development. The above concepts are for improved teaching in the classroom, providing students with real-world experiences through the engagement of internships and active learning in the classroom and laboratory. To achieve this, support and teacher professional development for teachers and faculty should be conducted. From the skill gap analysis (assessment) and the identification of subject matter that needs to be taught (content), the final step in the model is agriculture teacher development.

To achieve agriculture teacher development focus on workshops and train-the-trainer professional development should address the teaching and learning needs of teachers of agriculture. Consideration of professional development to address the following:

1. Provide workshops and programs for teachers that address how to utilize active learning strategies in teaching personal and leadership development competencies, real-world internship experiences, and technical subject matter content.
2. Work with faculty and instructors to develop instructional materials that support the teaching of skills and competencies.
3. Assist teachers and faculty in developing curricula, courses, and lesson plans.
4. Develop plans for using school farms and facilities for practical skill training.
5. Assess progress and refine teaching strategies, lesson plans, and instructional materials.

When offering professional development programs for agriculture faculty, two primary recommendations are made (Roberts et al., 2008).

· Know the audience, learn the culture, and understand the local situation
· Plan well and be prepared to alter plans as the activity or program progresses.

SUMMARY

This chapter focused on the holistic approach to understanding the model of agricultural education for development in Arab counties. Three components are outlined within the model. Assessment addressed the assessment of the program and seeking input from industry stakeholders and student competences to develop identify needed skills through the skill gap analysis. Content addresses educational components that make up the Agricultural Technical School (ATS) curriculum. Process proposes actionable items to develop teachers to carry out the process to develop a well-prepared agricultural workforce in Arab countries. To achieve this, four objectives were addressed: 1) Describe a model of Agricultural Education for Development in Arab countries. 2) Describe components for assessment of student competence and agricultural industry needs. 3) Identify content focus areas for curriculum development. 4) Propose a process to enhance teaching and learning.

REVIEW QUESTIONS

1. What are the three main components of the model that lead to a well-prepared agriculture workforce?
2. What is the importance of conducting a skill-gap analysis?
3. How does a skill-gap analysis inform needs for technical skills?
4. What is the outline for the five-step process in teaching decision-making skills?
5. Why are experiential learning opportunities, such as internships, important for the development of a well-prepared agriculture workforce?
6. Why is it important to outline responsibilities of the intern, on-site supervisor, and teaching supervisor prior to an internship beginning?
7. Who should make up an advisory committee?
8. What are three leadership skills that can be taught either through an internship or within classroom active learning?
9. According to Rosenshine and Furst (1971), what are the top five teaching behaviors for effective teaching and enhanced student achievement?

10. What are Bloom's (1956) three domains of learning and what are examples of how they can be utilized in an agricultural education setting?

REFERENCES

Armstrong, M. (2002). *Employee reward.* CIPD Publishing.

Barrick, R. K., Samy, M. M., Gunderson, M. A., & Thoron, A. C. (2009). A model for developing a well-prepared agricultural workforce in an international setting. *Journal of International Agricultural and Extension Education, 16*(3), 25–32. doi:10.5191/jiaee.2009.16303

Bloom, B. S. (1956). *Taxonomy of Education Objectives: The Classification of Education Goals.* Edwards Bros.

Chickering, A. W., & Gamson, Z. F. (1987, March). *Seven principles for good practice in undergraduate education.* Washington, DC: AAHE Bulletin.

Dale, E. (1969). *Audiovisual methods in teaching* (3rd ed.). Dryden Press.

DiBenedetto, C. A., Willis, V. C., & Barrick, R. K. (2018). Needs assessments for school-based agricultural education teachers: A review of literature. *Journal of Agricultural Education, 59*(4), 52–71. doi:10.5032/jae.2018.04052

Ebner, P., Ghimire, R., Joshi, N., & Saleh, W. D. (2020). Employability of Egyptian agriculture university graduates: Skills gaps. *Journal of International Agricultural and Extension Education, 27*(4), 128–143. doi:10.5191//jiaee.2020.274128

El Ashmawi, A. (2015). *The skills mismatch in the Arab world: A critical view. British Council Cairo Symposium.*

Garton, B. L., Miller, G., & Torres, R. M. (1992). Enhancing student learning through teacher behaviors. *The Agricultural Education Magazine, 65*(3), 10–11.

McKinley, B. G., Birkenholz, R. J., & Stewart, B. R. (1993). Characteristics and experiences related to the leadership skills of agriculture students in college. *Journal of Agricultural Education, 34*(3), 76–83. doi:10.5032/jae.1993.03076

Othayman, M. B., Mulyata, J., Meshari, A., & Debrah, Y. (2022). The challenges confronting the training needs assessment in Saudi Arabian higher education. *International Journal of Engineering Business Management, 14*, 1–13. doi:10.1177/18479790211049706

Puckett, J., Hoteit, L., Perapechka, S., Loshkareva, E., & Bikkulova, G. (2022, January). *Fixing the global skills mismatch.* Boston, MA: Boston Consulting Group.

Rasmussen, C., Pardello, R. M., Vreyens, J. R., Chazdon, S., Teng, S., & Liepold, M. (2017). Building social capital and leadership skills for sustainable farmer associations in Morocco. *Journal of International Agricultural and Extension Education, 24*(2), 35–49. doi:10.5191/jiaee.2017.24203

Roberts, T. G., Thoron, A. C., Barrick, R. K., & Samy, M. M. (2008). Lessons learned from conducting workshops with university agricultural faculty and secondary school agricultural teachers in Egypt. *Journal of International Agricultural and Extension Education, 15*(1), 85–87. doi:10.5191/jiaee.2008.15108

Rosenshine, B., & Furst, N. (1971). Research on Teacher Performance Criteria. In B. Othanel Smith (Ed.), *Symposium on Research In Teacher Education* (pp. 37-72). Englewood Cliffs, New Jersey: Prentice-Hall, Inc.

Swanson, B. E., Barrick, R. K., & Samy, M. M. (2007). *Transforming higher education in Egypt: Strategy, approach and results. Proceedings of the 23rd Annual AIAEE Conference*, Polson, MN.

Swanson, B. E., Cano, J., Samy, M. M., Hynes, J. W., & Swan, B. (2007). Introducing active teaching-learning methods into Egyptian agricultural technical secondary schools. *Proceedings of the 23rd Annual AIAEE Conference*, Polson, MN.

United States Department of Education (2008). *Building an effective advisory committee. Mentoring Resource Center Fact Sheet No. 21.* Building an Effective Advisory Committee (Fact Sheet) (educationnorthwest.org)

Vreyens, J. R., & Shaker, M. H. (2005). Preparing market-ready graduates: Adapting curriculum to meet the agriculture employment market in Egypt. *Proceedings of the 21st Annual AIAEE Conference*, San Antonio, TX.

Wiggins, G., & McTighe, J. (1998). *Understanding by Design.* Association for Supervision and Curriculum Development.

World Economic Forum. (2017, May). *The future of jobs and skills in the Middle East and North Africa.* World Economic Forum, Geneva, Switzerland.

Chapter 2
Establishing and Managing Successful Agricultural Education Programs at Universities and Secondary Technical Schools in the Arab Countries

Mohamed Samy
https://orcid.org/0000-0003-3522-0989
Independent Researcher, USA

ABSTRACT

Providing quality and innovative agricultural education programs (AEPs) to the rural communities should be a vital goal for the Arab countries to be able to accelerate rural development and achieve acceptable level of food security. A successful and effective AEP includes four overlapping, interactive components: 1) positive and encouraging learning environments, 2) effective supervised hands-on practical experiences, 3) supportive career development and placement services, and 4) enabling partnerships with agribusiness and community leaders. Effective program leaders are essential to the success of planning and implementing AEPs. Successful implementation of these programs will result in significant improvement of students' academic achievement and skill acquisitions, and increase the overall employability skills of agricultural graduates to allow them to actively participate in serving their communities; and assist in achieving acceptable levels of food security, increasing agricultural exports, and wisely managing the environment and natural resources, especially water.

DOI: 10.4018/978-1-6684-4050-6.ch002

INTRODUCTION AND PROSPECTIVE

As the global food crisis continues to intensify because of pandemics, political conflicts, and natural calamities caused by climate change, the pressure on rural communities in the Arab countries continues to produce more food and fibers (World Food Program, 2023). These Arab rural communities are assuming the responsibility of achieving acceptable levels of food security and boosting agricultural exports for their countries. In spite of this enormous and critical responsibility, these Arab rural communities have been experiencing a range of serious and severe economic and social challenges hindering their efforts of strengthening the agricultural production and export capacities of their countries. These challenges include widespread poverty, high rates of youth unemployment, lack of skilled labor, inadequate infrastructure and facilities, and very limited business opportunities (The World Bank, 2008; Elmenofi, et al, 2014; Abu-Ismail, 2018). These rural communities are a significant part of the Arab population, representing about 40 percent of the total population in the Arab countries in 2021. They range between 66 percent in Sudan to 15 percent in Saudi Arabia of the total population (The World Bank. Data, 2023). Figure 1 shows that more than 55 percent of the population in Yemen and Egypt lives in rural areas; while in Tunisia and Morocco, rural population represents between 30 to 35 percent of the population (The World Bank. Data, 2023).

Figure 1. Rural population in selected Arab countries in 2021

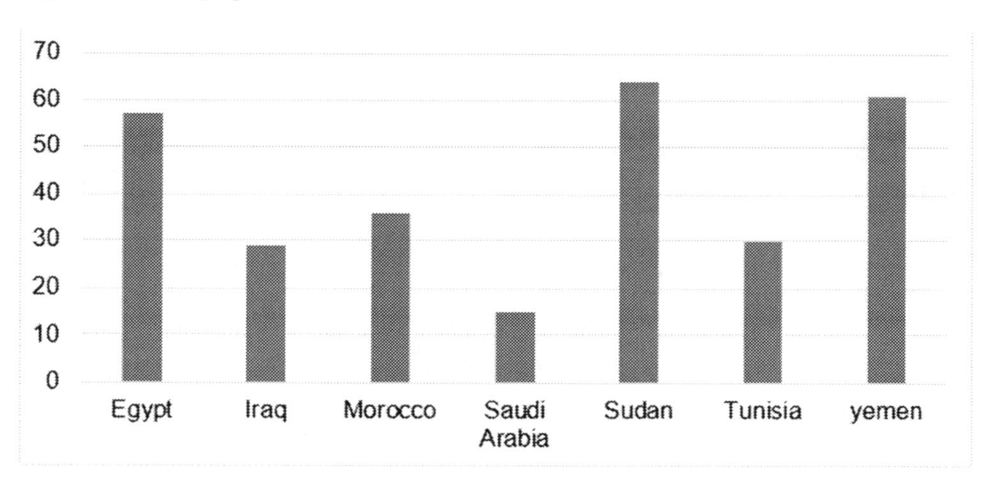

The agricultural sector provides a significant proportion of employment in most Arab rural communities, however its contribution to GDP is limited, around 12 percent

(The World Bank. Data, 2023). At the same time, these Arab rural communities possess the natural and human resources necessary to effectively participate in agricultural development, contribute to food security, and become important suppliers of high value agricultural products to global markets. Recognizing the critical importance of both agricultural development and quality education in accelerating economic growth in developing countries, the United Nations (UN) included both in its Sustainable Development Goals (SDGs) for 2030 (United Nation, 2022). In addition, the UN SDGs for 2030 emphasize that:

Quality education and lifelong learning opportunities for all are central to ensuring a full and productive life to all individuals and to the realization of sustainable development. (UN, 2023)

To be able to participate effectively in achieving these SDGs for 2030, the Arab countries should have the wisdom of focusing on building agricultural human capital through innovative and effective education programs that recognize the importance of the potential to transform people, the educational process and the economic foundation undergirding a progressive agriculture industry in their countries. Providing quality agricultural education and innovative training programs to the rural communities should be a vital goal for the Arab countries if rural development is to take place, agricultural competitiveness strengthened, and agricultural industry developed to the degree that it can capitalize on opportunities to help reduce poverty and unemployment in rural communities.

Starting in the 1990's, a significant number of commercial farms were established in Egypt, Morocco, United Arab Emirates (UAE), Saudi Arabia and other Arab countries, geared more toward high value horticulture crops for export markets and livestock products for domestic markets, which created an increasing demand for a skilled labor force that could contribute to advanced, large-scale, production enterprises. In spite of the increasing demand on skilled labor, most Arab countries suffer from shortage of a well-trained agricultural workforce. The World Bank indicates that the shortage of skilled labor is a major constraint to rural development and the economic growth in Egypt, Morocco, Saudi Arabia and other Arab countries (The World Bank Group. 2015; The World Bank Group. 2020; and Moshashai & Bazoobandi, 2020). In addition, the rate of unemployment in the Arab rural communities reached over 32.0 percent during the 2010's, creating more economic, social, and political problems for the Arab countries. The World Bank indicates that the share of youth not in education, training, or employment in the Arab countries range from 21.0 percent in Algeria to 44.8 percent in Yemen in 2021 (Figure 2) (The World Bank. Data, 2023). Figure 2 also shows that youth unemployment reached more than 30 percent in several Arab countries such as Sudan, Iraq, Egypt, and Tunisia in 2021.

Figure 2. Share of youth not in education, training, or employment in selected Arab countries

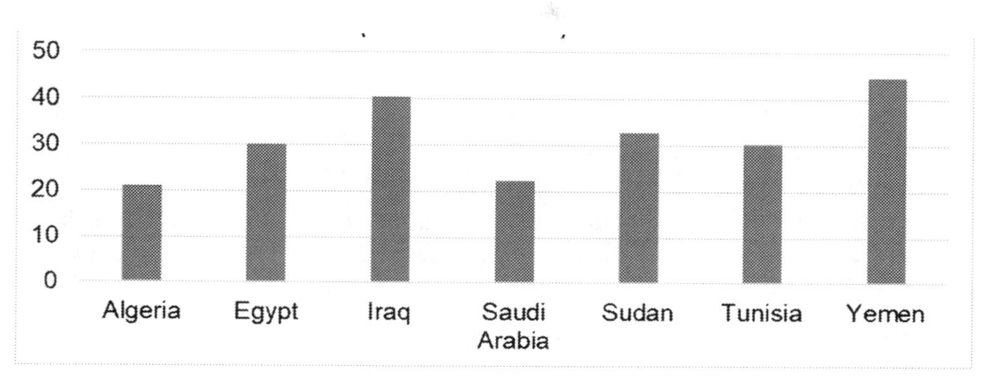

Swanson, et.al. (2007) indicated that engaging ATS students in various practical training activities has not been a priority in Egypt

The dramatic growth in student numbers and insufficient expenditures in most Arab countries have not delivered positive returns in quality of education or contributed to economic growth (Alaoui & Springborg. 2021). Most of the graduates of the Arab agricultural technical and secondary schools and colleges of agriculture are among the unemployed rural youth; these graduates lack employable skills and knowledge required by commercial farms and global food supply chains. For example, in Saudi Arabia, the education system at all levels faces many challenges and does not equip students with the necessary skills for labor market (Moshashai & Bazoobandi, 2020). In Egypt, Agricultural Technical Schools (ATSs) teachers had few teaching aids or materials, other than simple textbooks and a chalkboard in each classroom. As a result, student interest was low and the drop-out rate among third year students was high, about 50% (MUCIA, 2006). Typically, ATS graduates lack employable skills and knowledge required by the commercial farms and private sector firms (MUCIA, 2006). Swanson, et al. (2007) add that providing practical training to ATS students in Egypt was not a priority. Furthermore, Daghur (2018) points out that colleges of agriculture in the Arab world need a fundamental reform to support improvements in food security and environmental sustainability. Adam (2019) indicates that the quality of agricultural graduates in Sudan is below the expectations; Ben Haman (2020) states that lack of quality holds back Moroccan schools and universities from achieving the desired outcomes.

OBJECTIVES

The need for a skilled agricultural workforce continues to grow in the Arab countries not only to accelerate rural development and increase agricultural production, but also to sustainably manage the environment and natural resources, especially water. The rural communities in the Arab countries urgently need efficient managers and skilled technicians, these skilled workforces will work in different functions of the field crop, horticulture and livestock supply chains and in natural resource management. Innovative and effective agricultural education programs (AEPs) are urgently needed to support the development of rural communities and economic growth in the Arab countries. A new and innovative generation of AEPs have to be initiated and successfully managed in Arab rural communities to prepare rural students and youth to participate effectively in developing their communities, achieving acceptable levels of food security and sustainably managing natural resources for their countries.

This chapter describes the current situation of formal secondary/technical and university AEPs in the Arab countries; and identifies challenges hindering these programs from providing effective educational and training services to rural communities. The objective of this chapter is to introduce a new model for providing effective formal AEPs at the secondary/technical school level as well as at the university level in the Arab countries. Furthermore, this chapter provides several successful examples of how to implement various components and elements of this model to help leaders of agricultural education in the Arab countries successfully establish, manage and evaluate the effectiveness of AEPs.

FORMAL AGRICULTURAL EDUCATION PROGRAMS (AEPs) IN THE ARAB COUNTRIES

The AEP is a formal learning activities and services designed by the Ministry of Education (MoE) or a university, engaging mainly in the provision of agricultural education and determining the learning progress of each student in all stages. The administration of AEPs at the secondary/technical school level is part of the MoE in most Arab countries, including Egypt, Saudi Arabia, Morocco, Tunisia, Jordan, Iraq, Sudan, Oman, Yemen, and UAE. In addition, major universities in the Arab countries that offer degrees in agricultural sciences include Cairo University, Alexandria University, Assiut University and Zagazg University in Egypt, King Saud University and King Abdelaziz University in Saudi Arabia, University of Sfax in Tunisia, The University of Jordan in Jordan, University of Sana'a, in Yemen, and UAE University in UAE. At the present time, university and secondary technical school

AEPs in most Arab countries do not provide sufficient or appropriate education or training to meet the skilled workforce needs of the agricultural industry and global economy. For example, in Egypt, Sudan, Morocco, Saudi Arabia and other Arab countries, the skill capabilities of agricultural students at all educational levels have not met the needs of business and industry (Al-Zaidi,1982; Swanson, et al, 2007; Campbell, et al, 2008; Daghur, 2018; Adam, 2019; Moshashai & Bazoobandi, 2020; and Ben Haman, 2020).

These formal AEPs are producing unskilled graduates, adding to the rural unemployment problem and increasing the number of people living under the poverty line in the Arab countries. Most of these AEPs have provided education based mainly on rote memorization, teacher-centered instruction, minimal private sector and community input, and the use of low-level instructional technologies. In addition, in-service professional development opportunities for teachers, faculty members, administrators and program leaders have been insufficient or non-existent in most programs (MUCIA, 2006). Furthermore, linkages of the university and agricultural technical school education programs, and business and industry appear to be very limited because most students have not had the benefit of active practical hands-on training (e.g., internships), experience-based education, use of computers and career development opportunities. Moreover, formal AEPs had minimal input from business and industry through the use of external stakeholder advisory councils or practical training on commercial farms. In summary, the skill capabilities of students at all levels of formal agricultural education have not met the needs of agribusiness and industry in most Arab countries.

For example, the majority of agricultural graduates in Egypt, Morocco, Saudi Arabia and other Arab countries over the last two decades are either unemployed or not working in agriculture. At the same time, most commercial farms, agribusiness firms and food processing companies have reported that agricultural graduates lack the skills required to start a career in food security or global food supply chains (MUCIA, 2008; Moshashai & Bazoobandi, 2020; The World Bank Group, 2020). This lack of employable skills is a direct result of the ineffectiveness of these formal AEPs and an indicator that these programs are not responsive to the needs of the job market, and the local and global economy. Furthermore, these AEPs give little attention to problem-solving, critical thinking, communication skills, and computer skills (Campbell et al, 2008; Moshashai & Bazoobandi, 2020). In addition, students in these programs have very few resources for "hands-on" practical training in school farms and almost no opportunities exist for practical training on commercial farms or agribusiness firms.

The development and improvement of formal AEPs in the Arab countries must continue if these programs are to meet the needs of students and their communities in global economy (Krueger & Mundt, 1991). Strengthening the institutional capacity of the universities and secondary technical schools in the Arab countries to provide quality and effective AEPs will make their graduates more employable and active participants in serving their communities.

THE ROLE OF AGRICULTURAL EDUCATION PROGRAM LEADERS (AEPLs)

The role of the AEPLs is essential to the success of the AEPs at the secondary/ technical school level as well as at the university level. Program leaders at the secondary/ technical school level include minsters of education and her/his assistants, directors, managers, headmasters, supervisors. At the university level, program leaders include university presidents, deans, department heads, directors and managers. The AEPLs are responsible for setting up policies, planning, directing, managing, supervising, coordinating, and developing all human and physical resources, including the teaching-learning activities, allocating educational resources, and supervising school organization and accountability that holds teachers/faculty members and administrators responsible for the success of all students. A successful leader must have deep knowledge of program management, and well-articulated vision and goals that fully integrate education into the activities of the education institute (Amadi, 2008). The leader should also be able to bring teachers/faculty members, and administrators together to develop and implement various educational and logistic activities, especially teaching-learning activities. In addition, the program leader is responsible for selecting and supervising staff to manage certain components and activities, overseeing the program's budget, and establishing indicators for evaluating progress to meet program goals.

Successful execution of the responsibilities and duties of the AEPLs will result in significant and positive improvement in the quality and effectiveness of the AEPs, and in turn will boost students' academic achievement and skill acquisitions and increase the overall employability skills of agricultural graduates to allow them to actively participate in serving their communities, and assist in achieving acceptable levels of food security.

EFFECTIVE AGRICULTURAL EDUCATION PROGRAMS

The present situation of the rural communities and rural development in most Arab countries is calling for a new and innovative approach to the AEPs based on market and community needs, stakeholder input, student-centered and experience-based education, use of new technologies, updated technical information, as well as training and empowering agricultural teachers, university faculty members and program leaders. A successful and effective AEP includes four overlapping and interactive components (see Figure 3). These components are:

1. Positive and engaging learning environment,
2. Effective supervised student experiences,
3. Inspiring and supportive career development activities and placement services, and
4. Enabling partnerships with agribusiness and community leaders.

Effective AEPLs should be able to manage the process of developing and improving the four components of AEPs in the Arab countries by improving learning environment and engaging students in their learning, developing students' technical skills and competencies, enhancing students' personal, career, entrepreneurial, and leadership skills and providing placement services, and partnering with agribusiness and community leaders in planning and implementing AEPs and activities.

Figure 3. Components of effective agricultural education program

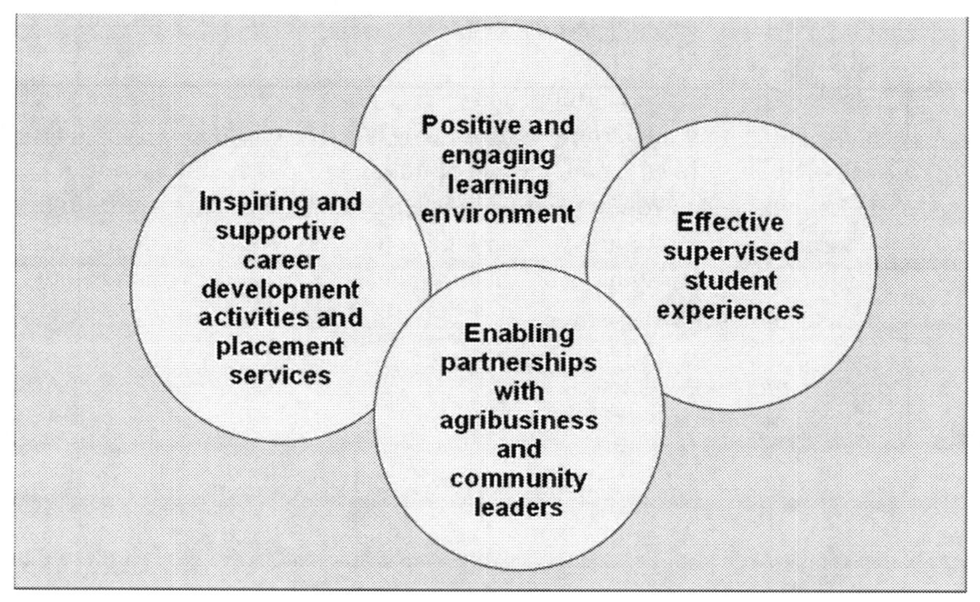

POSITIVE AND ENCOURAGING LEARNING ENVIRONMENT COMPONENT

Improving school learning environment and engaging students in their learning is a required condition to support students' development and growth. Irby refers to learning environment as all organizational structures and cultures, social interactions, and physical and virtual spaces that surround and shape students' experiences, perceptions, expectations, and learning (Irby, David M., 2018). At the present time, the learning environment of most AEPs in the Arab countries is less than optimal. Classrooms are overcrowded, interaction between teachers and students is very limited, teaching methods are teacher-centered and ineffective, course contents have become outdated, practical training is inadequate, educational resources are insufficient, and community and private sector inputs are minimum.

Improving the learning environment and engaging students in AEPS includes four elements (see Figure 4):

1. Identifying areas of improvement and development by assessing academic programs and conducting student skill-gap analysis (SGA),
2. Transforming the current curricula into market-responsive education programs that are more inclusive of "competency-based instruction",
3. Improving teachers' skills to be able to use active-learning strategies and engage students in their learning, and
4. Enhancing experiential learning by integrating practical training with classroom instructions and strengthening the use of training resources and facilities.
    - **Learning Environment:** Learning environment is all organizational structures and cultures, social interactions, and physical and virtual spaces that surround and shape students' experiences, perceptions, expectations, and learning (Irby, 2018).

Figure 4. Elements of improving learning environment

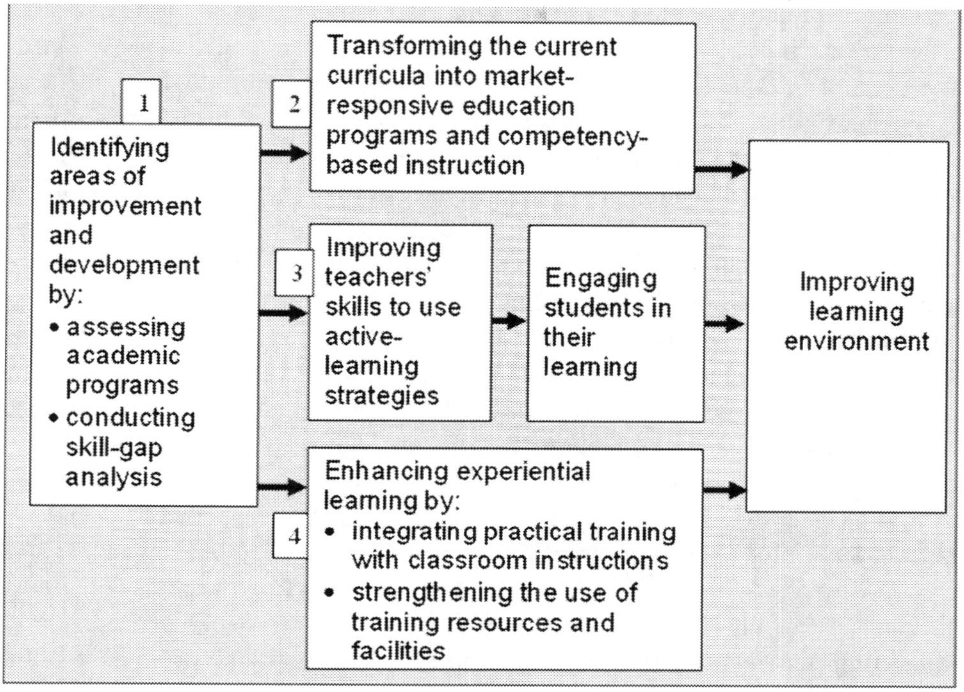

Identifying Areas of Improvement and Development by Assessing Academic Programs and Conducting Student Skill-Gap Analysis

The first element of improving the learning environment in a formal AEP is by assessing and understanding the current situation of human and physical resources, curricula and courses, teaching methods and student engagement, and involvement of agribusiness and community leaders. By conducting a needs assessment and carrying out skills-gap analysis study, the AEPLs will be able to pinpoint specific skill shortfalls in recent graduates and develop a clear picture of high-priority curricular improvements, teaching-learning strategy enhancements and educational resource enrichments.

A capacity-building team of academic program specialists, representing different agricultural and agribusiness areas should be assembled to conduct the needs assessment and skill-gap analysis study. Team members should be faculty members and teachers with excellent teaching skills and significant experience in curriculum development and educational policy. The team members start the needs assessment by analyzing existing curriculum and degree requirements and then assessing teaching-learning methods, educational resources, including library, laboratories and farm resources, and information technology, for technical adequacy and market relevance. The needs assessment should also recognize opportunities to integrate practical, hands-on workplace experience and internships into the education programs. At the end of their work, the capacity-building team members submit a needs assessment report with recommendations identifying specific weaknesses within agricultural technical education programs that are related to curriculum reform, inadequate and ineffective teaching methods currently used by teachers and faculty members (e.g., too much attention on rote learning and knowledge recall); lack of attention to problem-solving and decision-making skills; resource constraints (e.g., inadequate teaching materials and equipment, especially for practical skill training), professional development constraints for teachers and administrators, and lack of involvement with the agribusiness and community leaders, including the absence of supervised internship and experiential learning programs.

The capacity-building team will also carry out a skills-gap analysis study in each agricultural technical program area to identify the market-relevant skills and knowledge that graduates lack at different education levels. Rogers and Taylor (1998) distinguished learning outcomes among four general areas: knowledge, understanding, skills/competencies and attitudes. The focus of the skills-gap analysis is on the skills and competencies to reflect the ability to put knowledge to use. A skill/competency reflects an understanding of principles and theory as one applies knowledge (Vreyens & Shaker, 2005). Therefore, the skills-gap analysis is an outcome assessment tool designed to measure skills or competencies. It can measure the difference between expectations, or reality, and the needed ability or level of performance. Using the approach of Wiggins and McTighe (1998), the skills-gap analysis allows for planning of the curriculum reform in the AEPs (Vreyens & Shaker, 2005).

The survey instrument should be designed to measure the expectations of skills and abilities desired by employers in university and school graduates compared to the perceived competency and use of these same skills on the job by recent graduates. Two groups should be included in the study: agribusiness and private sector employers, and recent graduates, employed less than three years, who have successfully entered the agricultural workforce. The analysis, which should focus on critical areas such as food production, horticulture, livestock, farm-level production for small and medium enterprises, new irrigation techniques, market opportunities, and export development – will be an invaluable tool for identifying skill and knowledge gaps that must be addressed by curricular reform. The capacity-building team members develop the survey instrument to carry out the analysis, with the staff conducting structured interviews.

Example of Skills-Gap Analysis Study

A good example of skills-gap analysis is the study that was conducted in Egypt by the Linkages Project. This study was funded USAID- Egypt and implemented by the Midwest Universities consortium for International Activities (MUCIA) in 2004. This study was designed to prepare a baseline survey of the skill gap between university graduates' preparation to enter the labor force and the needs of prospective employers. The study surveyed 254 employers and 1,000 graduates from four colleges of agriculture in Cairo, Fayoum, Minia, and Assiut Universities. Employers ranked the importance of skills and level of competence of newly hired employees coming from the universities. Graduates ranked the importance of the skill in relationship to their current job as well as their perceived level of preparedness upon graduation.

Cross tabulations of the data revealed the skill gaps. An analysis of the mode scores revealed the most critical skills required on the job compared to the level of preparation or competence. The difference in employer expectations and graduates perceived preparation overlapped in several of the key skill areas identified using the national curriculum. There were skills, especially in the communication, management and computer skill sets that were not being addressed in the curriculum. The list of skills identified by both employers and graduates as essential or very important and reflecting a low level of preparation or competency was very similar. The results of this study were used for curriculum revision for all four colleges of agriculture to ensure market ready graduates enter the current agricultural labor force of Egypt with the skills required by the private sector (Vreyens & Shaker, 2005).

Developing the Current Curricula into Market-Responsive Educational Programs That Are More Inclusive of "Competency-Based Instruction"

The second element of improving the learning environment is the implementation of the results of the needs assessment and the skills-gap analysis study. These results should guide the process of developing and improving the current curricula into new market-responsive educational programs that are more inclusive of "competency-based instruction." A team of agricultural curriculum development and improvement specialists (e.g., teacher educators/senior teachers) should be assembled to transform the current curricula into market-responsive educational programs. The main tasks of the curriculum development and improvement team are:

1. to translate the results of the needs assessment and skills-gap analysis into new competency-based, market-responsive curricula by including the specific

knowledge and skills that are looked-for by employers and needed by students to be successfully employed within the agricultural sector; and

2. to train teachers and faculty members on how to use active teaching-learning strategies and effectively integrate the needed knowledge and skills into the contents of technical courses for all agricultural and natural recourse areas.

Curriculum is how the lesson is planned, designed, and constructed to address scientific standards and to identify learning outcomes and core competencies that students must demonstrate before advancing to the next level; while instruction is the way the curriculum is delivered to students. (Skyepack, 2021)

The process of agricultural curriculum development and improvement is a multi-step process of continuously making instructions better by creating and improving a set of courses taught at an agricultural technical school or university, based on the needs of students as well as the needs of the community and the agriculture industry. The exact curriculum development process varies from institution to institution, the broad framework includes defining program views and values; outlining the program grade-level and course goals; developing and sequencing of grade-level and course objectives; identifying resource materials to assist with implementation; and developing assessment tools to measure student progress (Connecticut's Official State Website, 2022). The curriculum development team members should carry out curriculum development workshops and conduct on the job training for program leaders, teachers and faculty members.

The effective curriculum should be:

An evidenced-based curriculum means it should be based on the best available scientific evidence, rather than norms, personal judgement, or other cultural influences (The Wing Institute at the Morningside Academy, 2023).

The evidenced-based curriculum consists of practices that have been evaluated and proven to be correct through rigorous scientific research. The new curriculum should be carefully developed after a thorough examination of recent scientific research results. Additional criteria for developing and improving new curricula include: defining levels of competency; determining scope and sequencing that lead to increasing levels of difficulty; requiring mastery-based instruction; specifying formative assessment; and identifying channels of providing feedback (The Wing Institute at the Morningside Academy, 2022). Feedback from teachers, faculty members, and administrators is critical in developing, implementing, assessing and modifying each course of the curriculum.

The contents of all general and technical courses must be updated and strengthened to meet the most recent scientific standards and to be consistent with the market and community needs, including the development of problem-solving, decision-making, and communication skills within each course. Course contents and materials include lectures, lecture notes and materials, syllabi, study guides, bibliographies, visual aids, images, diagrams, multimedia presentations, exams, web-ready content, and educational software (University of California Policy, 2012). In addition, supplemental instructional materials for all courses should be developed to enhance the teaching-learning process and facilitate the development of practical and problem-solving skills and to provide students with real-life learning experiences. These supplemental instructional materials include lesson plans, case-studies, worksheets and internet-based instructions.

Effective leaders of AEPs should continuously develop and improve the curricula based on the best available scientific evidence and the needs of students as well as the needs of the community and the agriculture industry. Moreover, effective program leaders should assist in designing the curriculum to facilitate the acquisition of skills and knowledge. They also should provide teachers, students, school administrators and community stakeholders with a measurable plan and structure for delivering quality education programs. In addition, effective leaders should use the curriculum as a road map for teachers and students to follow on the path to academic success (Gibb, 2016).

Improving Teachers' Skills to Use Active-Learning Strategies and Engaging Students in Their Learning

The third element in improving the learning environment and promoting high quality teaching and learning is to prepare teachers to use active learning strategies and engage their students in their learning activities. Teacher professional development is a type of continuing education for educators to improve their teaching skills and enable them to transform teaching methods into student-centered learning activities and engage their students in their learning; and in turn, boost student outcomes.

The leaders of AEPs are responsible for providing continuing and sustained professional development for teachers, faculty members and administrators. The program leader should also recognize that teacher professional development activities are essential to promote high quality teaching and learning, and it is a crucial and necessary element of the work of all agricultural educators throughout their entire careers. The primary goal of agricultural teacher professional development is to help agricultural educators grow professionally and become more effective teachers and/or administrators (National Education Association, 2022).

Agricultural educators in most Arab countries are in a greater need to continuing professional development; they need to be better equipped to teach students who have low academic performance, these students represent a significant number of the students enrolling every year in AEPs in the Arab countries. In addition, professional development activities help new agricultural teachers develop the skills they need to feel confident in teaching and in engaging students in their learning. The important types of agricultural teacher professional development are workshops, in-class observations, and seminars (Power School, 2022). In addition, there are several informal opportunities for agricultural teacher professional development that include individual research and peer observations and consultations.

An important part of effective teaching is the use of active teaching and learning strategies. Education research shows that incorporating active learning strategies into teaching significantly enhances student learning experiences (Freeman et al., 2014). Agricultural educators should be trained to motivate students to play an active role and get engaged in their learning about the subject matters being taught (Purdue University, 2022). A wide range of teaching-learning strategies and tools, which ask students to apply what they are learning, can be utilized to encourage active learning in the classroom, including problem-solving exercises, cooperative student projects, informal group work, simulations, case studies, and role playing. In addition, these tools can be blended with the technological and human resources available outside the classroom (Meyers & Jones,1993).

Active learning strategies ask students to engage in their learning by thinking, discussing, investigating, and creating. In class, students practice skills, solve problems, struggle with complex questions, make decisions, propose solutions, and explain ideas in their own words through writing and discussion. (Center for Teaching Innovation, Cornell University, 2022)

It is commonly known that students' attention waivers every 12-20 minutes (Promethean, 2022), so by utilizing active learning strategies throughout the lesson, teacher should be able to keep students more focused and engaged with the topics being taught. For example, think-pair-share, as a type of informal group work, is an effective active learning strategy to develop thinking skills and engage students in group work.

Think-pair-share works by providing students with a question or situation and giving them time to respond to it independently by writing down their ideas. After a set amount of time, students then share their responses with a partner for five to twelve minutes, discussing similarities and differences. Finally, each pair feeds back their ideas to the class to facilitate a whole group discussion. (Promethean, 2022)

Despite many studies confirming that active learning in classrooms improves student outcomes (Theobald et al., 2020), most Arab colleges of agriculture and

ATSs are very slow in using active learning strategies in classrooms. The teaching in most of ATSs and colleges of agriculture in the Arab countries is teacher-centered instruction, based mainly on rote memorization, and the use of low-level instructional technologies. The effective leaders of AEPs in the Arab countries should provide continuing and sustained professional development for teachers, faculty members and administrators to improve their teaching, stay current with new technical and educational technologies, and to improve interaction with students in class and online environments. In addition, the effective program leaders should motivate and encourage educators to use active learning strategies in their teaching and engage them in their learning.

Enhancing Experiential Learning by Integrating Practical Training with Classroom Instructions and Strengthening the Use of Training Resources and Facilities

The fourth element in improving the learning environment is to integrate classroom instruction with practical training and work experience through experiential learning. Experiential learning is defined as hands-on practical learning and may include: practical learning in an agricultural discipline, experimental activities, internships, and community service learning (Austin & Rust, 2015). Experiential learning is also referred to as:

Learning through action, learning by doing, and learning through experience. (Center for Innovative Teaching and Learning, Northern Illinois University, 2022)

In addition, Kolb refers to learning as "the process whereby knowledge is created through the transformation of experience" (Kolb,1984).

Enhancing experiential learning in agricultural education requires strengthening and redirecting the use of practical training resources and facilities (e.g., school farms and laboratories) at each school.

In experiential learning activities, the teacher should facilitate rather than teach. Learning opportunities should be planned to provide students with hands-on experiences in real problems in their community and their life; and students should work together and learn from each another (University of California, Davis, 2011). Experiential learning in agricultural education benefits students with special learning needs, especially students who need motivation, and students who learn by doing and hands-on examples (Cantor, 1995). Furthermore, experiential learning is considered as an excellent educational tool for agricultural educators to motivate students who have trouble learning within formal classrooms; and help them understand complicated agricultural problems and apply advanced technology. Through experiential learning activities, students can also develop communication skills and self-confidence and gain and strengthen decision-making skills by responding to and solving real problems (Center for Innovative Teaching and Learning, Northern Illinois University, 2022). In addition, experiential learning in agricultural education builds positive attitude towards future career in agriculture, and management of natural resources.

Effective leaders of AEPs should integrate classroom instructions with laboratory and farm experiences for all technical courses to provide students with meaningful, practical learning opportunities. Effective leaders also should guide educators to create exciting experiential learning opportunities for their students across the curriculum. The leaders should also encourage teachers and students to pursue co-learning strategy and apply technology to solve real agricultural problems in their community to increase agricultural productivity. In addition, effective program leaders should give more attention to experiential practical training on new irrigation systems, production and processing of major food crops and high-value export vegetables and fruits, and healthy organic products.

In summary, the role of the Arab AEPLs is central to the success of improving learning environment and engaging students in their learning activities. The program leaders need to conduct regular needs assessments and skill-gap analyses to start the development and continue the improvement process of the AEPs. They should also continuously develop and improve the current curricula based on the best available scientific evidence and the needs of students as well as the needs of the community and the agriculture industry. The program leaders are responsible for developing and improving the current curricula into market-responsive educational programs that are more inclusive of competency-based instruction and based on the needs of students and the community. In addition, the successful program leaders in the Arab countries should provide continuing professional development for educators to use active learning strategies and engaging students in their learning activities; and integrate classroom instruction with practical training and work experience through experiential learning.

EFFECTIVE SUPERVISED STUDENT EXPERIENCES COMPONENT

The second component of successful AEPs is the development of students' technical skills and competencies through effective supervised experiences. This can be done through: 1.) school-farm hands-on practical training, and 2.) supervised internship programs in partnership with the agribusiness firms, commercial farms, and community organizations; and can be organized at school-level or national- level.

School Farm Practical Training

The school farm is an important educational tool to provide students with hands-on experience of agricultural sciences and technology; it also provides the opportunity for skill acquisition and entrepreneurship development. Practical training lessons in the school farm should go beyond basic production skills and include business development, financial management and marketing skills. Practical training in the school farm can include food security and export skills such as soil and fertilizers, horticulture production, new irrigation techniques, animal breeding and feeding, operating simple machines such as irrigation pumps and small tractors, farm management, business plans and marketing. Effective practical training gives students the opportunity to perform and master skills needed to start a small farm or work on commercial farms, without the fear of failing.

Supervised Student Experiences

Supervised student experiences are carried out to develop students' skills and competencies and can be done through:

1. School farm training
2. Supervised internship training organized at:
    a. School-level
    b. National-level

Today, some of the agricultural students in the Arab countries are less likely to have grown up on a farm, which means there is an increased need for hands-on training opportunities during their formal education. AEPs, especially in the Arab cities and urban centers, should offer a range of on-farm training programs that provide learning opportunities for students, who are new to farming. These on-farm training programs include organic farming, fruits and vegetables production, and small animal production, to get hands-on experience in the field. For example, in Egypt most agricultural technical schools have farms ranging between 10-20 acres, these farms are mainly used to train students only on the production of field crops and livestock, with very limited resources for training on new irrigation systems, organic farming practices, and growing vegetables and fruits. A study of AEPs in South Nigeria revealed that quality of school farm practical training contributes directly to the quality of AEPs and to the employability of graduates (Amadi & Solomon, 2020). Program leaders in the Arab countries should provide adequate resources and facilities for successful implementation of practical training lessons in the school farm; these practical training lessons should include appropriate topics for growing healthy food, increasing productivity for small-scale farms, using new irrigation systems, and sustainably managing natural resources, especially water.

Supervised Student Internship Program

An effective supervised student internship program contributes significantly to the success of AEPs, and it is a direct and effective learning strategy to develop hands-on, practical skills and knowledge that students need to be successful employees at various food supply chain functions and services, or to become independent, successful farmers and producers. The supervised internship program includes placement in employment or in a non-paid internship in field crops, horticulture and livestock supply chains, food processing or agribusiness jobs, as well as entrepreneurship and personal ownership of an agriculturally-related enterprise. In addition, internship training can provide placement opportunities to students and introduce them to businesses in their community and in other regions of the country.

With the collaboration of commercial farms and agribusiness firms as well as community partners, students are offered training for 30-90 days to apply the concepts learned from classrooms, laboratories and school farms to actual situations. During this training, students learn about their strengths and interests, problem-solving skills, team work and organization skills, and abilities to deal with clients, as well as how to work under pressure. They also gain greater understanding of science-based agriculture and develop a positive attitude towards working in agriculture. Moreover, during their training students benefit from direct supervision of a manager or supervisor of the farm or food processing factory as well as their school teacher during their work experience in various functions of the food supply chains. The supervised internship program can be organized into two levels:

School-Level Supervised Internship Program

Establishing supervised internship programs at an agricultural education institute provides essential hands-on real work experiences for participating students. School-level supervised internships involve work experience on farms, agribusinesses, laboratories, food processing facilities, or related agricultural agencies in the geographic area of the school. This program is organized by each agricultural education institute in cooperation with commercial farms and agricultural businesses in the surrounding neighborhood. The school-level internship program is important to some students who cannot travel away from their homes. This type of internships allows students to develop new skills and stay in their homes without the need to travel away from their geographic area.

The program leader of each school plays an essential role in enhancing the effectiveness of the school-level supervised internship learning by:

- Appointing an internship teacher coordinator to oversee and coordinate program activities and contact commercial, private sector and government farms and agribusiness firms to arrange for student internship training.
- Establishing objectives and guidelines required to implement a successful internship program.
- Advising students about the requirements for participating in supervised internships,
- Training fellow teachers on how to monitor and evaluate student performance during the supervised internships
- Coordinating for each intern to be monitored by a single teacher on a one-to-one basis. This teacher is responsible for overseeing the internship experience and for ensuring that both employer and student follow the guidelines and objectives of the internship program.
- Conducting start-up training (seminars and workshops) for teachers, students and private sector and community partners participating in the internship programs. and
- Establishing collaborative arrangements with agribusiness firms, local farms, and NGOs located within the school geographic region to provide internship training opportunities.

National-level Internship Program

The national-level supervised internship program provides internship training opportunities to students on tasks and functions of field crop, horticultural and livestock supply-chains, that are not available in their geographic region. This program is managed by the leaders of the AEP at the national level, not the school level. These national programs offer students 30-90 day internship training opportunities, whereas students leave their geographic region and travel to a large-scale commercial, export farm, food processing facility or agribusiness firm to be trained on skills related to various functions and services of food supply chains. These programs help students get acquainted with the real work environment as well as the standards required for working in export and large-scale commercial farms and processing facilities. The commercial farm and agribusiness firm may also provide accommodations and meals. By offering these internship training opportunities, the commercial farms and agribusiness firms gain firsthand knowledge about the abilities of potential and future employees, and they can build a pipeline of skilled, young workers.

A good example of a successful National Supervised Student Internship Program was the program that was organized by the Value-Chain Training Project (The Value Project) to support Egypt's Ministry of Education in improving practical training in 5 agricultural technical schools in Qena, Upper Egypt. The Value Chain Project was funded by USAID- Egypt and implemented by the Midwest Universities consortium for International Activities (MUCIA) during 2007-2012. The Value Supervised Internship Program (VSIP) was a new initiative, applied for the first time as an educational strategy to complement the AEPs at Upper Egypt agricultural technical schools (ATSs).

The VSIP was designed to deal with two challenges. First, students in ATSs in Egypt are required to complete a 20-day summer training assignment on school farms (when very limited farming activities are taking place); most students perceived this assignment as forced labor rather than a learning opportunity to enhance their skills and employment options. Second, these ATSs had inadequate practical training resources, therefore school farm practical training activities were very limited and ineffective during the school year. The VSIP was also designed to provide students with practical, on-the-job training to develop skills for jobs in the global horticultural and livestock supply chains (e.g., input supply, production, post-harvest, handling, transportation, processing and marketing and exporting) that helped build a highly qualified labor pool. The VSIP provided training opportunities to students on tasks and functions of horticultural and livestock supply chains in commercial farms in the Delta area near Alexandria. These hands-on training opportunities did not exist in Upper Egypt at that time. Training opportunities included:

- Installing and maintaining drip and spray irrigation systems
- Applying post-harvest techniques
- Producing horticultural fruits and vegetables for export markets
- Grading, packing and storage of horticultural products
- Operating pre-cooling tractors
- Working in a clod store

Each student was part of a group of 20 students and a teacher, who was the leader and supervisor of the group. Each student group received practical, hands-on training and learning technical skills on the farm for 30-90 work days under the supervision of the farm manager and the supervisor teacher. Each student received a wage, and accommodation and meals that were provided by the commercial farm or agribusiness firm hosting the student group. By participating in the VSIP, students gained more than technical skills; they were also introduced to real life work conditions, work ethics and team work, and developed positive work experience during their internship training. Students learned about their strengths, interests, gained problem-solving skills, and abilities to deal with clients. They also gained greater understanding of science-based agriculture and developed a positive attitude about working in agriculture.

One of the students from Dan Dara ATS in Qena, who completed his VSIP in CELF Export Farm 60 kilometers south of Alexandria, described the benefits of his training by saying that he and his fellow students could now grow, harvest and pack colored peppers in greenhouses. He and his group also learned technical skills related to producing, harvesting and packaging horticultural products, maintaining new irrigation systems, driving tractors, and cultivating grapevines. The VSIP was a useful and effective mechanism to train and connect qualified students to private employers. Commercial employers were in dire need of skilled employees and ATS students require adequate training if they are to be employed after graduation. The VSIP was filling both requirements by bridging the gap between ATSs and the world of work. As the manager of CELF Export Farm noted, by providing these internships, the farm gained firsthand knowledge about the abilities of potential future employees, and that the internship program was an effective way to build a pipeline of skilled, young workers. The program leader of Dan Dara ATS stated that organizing supervised internship training programs for students in fruit and vegetable farms in other geographic regions in Egypt, during the summer provided a useful and viable alternative for the absence of farming activities during the school summer training assignment and for the lack of training resources at school farms (MUCIA, 2012).

INSPIRING AND SUPPORTIVE CAREER DEVELOPMENT ACTIVITIES AND PLACEMENT SERVICES COMPONENT

The third component of a successful AEP is the career development activities and placement services (CDAPSs). These activities and services enhance students' personal, career, entrepreneurial and leadership skills to prepare themselves for successful career in agriculture. This CDAPSs component represents the connection between the classroom instructions and supervised experience programs, offering students real-life learning opportunities to apply what they learn in the classroom and school farm to their out of school activities. It is considered an essential educational tool for the application of technical skills and lessons in career, occupation, business, entrepreneurship and leadership. At each education institute, the CDAPSs are managed and supervised by teachers who assist students in conducting career development activities. These teachers are selected and trained to become advisors to students and their respective groups and clubs. The CDAPSs also provide placement service to assist qualified students in obtaining employment in agribusiness firms and commercial farms. Successful CDAPSs should provide two types of activities, the career development activities and placement services.

Career Development Activities

The career development activities are planned and carried out to promote student career involvement, personal and professional growth, and entrepreneurial and leadership skills through agriculturally-related experience; and to complement and strengthen the classroom instructions. The career development practical learning opportunities promote personal and professional growth, career success and entrepreneurial and leadership development. Moreover, participation of students in supervised individual and group career development activities and incentive award competitions enables them to develop positive attitude towards successful career in agriculture.
The objectives of agricultural career development activities include:

- increasing awareness of the functions, requirements and job opportunities of the food security and global food supply chains,
- strengthening the confidence of agriculture students in themselves and their work,
- promoting informed choices and the establishment of an agricultural career,
- developing interpersonal skills required in teamwork, communications, human relations and social interaction,
- developing the ability to perform effectively in a competitive global job market,
- encouraging achievement in supervised agricultural experience programs,
- promoting cooperation and cooperative attitudes among all people,
- developing competent and assertive agricultural leadership,
- encouraging wise management of economic, environmental and human resources of the community, and
- promoting healthy lifestyles.

In planning career development activities, advisors integrate these activities with school curriculum and develop them based on the needs of the students and the agricultural activities in the school region. The career development advisors organize practical learning opportunities, such as training courses, workshops and incentive award competitions to support student personal growth, encourage them to develop technical and financial skills for success in their future careers, and help them demonstrate leadership skills.

Another type of career development activities is the supervised ownership/ entrepreneurship activity. In organizing this activity, the career development advisor prepares students for planning and operating agriculturally related enterprises or agribusinesses (projects) under the supervision of the advisor. The advisor provides assistance to students in: 1) identifying market opportunities (products or services), 2) organizing resources, and developing business plans and 3) creating a business organization to implement the opportunity. These enterprises can be individual or group projects. Examples include producing and marketing vegetables and fruits; nursery plants or livestock products; providing services such as tractor and machine repair; processing agricultural products; and the assembly of agricultural equipment.

In addition, the career development advisors should offer activities that give students the opportunity to practice new social skills such as working in teams and groups, communicating effectively and clearly, and assuming leadership roles. Furthermore, career development advisors should encourage students to participate in these activities to develop positive attitude towards agriculture and the environment. Possession of such skills and attitudes make students more employable and more competitive in the job market.

Examples of Career Development Training Activities

- Communication and presentation skills.
- Interview skills and CV writing.
- Food supply chain charts.
- Setting goals and career choices.
- Productivity and job performance.
- Conflict resolution and stress management.
- Integrity and work ethics.
- Teamwork and Interpersonal skills
- Personal finance and organization
- Time management
- Customer services
- Leadership development
- Business management, finance and marketing
- Information technology management
- Social media management
- Raising awareness about healthy life and the harms of smoking.

Placement Services

The second type of CDAPS activities is placement services. These services are provided to qualified students to obtain employment in agribusiness firms and commercial farms. The placement services include connecting students to employment opportunities by facilitating the matching of qualified agricultural graduates and employment opportunities, and tracking and supporting the employment of agricultural graduates and offer additional training and advice. Effective placement and tracking services increase morale and efficiency of the students and builds a good relation with agribusiness employers (Economic discussion, 2022). Another type of services to help students find employment is to sponsor career and job fairs. The key purpose of a career and job fair is to help both students looking for employment, and employers who are looking for employees. Agribusiness firm and commercial farm representatives at career fairs inform students of job openings and their hiring processes.

Effective leader of AEPs should not only provide career development learning opportunities to promote student personal and professional growth, and entrepreneurial and leadership skills through agriculturally-related experience; but also encourage students to take advantages of these opportunities in supervised individual and group activities and incentive award competitions. In addition, program leader should organize placement services to assist qualified students in obtaining and keeping their employment in agribusiness firms and commercial farms.

Example of Successful Career Fair

A good example of a successful career fair was the Agribusiness Career Fair that was organized by the Value project, April 2012, Cairo, Egypt. Participants in this Agribusiness Career Fair included 37 agribusiness firms, commercial farms and private companies and 2,000+ agricultural technical school graduates from 50 agricultural technical schools in Egypt; these students were trained by the Value project in the national supervised internship program in large-scale export farms and food processing facilities. Many of the participating students received employment offers from the agribusiness firms, commercial farms and private companies during the fair (MUCIA, 2012).

ENABLING PARTNERSHIPS WITH AGRIBUSINESS AND COMMUNITY LEADERS' COMPONENT

The fourth component of a successful AEP is an enabling and effective partnership with the agribusiness and community leaders in planning and implementing of educational activities. The United Nation's Sustainable Development Goals for 2030 have clearly stated that the partnership between the public and private institutions is a necessity to achieve the goal of providing quality education in the developing countries (United Nations, Sustainable Development, 2021). To achieve this goal of providing quality AEPs in Arab countries, the agribusiness community, private sector companies, and local non-governmental organizations (NGOs) have to work with the public education institutions in modernizing AEPs and supporting agricultural development.

Improving the quality of AEPs and producing qualified skilled labor will considerably benefit the agricultural sector in the Arab countries, including the agribusiness community, private companies and NGOs. Therefore, the leaders of the agribusiness community and private companies are looking for opportunities to become active partners in improving the quality of AEPs; and they are ready to engage in modernizing AEPs in the Arab countries. The agribusiness community and private companies possess a wide-range of expertise and resources, and they are willing to share their expertise and resources with the public agricultural education institutes. The agribusiness communities and private sector companies in the Arab countries can support the modernization of AEPs in three major aspects:

Assist in the Creation of External Advisory Councils at Agricultural Education Institutes, and Participate in Their Activities

These external advisory councils include volunteer members from agribusiness and private sector companies, and NGOs in the locations of the education institutes. The main function of these councils is to facilitate the communication and cooperation between the agribusiness community, private companies, local NGOs and the education institutes. Agribusiness and local leaders in these councils bring perspectives from the agricultural industry, private companies, research institutes, and global supply chains. In addition, agribusiness leaders will provide strategic guidance, advice, and direction to the education institutes and assist in developing market-responsive curricula, recommending new technical and business courses, and improving technical and business training programs.

The External Advisory Council (EAC) at Faculty of Agriculture, Cairo University (2005-2008)

A Success Story

The Faculty of Agriculture at Cairo University established its first External Advisory Council (EAC) in 2005, with the participation of faculty leaders and private sector leaders in the greater Cairo region. It was the first time such organizations have been formed in Egypt. This activity was supported by the AERI Linkages Project, funded by USAID/Egypt and implemented by MUCIA. The EAC, with a majority of private sector members, 8 out of 15 members, met twice every year from 2005 to 2008 to review and discuss curriculum changes, internship programs, and other improvements to be introduced to improve the teaching-learning process.

Private sector leaders started for the first time to provide input and feedback on how various academic programs could be made more responsive to employment needs of the agricultural sector, especially agribusiness firms that are engaged in input supply, production, processing, and marketing of agricultural products. In addition, private sector members of the EAC identified resources for internship training and provided internship training opportunities to students. They also shared resources with the faculty to provide career development programs. The EAC has been proven to be a useful mechanism for faculties of agriculture in providing needed feedback, supporting and identifying resources for improving academic programs, internship training and career development.

Source: MUCIA (2008).

Provide Supervised Learning Experiences and Internship Programs to Agricultural Students

The agribusiness community and private companies can also provide students with training opportunities to practice real-world skills. Students benefit from this direct work experience related to production, processing and marketing of major crops (e.g., input supply, production, post-harvest, handling, processing, marketing and exporting). Students can remain on the private sector training venue/facility for 30-90 work days, receiving practical, hands-on training and learning business skills under the supervision of a manager or supervisor. During this training, students learn about their strengths and interests, problem-solving skills, team work and organization skills, and abilities to deal with clients, as well as how to work under pressure. They also gain greater understanding of science-based agriculture and develop a positive attitude about working in agriculture. The commercial farm and agribusiness firm may also provide accommodations and meals. By providing these internship training opportunities, the commercial farms and agribusiness firms gain firsthand knowledge about the abilities of potential and future employees, and they can build a pipeline of skilled, young workers.

Assist in Career Development and Placement Service Programs to Develop Students' Personal and Professional Skills and Connect Qualified Graduates with Employment Opportunities.

The agribusiness, private sector and local leaders can also assist in planning and implementing Career Development Activities and Placement Service Programs (CDAPSPs) in the education institutes. These programs offer opportunities to enhance personal and professional skills, improve career opportunities and deepen knowledge of science-based agriculture. These CDPSPs support the close collaboration among agribusiness and community leaders, educators and program leaders to enhance student career development by providing the following services:

- Sponsor events and activities that help career development of agricultural students and graduates
- Facilitate the matching of qualified agricultural graduates and employment opportunities
- Expand training, internship, employment, and ownership opportunities for female students
- Track and support the employment of agricultural graduates

An effective program leader should encourage agribusiness and community leaders to participate in planning and implementing agricultural education activities. Most of the agribusiness, private sector, community leaders in the Arab countries will welcome the opportunity to be active partners and they are ready to engage and share resources with the public universities and ministries of education in modernizing the AEPs. However, successful partnership depends on the existence of a sound enabling environment and the right mix of public and private sector skills (FAO, 2018).

INDICATORS OF THE SUCCESS OF AEPs AND ITS LEADERS

The success of AEPs and their leaders is measured based on short-term and long-term indicators of all the schools or the education institutes included and/or participating in the program. The short-term indicators include rates of attendance and suspension, and test scores, whereas the long-term indicators include graduation rates, market readiness and rates of employment. In addition, the success of each school or education institute in the AEP is measured by indicators such as academic achievement and progress, language proficiency, rates of suspension, rates of drop-outs, and rates of graduation. The success of the students is measured by their academic achievement and skill acquisitions. The rate of success of the students is an indicator of the success of the school or agricultural education institute and the AEP (Ed100, 2022).

CONCLUSION

The Arab countries should have the vision to build agricultural human capital through innovative and effective AEPs that recognize the importance of the potential to the educational process and the economic foundation undergirding a progressive agriculture industry in their countries. The need for skilled agricultural workforces continues to grow in these Arab countries to accelerate economic growth, rural development and sustainably manage natural resources, especially water. Providing quality and innovative agricultural education and training programs will help prepare rural students and youth for employment in different functions and services of the food and fiber supply chains and in sustainably managing natural resources. A successful and effective model for providing a formal AEP includes four overlapping, interactive components: 1.) positive and encouraging learning environments, 2.) strong and effective practical, hands-on technical training, 3.) inspiring and supportive career development and placement services, and 4.) enabling partnerships with agribusiness and community leaders.

Efficient AEPLs are essential to the success of establishing and managing formal AEPs in the Arab countries. Program leaders should be able to manage the process of developing and improving the four components of AEPs by improving learning environment and engaging students in their learning, developing students' technical skills and competencies, enhancing students' personal, career, entrepreneurial, and leadership skills, and partnering with agribusiness and community leaders to share plans, feedback, expertise and educational resources. Successful program leaders integrate classroom instructions with laboratory and farm experiences for all technical courses and guide educators to create exciting experiential learning environment for their students across the curriculum to provide students with meaningful, practical and useful learning opportunities. In addition, program leaders should encourage educators and students to pursue co-learning strategies and apply technology to solve real agricultural and environmental problems in their community.

REFERENCES

Abu-Ismail, K. (2018). *Extreme poverty in Arab states: a growing cause for concern*. The Forum ERF Policy Portal. https://theforum.erf.org.eg/2018/10/16/extreme-poverty-arab-states-growing-cause-concern/

Adam, A. H. M. (2019). The Future Perspectives of Agricultural Graduates and Sustainable Agriculture in Sudan. *Journal of Agronomy Research*, *1*(4), 36–43. doi:10.14302/issn.2639-3166.jar-19-2732

Al-Zaidi, A. (1982). *Adequacies of Curriculum and Training in Agriculture Provided at Three Saudi Institutions as Assessed by Administrators, Instructors, Senior Students, and Regional Directors*. [Unpublished doctoral dissertation. Oklahoma State University, Stillwater, Oklahoma, United States].

Alaoui, H., & Springborg, R. (2021). *Education in the Arab World: A Legacy of Coming Up Short*. The Wilson Center. https://www.wilsoncenter.org/article/education-arab-world-legacy-coming-short

Amadi, E. C. (2008). *Introduction to Educational Administration: A Module*. Harey Publications.

Amadi, N. S., & Solomon, U. E. (2020). *Issue 3 Ser* (Vol. 10). Assessment of Quality Instruction Indicators in Vocational Agricultural Education in South –South Universities Nigeria. Journal of Research & Method in Education (IOSR-JRME). May - June.

Austin, M. J., & Rust, D. Z. (2015). Developing an Experiential Learning Program: Milestones and Challenges. *International Journal on Teaching and Learning in Higher Education*, *27*(1), 143–153.

Ben Haman, O. (2020). The Moroccan education system, dilemma of language and think-tanks: the challenges of social development for the North African country. *The Journal of North African Studies*, *26*(4), 709-732.

Campbell, J., Martin, R., & Diab, A. (2008). *Impact Assessment Report: The AERI Linkage Project in Retrospect and Prospect*. MUCIA.

Cantor, J. A. (1995). ASHEERIC Higher Education Report: (Vol. 7): *Experiential Learning in Higher Education. Washington, D.C.*

Center for Innovative Teaching and Learning, Northern Illinois University. (2022). *Experiential Learning*. https://www.niu.edu/citl/resources/guides/instructional-guide/experiential-learning.shtml

Center for Teaching Innovation. Cornell University. (2022). *Active Learning*. https://teaching.cornell.edu/teaching-resources/active-collaborative-learning/active-learning

Connecticut's Official State Website. (2022). *A Guide to Curriculum Development: Purposes, Practices,* Procedures. https://portal.ct.gov/-/media/SDE/Health-Education/curguide_generic.pdf

Daghir, N. (2018). Higher Agricultural Education in the Arab World: Past, Present, and Future. In A. Badran, A. Baydoun, & R. Hillman (Eds.), *Universities in Arab Countries: An Urgent Need for Change* (pp. 209–224). American University of Beirut. doi:10.1007/978-3-319-73111-7_12

Ed100. (2022). *Measures of Success.* Ed100. https://ed100.org/lessons/measures

Economic discussion. (2022). *Placement: Meaning, Definition, Importance, Principles, Benefits, Problems.* Economic Discussion. https://www.economicsdiscussion.net/human-resource-management/placement/placement/ 32361.

Elmenofi, G., El Bilali, H., & Berjan, S. (2014). *Contribution of extension and advisory services to agriculture development In Egypt. Sixth International Scientific Agricultural Symposium Agrosym*, Egypt. doi:10.7251/Agsy15051874e

Food and Agriculture Organization of the United Nations (FAO). (2018). Agricultural Development Economics [Rome, Italy.]. *Policy Brief*, 7.

Freeman, S., Eddy, S. L., McDonough, M., Smith, M. K., Okoroafor, N., Jordt, H., & Wenderoth, M. P. (2014). Active learning increases student performance in science, engineering, and mathematics. [NATL ACAD SCIENCES. United States.]. *Proceedings of the National Academy of Sciences of the United States of America, 111*(23), 8410–8415. doi:10.1073/pnas.1319030111 PMID:24821756

Gibb, N. (2016). knowledge-rich curriculum to the Social Market Foundation [Paper presentation]. The Association of School and College Leaders (ASCL) Event 'Taking ownership of your curriculum: a national summit' in 2016. London, UK.

Irby, D. M. (2018). *Improving Environments for Learning in the Health Professions. Proceedings of a conference on Improving Environments for Learning in the Health Professions*. Atlanta, Georgia. United States.

Kolb, D. A. (1984). *Experiential learning: Experience as the source of learning and development* (Vol. 1). Prentice-Hall. United States.

Krueger, D., & Mundt, J. (1991). Change: Agricultural education in the 21st century. *The Agricultural Education Magazine.*, *64*(1), 7–9.

Meyers, C., & Jones, T. B. (1993). *Promoting Active Learning. Strategies for the College Classroom*. Jossey-Bass Inc., Publishers.

Moshashai, D., & Bazoobandi, S. (2020). The Complexities of Education Reform in Saudi Arabia. *The Gulf Monitor*. https://castlereagh.net/the-complexities-of-education-reform
-in-saudi-arabia/

MUCIA. (2006). *Request for Second Amendment to the AERI Linkage Project*. Unpublished Manuscript.

MUCIA. (2008). *Final Technical Report: AERI Institutional Linkage Project*. Unpublished Manuscript.

MUCIA. (2012). *The Second Quarter Report, Year 4:* The Value Chain Project. Unpublished Manuscript.

National Education Association. (2022). *The Teacher Professional Growth*. NEA. https://www.nea.org/professional-excellence/professional-learning/teachers

Power School. (2022). *The Three Most Common Types of Teacher Professional Development and How to Make Them Better*. Power School. https://www.powerschool.com/blog/the-three-most-common-types
-of-teacher-professional-development-and-how-to-make-them-be
tter

Promethean. (2022). *12 active learning strategies in the classroom.* Promethean World. https://www.prometheanworld.com/gb/resource-centre/blogs/12-active-learning-strategies-in-the-classroom/)

Purdue University. (2022). *Active Learning Strategies*. Purdue University. Error! Hyperlink reference not valid.

Rogers, A., & Taylor, P. (1998). *Participatory Curriculum Development in Agricultural Education: A training guide*. Food and Agriculture Organization of the United Nations.

Skyepack. (2021). *Curriculum Development: Complete Overview & 6 Steps*. Skyepack. Error! Hyperlink reference not valid. curriculum-development.

Swanson, B. E., Cano, J., Samy, M. M., Hynes, J. W., & Swan, B. (2007). *Introducing active Teaching–Learning methods and materials into Egyptian agricultural technical secondary*.

Swanson, B. E., Cano, J., Samy, M. M., Hynes, J. W., & Swan, B. (2007). Introducing active teaching-learning methods and materials into Egyptian agricultural technical secondary schools. *Proceedings of the Association for International Agricultural Extension and Education Research Conference*. Polson, MT.

The Wing Institute. (2203). *Evidence-Based Education*. The Wing institute at the Morningside Academy. https://www.winginstitute.org/evidence-based-education

The World Bank. (2008). *Sector Brief: Agriculture & Rural Development in MENA*. The World Bank. http://go.worldbank.org/WMLZXRV380

The World Bank. (2023). *Data*. World Bank. https://data.worldbank.org/indicator/SP

The World Bank Group. (2015). *Egypt PROMOTING POVERTY REDUCTION AND SHARED PROSPERITY: A Systematic Country Diagnostic*. (P151429). World Bank. https:// documents1.worldbank.org/curated/en/853671468190130 279/pdf/99722-CAS-P151429-SecM2015-0287-IFC-SecM2015-0142-MI GA-SecM2015-0093-Box393212B-OUO-9.pdf

The World Bank Group. (2020). *Morocco: A case for building a stronger education system in the post Covid-19 era*. World Bank. https://www.worldbank.org/en/news/ feature/ 2020/10/27/a-case-for-building-a-stronger-education-system-in-the-post-covid-19-era

Theobald, E. J., Hill, M. J., Tran, E., Agrawal, S., Arroyo, E. N., Behling, S., Chambwe, N., Cintrón, D. L., Cooper, J. D., Dunster, G., Grummer, J. A., Hennessey, K., Hsiao, J., Iranon, N., Jones, L. II, Jordt, H., Keller, M., Lacey, M. E., Littlefield, C. E., & Freeman, S. (2020). Active learning narrows achievement gaps for underrepresented students in undergraduate science, technology, engineering, and math. *Proceedings of the National Academy of Sciences of the United States of America*, *117*(12), 6476–6483. doi:10.1073/pnas.1916903117 PMID:32152114

United Nations. (2022). *Sustainable Development*: UN. https://www.un.org/ sustainable development/

United Nations. (2023). *Quality Education*. Department of Economic and Social Affairs. Sustainable Development. https://sdgs.un.org/goals/goal4

University of California. Davis. (2022). *Experiential Learning*. UCal. https://www.experientiallearning.ucdavis.edu/ module1/el1_40-5step-definitions.pdf

University of California Policy. (2012). *Ownership of Course Materials*. Ucal. (https://policy.ucop.edu/ doc/2100004/CourseMaterials)

Vreyens, J., & Shaker, M. H. (2005). *Preparing Market-Ready Graduates: Adapting curriculum to meet the agriculture employment market in Egypt. Proceedings of the 21st Annual Conference of the Association for International Agricultural and Extension Education*, Egypt.

Wiggins, G., & McTighe, J. (1998). *Understanding by Design*. Association for Supervision and Curriculum Development.

World Food Program. (2023). *Global Food Crisis*. WFP. https://www.wfp.org/ emergencies/ global-food-crisis

Chapter 3
Agricultural Education in Egypt:
A Case Study

Sherine Fathy Mansour
https://orcid.org/0000-0002-6878-7255
Desert Research Center, Egypt

Mohamed Fawzii Shahien
Desert Research Center, Egypt

ABSTRACT

Developed countries attach great importance to agricultural education of all kinds and stages. Agricultural education plays its role in creating skilled human forces capable of giving excellence, which supports the agricultural sector and enhances its economic and competitive standing in those countries. However, the philosophy and characteristics of agricultural technical education vary from state to state; as well as the role of the state and the private sector in shaping its policies and defining its objectives. Despite these differences, which are attributable to the nature of each state, practical practices have proven to be successful. However, there are some problems that impede progress and development in agricultural education in particular, and this chapter tries to highlight the features of these problems.

DOI: 10.4018/978-1-6684-4050-6.ch003

INTRODUCTION

The results of economic and social studies emphasized the strong and effective impact of education in increasing people's productivity.

This has a positive impact on income at both the individual and national levels, that is, to increase gross domestic product (GDP). Thus, education's association with the economy is undoubtedly close and certainly contributes to economic growth.

Education also has important social dimensions, expanding the understanding and orientation of emerging generations in a sound and ambitious scientific direction so as to unlock the forces of creativity. In addition to the importance of education in bringing the children of society closer together through the dissolution of many customs and traditions and their son-in-law into a single crucible and the resulting similar behavioral and social patterns that make them closer and more attractive to current socio-economic dynamic variables in addition to the important role that education plays in the refinement of social behavior and freedom from customs and traditions that stand in the way of development.

In view of the contribution of the agricultural sector to increasing national output as one of the leading sectors of economic development, and the employment of large proportions of F employment, agricultural development needs to be accelerated so that maximum productivity can be achieved from the use of available agricultural resources, especially educated agricultural labor, "agricultural graduates," as the cornerstone of agricultural production development. Since agricultural graduates and the agricultural workforce are the main pillars of sustainable agricultural development, it is their responsibility to cope with developments, remove obstacles and focus on the vital role in agricultural development by maximizing the return of agricultural resources.

Education is a product and developer of human skills, and thus a product of human capital as an investment commodity and as an instrument of social policy. Moreover, education can create possibilities for discovering new goods, modern technology, and new tools of social policy. No other type of capitalist formation possesses such characteristics. The paramount importance of education in any labor market is essentially crystallized in its ability to produce an educated and flexible workforce through quality education. As a result, with economic development new technology can be applied in production, increasing the demand for e workers and improving the quality of education. Education at all levels also contributes to economic growth by improving health and income and possibly by contributing to political stability.

Educational institutions are primarily and, among their core tasks, contributing to research. Studies in the different fields of agriculture, providing extension services and technical consultations to the competent authorities, especially that developing societies are in urgent need of university cohesion with society and to make much change in the methods and development of rural living patterns in order to keep pace with the development of the required production rates. It can also do an important and dangerous job in making the extension service a success wherever it is located or where rural communities are in dire need of new assimilation in agriculture, economy, and rural life.

Given that there are some problems impeding progress and development in agricultural education in particular, the main objective of this chapter is to highlight the features of those problems that impede agricultural education in Egypt by reviewing the most important concepts of agricultural education, The development of agricultural education in Egypt and the importance of agricultural education in Egypt ' In addition to presenting the most important problems facing agricultural education, offering some machinery to confront and solve these problems.

Concept and Importance of Agricultural Education

A number of researchers provided several different definitions of technical education.

Technical education refers to the type of formal education that includes educational preparation and the acquisition of degrading oryx and knowledge, carried out by formal educational institutions at the secondary level for the purpose of preparing skilled workers in various industrial, agricultural and commercial disciplines (Ismail,1991)

It is also known as education aimed at the preparation of a category of professionals in the fields of industry, agriculture, administration and services, the development of technical qualifications of scholars, and the duration of study three years after the preparatory stage (Ismail, 1991).

Agricultural education has also been known and elements of the definition have been identified in the following areas:

1. Agricultural education is an integral part of general education.
2. It is a means of preparing for the areas of work and active participation in the world of work.
3. It is a form of continuous education for life and the preparation of a responsible citizen.
4. It is a way to promote and develop the environment. (Ahlam,1983)

After a review of previous tariffs, the specific elements of the concept of agricultural education can be found as follows:

1. A type of technical education that engages in the graduation of skilled artistic labor that benefits the labor market in all disciplines; To provide the various

production sector and services, public and private bodies, and institutions with all the needs of their workers and skilled technical personnel.

2. A type of technical education aimed at gaining the individual a certain amount of culture, technical information and practical skills through applied training that enables him to perform agricultural operations efficiently and effectively in order to prepare the human strength necessary to work in the agricultural sector. Due to the importance of technical education in the preparation of an effective human component capable of implementing economic development plans, without which such plans and programs are stalled and unable to achieve their objectives. (Hassanin & Abdelaziz, 1994)
3. It is an education that helps to provide skilled labor.
4. It is education with a duration of three years, or five years after the prep stage.
5. It is an education that provides the student with professional experience, not just It is an education that prepares the individual for entry into the labor market and is equipped with certain skills that help him to promote society and contribute to its development. (Ministry of Education,1965)

The study of the previous definitions of **agricultural education** shows the number of such definitions depending on where agricultural secondary education is viewed education ", some are seen as a systematic education involving educational preparation, acquisition of skills and knowledge, preparation of the category of professionals, development of art or as a means of promoting and developing the environment, or the graduation of artistic work or the acquisition of an individual ' However, **Agricultural education** is a basis for sustainable development with all its natural and human resources, and its preservation to a maximum extent, without wasting it, taking into account all economic, social and environmental dimensions. knowledge, as it combines theoretical and practical study.

Importance of Agricultural Secondary Education

Education is the basis for inclusive development as a treasure within society and a lifeblood to eliminate unemployment and one of the capital components and assets affecting the economic and social situation in several dimensions, including:

- **Economic dimension:** Through qualified and trained human resources.
- **Social dimension:** Education develops an individual's intellectual and intellectual abilities and earns them behavioral patterns and values.
- **The security dimension:** Care for the education and training of individuals leads to the reduction of the unemployment rate contributes to the security stability of society.

- **Educational dimension:** Technical education provides scientific cadres capable of research, innovation, invention, development and technical progress in various areas of life and continuous improvement in means of living.

Technical education is a strategic element of educational policy because it is the basic component for acquiring the skills and knowledge needed by professionals in all sectors, which largely depends on technical education, because the prosperity of nations depends on modern and advanced technology. This requires a strong workforce with a scientific background, modern skills and continuous training to meet the challenges faced by society. Hence the importance of technical education to improve its quality in order to meet those challenges. The continuation of development and evaluation has become an essential feature of the times and an indispensable need to keep pace with future contemporary changes (Aliyah, 2007).

We find that technical education in most European countries is of great interest because it is an important source of technical employment trained on scientific and practical technological grounds and in this spirit, Attention was paid to technical education of all kinds in order to prepare the category of advanced technicians and technicians in the fields of Agriculture, industry, trade, management, services and the development of their technical skills (Al-Zeid, 1993).

Hence, agricultural education is an urgent need, especially in agricultural governorates, because of the important role played by agricultural secondary education institutions. Agricultural schools have been created in most Arab States with the aim of graduating agricultural technicians with knowledge, scientific experience and all that would develop agriculture; Contributing to students' awareness of the various problems experienced by the agricultural community and ways to address these problems The importance of agricultural education can be limited to:

1. Agricultural education is most important means of preparing the agricultural, trained and skilled workforce for development projects.
2. Agricultural secondary education has become a fundamental pillar in the establishment of agricultural society.
3. Agricultural education is the primary responsibility for preparing the human strength that contributes to the achievement of inclusive development.
4. Agricultural education is necessary to improve productivity levels in order to achieve the ability to compete locally and globally.
5. international conferences and international organizations, such as the United Nations Educational, Scientific and Cultural Organization (UNESCO), underscored its importance to development as a prerequisite for economic

and social development during his role in the preparation of the agricultural workforce.

Agricultural education is based on its effective role as a productive and economic sector in achieving the outcome and in pivotal aspects at the national and international levels. Generally speaking, the role and impact of agricultural education on economic development can be defined by the following variables:)Hanafi, 1996).

- **First:** To provide for the nutritional needs of individuals, agriculture provides the nutritional needs of cereals, vegetables, fruit and animal products such as meat of various kinds or dairy products. These foodstuffs may be consumed either directly or manufactured. The demand for foodstuffs has increased as a result of population growth, especially in developing countries.
- **Second:** Providing hard currencies. Some developing countries rely on the agricultural sector to obtain hard currencies. For example, Egypt exports cotton to countries around the world, Brazil exports coffee to the United States, Somalia exports bananas, and Sri Lanka exports tea, so these countries get hard workers. Developing countries exporting some goods such as rubber and fiber suffer from lagging demand for these goods as a result of the discovery of new alternatives such as petroleum products and the expansion of agricultural crop production in developed countries as a result of the use of technical means in their agricultural production, yet agricultural products.

It will continue to be a source of hard currency, so developing countries must pay attention to increasing the productive efficiency of agricultural crops in terms of quality so that they can compete with similar commodities in foreign markets and export agricultural commodities as manufactured. (Ministry of Education, 1963)

- **Third:** Agriculture provides employment for agricultural productive sectors, using mechanization and modern technology in agriculture. Labor diverts to other sectors such as industrial, commercial, and service sectors.
- **Fourth:** Food industrialization. Growth in food industries depends on growth in agricultural crops, for example, the sugar industry needs large areas grown with sugarcane or beets. cotton industries also need cotton crop and fiber crops. The fat and oil industry also depends on the volume of production of oil crops such as sunflower seed, sesame, soybeans, and other oil crops.
- **Fifth:** The agricultural sector is a market for agricultural goods. Agriculture has an indirect impact on the development of other economic sectors, for example, growth in the agricultural sector entails the use of modern technology such as the use of fertilizer, agricultural machinery, and chemical pesticides,

thereby expanding the fertilizer industry, agro-machinery industry, chemical industry and other industries (Aliyah, 2007).

Agricultural education is the basis of a technological renaissance. Technological progress and its applications in agriculture have undoubtedly made it a broken nation that is transformed into a nation of leaders, such as Japan, which, after emerging from the Second World War, became a country of technological progress, defeated, and shattered. It is the development of agricultural education at that time, Malaysia and its Government are great fans and imitators of the Japanese experience and its officials do not deny this admiration and the importance of agricultural education in generating inclusive development.

Agricultural education is one of the branches of technical education, so the developed countries have taken care of technical education as the education of the current century and any State cannot enter this century unless it possesses all the essentials of the economic revolution. In Sweden, about 74% go to technical education, 57% in France, 61% in Denmark, 62% in Austria, while in the Arab countries the enrolment rate falls to 0.3% in Saudi Arabia, in Kuwait 8%, in Jordan 1.6% and in the Arab Emirates 0.8% (Mohamed, 1996).

Major countries such as the United States, Germany, and Japan have also been keen on developing technical education. and an effective policy approach to economic development in particular, given that technical education is the cornerstone of economic miracle, they have therefore been keen to combine the theoretical and applied aspects of technical education and training in all practical skills using the latest technology in addition to the radical modification and change of technical education programs to meet the challenges of the times development process (Mohamed, 1996).

Aims of Agricultural Secondary Education

Egypt's Agricultural Secondary Education aims to prepare the first professional and technical category in the field of Las Ara who has acquired a degree of knowledge and irresponsibility that qualifies him to carry out his agricultural specialization, Article 30 of the Education Act No. 139 of 1981 stipulates that technical education shall be aimed at the preparation of a professional category in the field of industry, agriculture, trade, administration and services. As stipulated in article 38 on technical education of the five-year system, It aims to develop the first professional category, and to achieve this overall objective of technical education, the patterns of agricultural secondary schools in Egypt have varied. These patterns can be explained below (Ya'qub & al-Tourist, 1988).

Agricultural Secondary Schools' Three-year System

This pattern is the most prevalent in agricultural schools, with 114 schools in the school year 2004/2005 66% per cent of the total number of agricultural schools. The agricultural secondary school consists of three years of two divisions:

Agricultural Division

The purpose of this division is to prepare the category of agricultural technician in one of the specializations available in the school, and study in this division generally in the first and second grades. The student in the third grade teaches one of the following disciplines: animal production, agricultural manufacturing, fish production, land reclamation and agricultural mechanization, dough and baked goods.

Laboratory Secretaries' Division

The Division's objective is to prepare laboratory secretaries for school laboratories at all types and levels of schools, and to study in the Division in general during the three grades (Alani and Saadallah, 1986).

Considering the realities of the three-year system of agricultural schools, it is noted that:

- Agricultural secondary school is the three-year system of the main pattern of agricultural secondary education in Egypt, and other patterns have been introduced from which it has embarked. Some schools have transformed the three-year system into other patterns.
- The Agricultural Division is the main division in schools with a three-year system, with all schools, while the Laboratory Trustees' Division is located in a limited number of schools.
- Multidisciplinary in the Agricultural Division, where there are five agricultural disciplines, one of which is selected by the student in the second grade, making the school's potential spread between five disciplines, thus weakening its ability to provide the necessary possibilities for good training that can be achieved if the school specializes in only one agricultural specialty (Alani and Saadallah, 1986).

Agricultural Secondary Schools' Five-year System

This pattern includes two schools at the Republic level, as well as the Livestock Production Division attached to the Three-Year System School, which can be explained below: (Ali, 1996)

Advanced Technical School for Food Manufacturing

The School was established by Ministerial Decree No. 87 of 20/4/1978, by transforming the Agricultural High School into a three-year system in Mashrad into an experimental five-year technical school, called the Technical Experimental School for Food Manufacturing in Mashrad. The School comprises the following specializations: The sugar industry, oil extraction industry, canned industry, dairy, slurries industry, the decision also stipulated the possibility of creating other disciplines or abolishing old disciplines, in accordance with the development needs and requirements of the environment and the manufacturing plan.

Advanced Technical School for Agricultural Mechanization and Land Reclamation

The school was established in Ismailia in accordance with Ministerial Decree No. 35 of 1980. The school aims to prepare a student for integrated preparation in the field of agricultural mechanization and land rehabilitation. The student studies general subjects, including theoretical and practical, as well as technical subjects.

Division of Preparation of the First Technician in the Field of Livestock Production

The Division was established pursuant to Ministerial Decree No. 80 of 1990. It is currently attached to the Damanhur Agricultural Secondary School. It has a three-year system. The duration of its study is two years. The three-year system is accepted by the student with high collections in the Agricultural Secondary School Diploma.

Professional Agricultural Secondary Schools

This type of school was introduced in the last decade of the twentieth century to accommodate the student with a certificate of completion of basic education (vocational preparatory). These schools are currently attached to agricultural schools with a three-year system. These schools are generally studied in the first and second grades. In the third grades, the student teaches one of the following disciplines:

Horticulture, Veterinary Health, Beekeeping and Farm Protection, Dairy and Food Industries, Hobbies and Baked Goods, Successful Third Grade Graders are awarded a Diploma in Agricultural Secondary Schools three-year professional preparation system. (Ministry of Education,1965)

The continuous development of the objectives of agricultural secondary education has become a necessity in modern times. in order to align these goals with changing labor market needs, this may be consistent with a pragmatic view that does not believe in any definitive, constant or unchanged goals and as long as we live in a changing world and the future ahead of us is uncertain, direct experience is subject to change. We therefore need to restructure our goals to meet the needs of our dynamic environment. The objectives of agricultural secondary education must therefore be linked to the requirements of the renewed labor market, and the primary objective to be pursued should be to prepare students who can run a small food industry enterprise, participate in reclamation processes, to create and land, and be able to understand the real role of agricultural mechanization in improving the process (Elhami, 2000).

Implementation of the Economic Development Plan in general or agricultural development in particular will undoubtedly be achieved only through the improvement of the outputs of medium agricultural education, whose responsibilities are limited in the creation of the category of assistant supervisor or technical assistant, the fourth episode of the employment structure in Egyptian farming, who provide them with appropriate education and practical training in one or more agricultural fields.

In the light of this, the objectives of the Agricultural High School have been set, so that the preparation of its members will give them the following skills:

- Guide the work to graze according to its correct arrival.
- Undertaking specific agricultural project burdens in the following respects: a) free agricultural exploitation, and (b) successful agricultural management.

Covering the needs of the public and private sectors in the areas of agricultural production and services of this level of employment (Ya'qub and al-Tourist, 1988).

- Teaching agricultural subjects in various schools after obtaining an educational qualification and permitting the need for appointment without the educational qualification.
- Agricultural Extension: Graduates can enroll in farming colleges in Egyptian universities on the terms of a decision issued by the Higher Council of Universities (Salim and Hassan, 2005).

The levels required for this phase have been set in various areas on the basis that the graduate is preparing to work in the following areas (Alani and Saadallah, 1986):

- Plant production
- Land reclamation
- Animal production
- Plant prevention and pest resistance
- Food manufacturing
- Slimming and silkworm husbandry

Most of these goals are usually not achieved because the majority of graduates work in fields that are disproportionate to the nature of their studies, as a result of the increased demand for government jobs, the lack of demand for agricultural projects because of their risks, and their preference for life in the city than in the countryside (Abdel, 1997).

A number of agricultural education's objectives in the area of agricultural development are generally accepted by agricultural practitioners, which are summarized as follows.

- **First:** Provide qualified and managerial national agricultural cadres to carry out planning and supervisory tasks on agricultural production programs in a way that can meet the needs of agricultural enterprises from trained manpower in all fields.
- **Second:** Conducting agricultural research for the purpose of finding the same ways to solve the problems of production of grazing and developing grazing and enriching agricultural information.
- **Third:** Transfer of agricultural information and knowledge developed to those who need to apply it as farmers; this is known as the indicative process.
- **Fourth:** Raising the community's awareness of the importance of agriculture for its well-being and stability. Agriculture provides the nutritional needs of cereals, vegetables, fruit and animal products such as meat of various kinds or dairy and their derivatives.
- **Fifth:** Raising the level of agricultural production. Agricultural education plays an active role in enhancing productive efficiency in agricultural fields by rationalizing the use of elements of production from land, labor, and capital. Since agricultural education is an important dimension of economic development because of its effective development cycle, agricultural education thus occupies its important place within the structures of different countries' agricultural systems (Hanafi, 1996).

Some Problems of Agricultural Secondary Education

Agricultural secondary education faces many problems in its structure, functions, students, teachers, and overall education inputs. This is reflected in the outcomes of education and its impact on society, its development and development. However, agricultural education problems can be included under the following categories:

Technical problems include curricula, textbooks, teachers' training, training, specific means, etc. These problems are described below.

Problems Associated with Curricula

Current curricula lack some important study aspects such as environmental studies and social sciences. The study found that continuous changes in the labor market require a similar development in the agricultural secondary education curriculum to meet the changing needs of the agricultural labor sector and eliminate existing disciplines. One of the features of this development may be the introduction of new disciplines according to the actual needs of the labor market, the integration of information technology into the content of the curriculum and the inclusion of adequate guidance to help the teacher to use educational software in teaching, considering that teaching content using information technology has become a necessary requirement to help the student to learn effectively. This is helped by the allocation of a separate course aimed at gaining the student a culture of dealing with computer and information networks. The following is a presentation of some problems associated with the curriculum.

- The training possibilities and currencies available to schools do not rise to the level necessary to train students in reality in the fields of production and services.
- The absence of sports, cultural and music classes in technical schools and the absence of an educational guide in some schools, which has led the student to direct their energy to other familiar activities.
- The curriculum is not linked to the developments and needs of the market in accordance with the foundations of modern technology, since most curricula have not been developed and therefore the student's learning is separate from the labor market.
- Teaching the curriculum of the teacher depends on the old style of indoctrination without developing the same educational attitude because of the lack of appropriate educational means for teachers.

- Teaching by some teachers of subjects not specialized in addition to their poor use of audiovisual means in the educational process, as well as the burden of increasing quotas.
- The curriculum and its contents do not adequately take into account the student's needs and development requirements because some subjects do not fit the actual realities and requirements (Ahmad,1993).

The absence of agricultural methodological books prompts teachers to prepare special memoirs for students in the primary grades and the efficiency of the memoir depends on the teacher's level and experience. These factors vary from subject to subject and are reflected in the student's achievement.

- In most cases, agricultural books do not represent agricultural reality.

The school lacks a comprehensive library, although agricultural schools have a library, but books with old books do not keep pace with today's needs.

Problems Related to Teachers and School Administration

The teacher is the cornerstone of the educational process in general and agricultural education in particular, as it is important for students' qualification s quality is an indicator of the nature of this education because of its multiple tasks. A currency that not only depends on the preparation of a skilled, observed or artistic worker, but goes beyond a larger goal of raising him as a good, productive citizen with conscious, positive experience and the ability to improve conformity with the community's directions in a sound manner (Mansour, 2012).

- Schools of education do not contribute to the preparation of agricultural teachers, as there are no academically prepared and professionally qualified agricultural teachers.
- Polarized groups to serve as teachers who are graduates of agricultural colleges need prior qualification before practicing their teaching profession but under the weight of need to fill the shortage are accommodated without preparation.
- The migration of old agricultural teachers to agricultural sites due to the search for higher pay and better social status. The salary of the agricultural engineer in the Ministry of Agriculture is different from that of the agricultural engineer in the Ministry of Higher Education.

- Lack of educated teaching staff in technical schools and recruitment of non-educationally qualified teachers and education professionals from the various non-pedagogical qualification campaign.
- There is a deficit in agricultural technical education teachers for theoretical technical subjects and their need for continuous training.
- The low level of performance of the school administration at the level of some technical education schools and the weak and low efficiency of the guidance system for those schools.
- The school management method is carried out more in the sense of management and control than in the concept of direction and participation in responsibility, which makes the administrative system a burden on the teacher and the widening of the gap between the administrative system and the educational system within different levels of education.
- The low level of education for teachers and their lack of acceptance of it because of the lack of specialized training programs for professionals in technical education of its different kinds, which results in a weak level of academic and educational enrolment of the teacher (Mansour, 2012).

Educational Issues

Teaching means All appointments used by the teacher to explain and clarify theoretical lessons and their practical applications are:

- Poor school equipment, equipment and mechanisms that do not keep pace with the developer's production methods and lack of efficiency in performance.
- Weak and minimized means of explanation used in teaching.

Technical Education Student Problems

The low academic level of students enrolled in agricultural secondary education institutions as a result of admission policies based on the grades obtained by the student. Agricultural schools accept students who are fortunate to continue the educational process, so it is difficult to communicate information to them and the weak level of graduates. The higher the academic level of agricultural education students, the more satisfactory the outputs.

Student intensity in the classroom is a major problem, since lack of density increases the cost of education and increases density, resulting in incomprehension.

Agricultural secondary schools suffer from difficulties related to the primary level of students' reluctance to attend because of the lack of a post-graduation work guarantee, poor material benefits and incentives for students, the existence of an outdoor residency system in some schools and the low level of insurance for teaching requirements (Ali, 1996).

This underscores the need for curriculum adjustment and the creation of a system for what they called a multidisciplinary school.

Schools, Equipment, Educational and Training Capabilities

- The inadequacy and inadequacy of school buildings for pupils, resulting in higher class intensity, in some governorates, reaching more than 50 students, to varying degrees, in both rural and urban areas, in addition to the student's lack of full assimilation (Al-Absi, 2017).
- Lack of equipment and non-Tu Afar educational facilities from laboratories, ores, farms, libraries and places to carry out educational and training activities in a large number of schools (Ali,1996).

Custom receipt system that converts equipment from technical to acquisition affecting its use and training as equipment that can be corrupted or damaged by use.

- Lack of teaching materials, photographs and illustrations in the relevant textbooks for explanation and teaching.
- The limited space of certain departments and laboratories and the lack of materials and tools used, as it limits the use of both students and students and the development of multiple skill aspects for them.
- The poor condition of certain buildings and facilities in some technical schools and their need for comprehensive maintenance or replacement and renovation.
- The existence of certain technical machinery, equipment and tools for the educational and training process, which require maintenance and maintenance services. This may not be possible owing to the absence of maintenance contracts, in addition to the lack of the availability of spare parts for them, which makes them unavailable and thus the student is trained in such machinery and equipment theoretically rather than scientifically (Habib, 2014).

Institutional and Organizational Problems

Inadequacies in educational institutions in terms of location, construction, equipment and weak institutional and organizational aspects within agricultural education institutions and departments include administrative problems, the organization of work and the relationship between them and research centers. These problems include:

1. Poor choice of location of agricultural educational establishments in terms of location. Most agricultural schools are away from basic services for both students and teachers. If these services are not available, there is difficulty in accessing these schools.
2. Some existing school buildings are poor due to discontinuation of maintenance
3. Lack of agricultural land to train students within the agricultural school, many agricultural schools find it disproportionate to the size of the farm. This is contrary to article 139 of Act No. 1981 of 31, which stipulates that each agricultural school shall attend its students' farm where its size is commensurate with its number and type of study.
4. Weak linkage between the Institute of Agricultural Education and the Ministry of Agriculture limits the resolution of graduates' problems.
5. Lack of coordination and linkage between the Institute of Agricultural Education and agricultural research and extension agencies limits the graduates' contribution to solving the problems and issues of agricultural production that have become complicated by the day.
6. While agriculture is important, society's perception of it is negative and underappreciated.
7. Lack of financial provision available to schools, resulting in the preparation and modernization of schools over a period of years. Problems and constraints include:
    a. The existence of unemployment as a result of the graduation of thousands of academics in excess of their need, which results in unbalanced growth in graduates.
    b. Low growth in agricultural education compared to other specializations due to wages and conditions of absorption.
    c. Women's low contribution to agricultural development owing to the fact that agricultural education is more dependent on males than females.
    d. Increasing the educational system's outputs to a magnitude beyond the capacity of the labor market to absorb them.
    e. The imbalance between supply and demand where there are many specializations and the labor force that seek employment from the holder of scientific qualifications and disciplines that the labor market is unable to absorb.

In general, the problems affecting the realities of agricultural secondary education can be summarized in different forms, which prevent the achievement of the following objectives:

1. Society's and students' ignorance of the value of technical education in general and its increasing role in the overall development process constitutes a major obstacle to the attendance at agricultural secondary schools, resulting in the reluctance of middle and higher students and their tendency to general secondary schools, which has an impact on the scientific level of graduates of such schools
2. Lack of accurate data on the needs of the production sector and the various services of agricultural secondary school graduates in both quantitative and qualitative terms, resulting in a lack of good planning for agricultural secondary education and a proper and realistic link to the labor force needs of the agricultural sector.
3. The seizure of some agricultural school farms or the reduction of their area so that it is insufficient to train the increasing number of students, which is detrimental to the scientific and training aspects of those schools.
4. The reluctance of many bodies, factories and production centers to train students in agricultural schools, given the increase in the number of students in these schools and the fact that the centers are the main sponsors of most schools.
5. Inadequacy of secondary school equipment to keep pace with technological advances in all agricultural facilities, owing to the lack of funds available (Habib, 2014).

Agricultural secondary education curricula remain inconsistent with the labor market because of their weak relevance. These curricula are not concerned with students' accustomed to logical thinking and have thus failed to prepare students to meet future challenges.

- The practical training of the student does not receive the deserved seriousness of both teachers and students, since the great interest is focused on obtaining the production of the sponsor without considering the student's professional and educational benefits (Al-Absi, 2017).
- The low contribution of agricultural secondary education to rural development, since the majority of the extension services by which agricultural secondary school can contribute to rural society are not realized.

The School Department of Agricultural Secondary Education has some problems affecting the achievement of its objectives. These include the lack of implementation of contemporary administrative trends, such as the management of overall quality and the inadequacy of the powers conferred on the Head of School, which are not commensurate with the responsibilities he carries out and the scarcity of preparation and training of the school's leadership at Agricultural Secondary School (Berl, 2001).

It is clear from the foregoing that there are many problems in the realities of agricultural secondary education, but efforts are being made to advance this reality. While there are many problems in agricultural education, there are also a range of mechanisms and means from which to start and build in the development process, including:

- The curriculum includes various subjects, some of which include topics on soil and its analysis, water sources, the general properties of Egyptian land and land reclamation methods.
- Modernizing the agricultural education plan and system so as to build institutional linkages between types of education (general, technical and vocational).
- The existence of joint subjects for all specializations at the secondary level taught by all students with an interest in technological sciences and foreign languages by integrating the types of education.
- Attention and expansion of agricultural technical education to meet the requirements of development plans for technical employment.
- Appropriate the size and space of the farm with schools and prepare the student and be close to agricultural schools so that each student can practice practical training at work sites.
- Attention to school buildings, classrooms and laboratories, as well as greenhouses and beaches, so that each student can perform practical tests and practical trainings and there are large numbers of students without classrooms, forcing them to damage most of the science in the school yard.
- Interest in the professional development of teachers during service by training teachers in agricultural workplaces.
- Work on the educational rehabilitation of graduates of agricultural colleges to improve their teaching skills.
- Administrative supervision and continuous follow-up and activation of its results.
- Raising teachers' productive efficiency and performance in their practical specializations.
- Accelerate the implementation of the cooperation plan between the Ministry of Education and the Ministry of War Production by supplying machinery and equipment for the workshops of agricultural schools.
- The establishment of technological complexes that include agricultural schools and advanced technological faculty and have the latest advanced equipment and equipment with foreign expertise to adopt the curricula and certificates obtained by the graduate so that he can work in any country.

- Allocation of a separate budget for agricultural secondary education for the agricultural education sector. The proposed concept of using this budget independently for disbursement provides for all investment projects required for agricultural education.
- Increasing the State's expenditure on education. The greater the expenditure by the State on education is indicative of the degree of sophistication and civilization of this nation. The educational process is productive and prepares the head of human capital without which material capital cannot be invested.
- Increase the proportion contributed by the State's budget for services to balance education so that it can be found in the proportion provided and disbursed for agricultural secondary education.

CONCLUSION

Education is an investment in human resources. The higher level of education leads to increase of the productivity, the higher the productivity of the individual, and leads to the optimum utilization of the available resources to society as a whole. However, there are some problems that impede progress and development in agricultural education in particular, and this chapter tries to highlight the features of these problems.

Agricultural education faces many problems in its structure, functions, students, teachers and overall education inputs. This is reflected in the outcomes of education and its impact on society, its development and development. However, agricultural education problems can be included under a number of classifications, including technical, institutional, organizational and other interrelated problems. It is also clear from the foregoing that there are many problems affecting the realities of agricultural education and preventing the achievement of its objectives as desired, but efforts have been made. To advance this reality, although there are many problems affecting agricultural education, there are also a range of mechanisms and means from which to start and build in the development process. One of the most important is to modernize the education plan and system so that it works to achieve institutional structure between types of education (general, technical and vocational), and the attention and expansion of agricultural technical education to meet the requirements of the development plans of technical employment.

REFERENCES

Abdel, G. A. (1997). *Comparative Education "Curriculum and its Application"*. Dar al-Thakr al-Arabi.

Ahlam, R. A. (1983) *Agricultural Education in Egypt in Light of Development Demands and Trends, Calendar Study,* [Doctoral Thesis, Faculty of Fayoum Education, Cairo University].

Ahmad, Z. H. A. (1993). *Technical Education and Development Requirements in Saudi Society "Applied Study on Samples of Students, Graduates and Educators of the Taif Industrial Institute"* [Master's Thesis, Institute of Arab Studies and Research].

Al-Absi, M. (2017). *The First Conference, The Future of Engineering and Technical Education*. MDPI.

Alani, T. A., & Saadallah, G. (1986). Vocational Education in the Arab World. Arab Educational, Cultural and Scientific Organization, Tunisia.

Ali, F. (1996). Some problems of vocational preparatory education in Egypt and ways to overcome them in the light of Japanese experience. *Journal of Contemporary Education, Egypt*, 41.

Aliyah, B. M. O. (2007). *Problems Faced in Agricultural Education in Areas of the Palestinian Authority from the Perspective of Agricultural School Teachers and Remedies,* [Master's Thesis, Islamic University, Gaza].

Berl, D. A. (2001). *Field study of some of the problems of school administration at the Agricultural High School in the Arab Republic of Egypt,* [Master's thesis, Bakfar el-Sheikh University of Tanta].

Elhami, M.G. (2000). Developing vocational education in Egypt in light of contemporary global trends. *Educational Sciences*, (1).

Habib, A. S. (2014). Technical Education in Egypt: Problems and Solutions, Administration (Consortium of Administrative Development Associations). Egypt, 51(1).

Hanafi, M. T. (1996). *A calendar study of Egypt's agricultural secondary school in light of contemporary trends.* Fourth Annual Conference (Future of Education in the Arab World between Regional and Global), Helwan University. h://portal.moe.gov.eg

Hassanin, E. H., & Abdelaziz, R. Z. (1994). *Study on the reality of agricultural secondary education and ways of developing it in the Arab Republic of Egypt.* General Directorate of Agricultural Education.

Ismail, Y. M. (1991). *The contribution of the second cycle of education mainly in preparing pupils for technical secondary education*, [Master's thesis, Faculty of Education, Ain Shams University, Egypt].

Mansour, H. M. M. (2012). *Developing Technical Education in Egypt in the Light of Malaysian Experience,* [Master's Thesis, University of Tanta]

Ministry of Education. (1965). Agricultural Secondary School Book. *Ministry Press.*

Ministry of Education: Act No. 233 of 1988 amending certain provisions of Act No. 139 of 1981.

Ministry of Education. (1963). Agricultural Education. *Ministry's Press.*

Ministry of Education - General Directorate of Agricultural Education - Report on Agricultural Education. (1890 -1970)

Salim, R., & Hassan, J. (2005) The experience of technical higher education in Egypt. General Directorate of Cultural Research of Egypt.

Ya'qub, S., & al-Tourist, O. (1988). Secondary and higher agricultural education and its role in meeting the needs and requirements of rural development in selected Arab countries. UNESCO Regional Bureau of Education in Arab States, Yundbas.

Chapter 4
Community Colleges and Universities in the Arab Countries

Mohamed Hassan Abdelgwad
https://orcid.org/0000-0002-3500-6395
King Saud University, Saudi Arabia

ABSTRACT

A community college, sometimes referred to as a junior college or technical college, is a taxpayer-supported two-year institution of higher education. The term "community" is at the heart of the mission of the community college. These schools offer a level of accessibility—in terms of time, money, and geography—that cannot be found at most liberal arts colleges and private universities. Community colleges are common in many different Arab countries, and just recently established within other countries. It is mostly responsible for graduating effective technicians in different disciplines. In addition, it plays an important role for students' manipulation for university enrollment in different Arab countries. Through this chapter there are many points and views relevant to concept, initiation, academic programs, employability, job destination, and benchmarking versus global community colleges.

INTRODUCTION

Most of Arab countries understood the importance of community colleges not only to meet the labor market's needs, but also to solve the growing demand problem. Community colleges tended to be state-owned in developing countries, with very few privately owned. Despite the issue of university enrollment, a wide gap between

DOI: 10.4018/978-1-6684-4050-6.ch004

universities and labor markets remains, and that is not only the case in Saudi Arabia, but also for most universities in the Arab World (Almannie, 2015). In Saudi Arabia and many Arab countries, colleges have faced several challenges related to internal programs and external links to the community, which have resulted in low participation of these colleges in the labor market. Alsolaimi (1996) and Fifi, (2013), in addition there is a weak English preparation. Its apparent that programs in community colleges do not necessarily apply in the field as required for globalized markets (Hamad, 1984).

Community colleges are one of the most prominent contributions and educational institutions that meet the needs of the labor market, according to the understanding and assimilation of its requirements from various scientific disciplines, as they receive large numbers of male and female students after completing secondary education, who seek to occupy some jobs and engage in the labor market without the need to receive University education in one of the undergraduate programs for a period of no less than four years.

Through this chapter many topics will be clarified for community colleges in Arab World countries and its inspiration for the new era of Industrial and technological development at labor market. This chapter objectives are:

1. Concepts & Foundation of community colleges in Arab world.
2. Community colleges academic programs in Arab world.
3. Questions and Answers (Q&A) for community colleges
4. Role of community colleges in relation to job & labor market.
5. Technology based community colleges in Arab world:
6. Community graduate privilege in Arab world.
7. Benchmarking community colleges in Arabs *versus* global experience

CONCEPTS AND FOUNDATION OF COMMUNITY COLLEGES IN ARAB WORLD

A community college is an educational institution that provides intermediate higher education, that is, education that begins after the secondary stage and does not extend to the end of the university stage but is limited to two academic years. It is thus similar to the Comprehensive Intermediate College and the Specialized Intermediate College, which existed before it and were originally intended for it, except that it was distinguished from these institutions by its direct connection to the community's needs for qualified manpower that increases in terms of specialization and depth at the level of secondary education, and does not reach the level of a university degree. It is also related to the desire of a wide range of graduates at the secondary level to obtain a short-term education that qualifies them to work and provides them with broader opportunities to obtain it. This direct connection with society prompted it to be called a community college. However, this distinction no longer exists accurately and clearly after the intermediate colleges and intermediate institutes took the same approach in relating to the needs of individuals and society (Arab Encyclopedia, 2023).

In the institutional and administrative organization of community colleges, there is a model that connects them to the university and makes them part of it and limits its role to teaching a set of university courses that are considered for the student if he continues his university studies after graduating from the community college. And to teach another group of technical and vocational courses that qualify the student to work if he suffices to study at the community college.

There is also a model of community colleges that are independent from the university in their management, curricula, academic system, and certificates they grant. On the other hand, community colleges may be comprehensive colleges that include several specializations in a manner similar to universities.

The history of community colleges at the international level goes back to the year 1901 AD, when the first community college was established in the United States of America and with the successes achieved and the role it played in qualifying the workforce, their number increased over the past decades until their number today reached (1280) community colleges in the states It has thousands of branches (Al Jazirah News Paper, 2008 G Issue 12981).

The importance of community colleges stems from its philosophy based on serving its local communities by providing comprehensive and diverse programs for specialized qualification directed at practicing professions and acquiring the skills required by the local labor market, and preparing to continue university education in a better, more qualified and effective manner.

There was no community college in Egypt until the present, and the focus was mainly on social service through a large number of social service institutes spread in many governorates in Egypt. In the modern time, during the past five years, the Ministry of Higher Education began establishing some community colleges in some Egyptian universities in cooperation with some distinguished American community colleges.

The Higher Institute of Social Work in Cairo is a private higher institute affiliated to the Ministry of Higher Education. It grants a bachelor's degree in social work. The institute was established in 1937 as the first higher institute in the Middle East and Africa with the aim of preparing social workers. It was then called the School of Social Work in Cairo. The first class of students graduated in 1940.

The Arab University College of Technology is a private intermediate university college, accredited with general and special accreditation, and is one of the first community colleges in Jordan. The college was founded in 1980 by a group of intellectuals and educators.

The Community College of the Kingdom of Saudi Arabia was established in the year 2000, under the supervision of King Saud University. Since that time, the college has sought to keep pace with the strategic premises and directions of King Saud University based on achieving global leadership and excellence in building a knowledge society.

COMMUNITY COLLEGES ACADEMIC PROGRAMS AND SPECIALIZATIONS IN ARAB WORLD

The community college includes two types of academic programs that enable high school students to enroll and getting medium Diploma with two years curricula or graduating at four years curricula with bachelor's degree in community or applied studies colleges. More detailed description and information for both types are listed below:

Transition Programs

These transitional programs are suitable for those students who did not have the opportunity to obtain university admission due to their low academic achievement in the secondary stage, so the student enrolls in one of the available specializations at the Community College for two years, after which he moves to the complementary department at the university.

King Saud University suffers from a high demand for university education from female students whose grades did not qualify them to join the university directly. Therefore, the university established the College of Applied Studies and Community Service, to undertake the provision of a number of transitional programs. The importance of the study is the experience of transitional programs in community colleges as an essential measure to increase the university's absorptive capacity, expand postgraduate studies and conduct research. The researcher used the descriptive analytical method for the study population of 368 students.

The results of the study showed that there are 18 problems faced by the students of the transitional programs, the most important of which is the lack of external spaces in the college building, the lack of several classes offered for the same course at different times, the lack of places to spend free time between lectures, the failure to clarify the names of the faculty members, there is no financial reward for the students (Al-Ghamdi, 2018).

Qualifying Programs

These are programs that qualify students in a number of specializations that are immensely popular in the labor market, and are studied for a period of two years, after which the student obtains the qualification of an intermediate college diploma or a participant's degree in some cases, which qualifies the student or individual for career advancement.

Arab University College of Technology (Arab Community College) is a private intermediate university college accredited by general and private accreditation in Jordan. The college was established in 1980 by a group of intellectuals and educators in Jordan. Its founders realized the rapid developments taking place in Jordan and the Arab world at the time, and the extent of the need for qualified technical cadres in various disciplines; To prepare and implement development plans, and all this called for several subjective and objective considerations, related to the diversity of the fields of specialization and their compatibility with the desires of students on the one hand and their response to the needs of the local and Arab market, and keeping pace with the requirements of development and progress in various fields on the other hand, it was necessary to establish a college that contributes to the effort And in preparing and training cadres in various fields of work, it was the Arab Community College whose founders chose this name to express an intellectual and educational content.

As for the intellectual content, it is represented in the fact that it is the totality of all the sons of the Arab world, regardless of their different countries and cities, and they are equal in them.

As for the educational content, it is represented in the new orientation in education, as it mixes theory and practice in higher education, with an emphasis on the practical aspect, rather than the theoretical aspect on which the traditional university study is based. Its specializations include computer science, medical equipment's, Nursing, car engineering, Interior Design, Aviation technology, The science of child development and retributive justice.

Applied Studies and Community Service

It is a specialization available in a number of Saudi universities. This specialization includes a number of different programs in which the student studies administrative sciences, accounting and marketing in order to graduate cadres capable of competing in the labor market significantly. It also aims to encourage scientific research in various applied fields.

The college aims to achieve many goals by studying there, and among the most prominent goals of the college are the following:

- Create appropriate scientific conditions for individuals and institutions.
- Providing educational services and spreading cultural awareness.
- Knowing the needs of society and social institutions and providing them.
- Contribute to community service and development by offering a number of applied and professional programs.
- Achieving quality and excellence in education.
- Encouraging scientific research in various fields of interest to society.
- Communication between government agencies and private entities.

King Saud University was established in 1377 AH and sought to achieve tremendous achievements, but it faced a set of problems and natural obstacles that any emerging society can go through (Al-Aqeel, 2005, p. 208). One of the most important problems that higher education suffers from in the Kingdom is the increase in societal demand for it, and that Saudi universities are still interested in the theoretical and literary disciplines that the labor market is saturated with, while the community suffers from a lack of scientific and applied disciplines (Al-Aqeel, 2005, p. 210). In the year 1403/1402 AH, King Saud University established the "Community Service and Continuing Education Center by a decision of the university rector to link the university with society, ministries and public and private institutions. The university also established the women's section in appreciation of the importance of the role of working women in society (Annual Report, 2000 AD, p. 2 And in 1421 AH, the Higher Council of the University issued a decision to transfer the name of the Center for Community Service and Continuing Education to the College of Applied Studies and Community Service (Student Guide, 2003 AD, pg. 7), which provides an opportunity for individuals to train and qualify, and keep abreast of what is happening from Develop in various scientific fields through programs and training courses.

The college has also contributed to creating educational opportunities and suitable alternatives for high school graduates, due to the increased societal demand for university education for girls, and the lack of professional specializations available for girls in higher education (Bubshait, 1997, p. 120).

His Excellency the Minister of Education, King Saudi Arabia, affirmed that the ministry seeks to transform community colleges into applied colleges whose outputs are in line with the requirements of the labor market, and to diversify their programs according to the skills of the 21st century, the requirements of the labor market, and the Fourth Industrial Revolution.

In a statement to the Saudi Press Agency, after inaugurating the Clinical Simulation Unit for Medical Training and Specialized Skills, and the Industrial Innovation and Robotics Center at the University of Tabuk, and inspecting a number of educational, research and health projects at the university: With the aim of developing curricula and educational fields in our universities to meet the specific needs of the market.

He added that the ministry supports professional diplomas in universities, in order to achieve spending efficiency in the admission process, respond to the needs of the labor market, and reduce admissions in theoretical disciplines, for example, if the labor market has a specific profession that originally does not require study for four years, and requires qualification in it. program for one or two years, so why should I charge the student and the community the cost of four years?

In addition, the President of Umm Al-Qura University issued an executive decision based on the decision of His Excellency the Minister of Education and Chairman of the University Affairs Council to transform the Community College and the College of Community Service and Continuing Education to become the Applied College.

QUESTIONS AND ANSWERS (Q&A) FOR COMMUNITY COLEGES

What is the Degree Offered by the Community College?

The college offers qualifying programs that end with the graduate obtaining an intermediate academic degree called the Associate Degree, which is equivalent to an "intermediate diploma", in addition to offering the tax accounting program with an intermediate diploma (modern program).

What Majors are Available in the Community College?

Department of Computer Science: It offers two programs:

1. Computer Science Program,
2. Web and Graphic Technology Programs, including two sub-specialties (website technology track, graphic technology track).

Department of Administrative Sciences: It offers two programs:

1. The Administrative Sciences Program, which includes five sub-specialties (Human Resources Management, Executive Secretarial, Marketing, Financial and Banking Management, and Insurance Business Management).
2. Tax accounting program.

The study plans for the programs available in the college programs at King Saud University are announced on the college's website at the following link:

https://rcc.ksu.edu.sa/ar/plans

What is the Duration of Study at a Community College?

The duration of study at a community college is not less than two years (four semesters) and most majors extend to five semesters in the normal range.

What are the Advantages of Studying at a Community College?

The short period of study at the college, the absence of tuition fees, a stimulating community and educational environment, qualifying the graduate to engage in the labor market in a record period.

Can a Community College Graduate Complete a Bachelor's Degree at the University?

Yes, it is possible to pursue a bachelor's study at the university after fulfilling the conditions approved by the university and for the specified and specified percentage, and that will be after completing the diploma study, bearing in mind that the primary role of the college is to prepare the graduate to engage in the labor market.

What are the Admission Requirements for a Community College?

Admission to the Community College is available to all high school graduates of all kinds, in addition to graduates of scientific institutes and those who are educationally similar. The Community College, like any college of the university, requires that there be a compatibility between the student's specialization in the secondary stage and the specialization in which he wishes to join the college, and some admission requirements differ from one department to another in the college. For more information about the college admission requirements, you can visit the following link:

https://rcc.ksu.edu.sa/ar/registeration_

COMMUNITY COLLEGES ACADEMIC PROGRAMS IN ARAB WORLD

The Community College includes a variety of specializations that largely target the needs of the labor market, the most popular of which are:

1. Computer science.
2. Medical equipment.
3. Nursing.
4. Car engineering.
5. Interior Design.
6. Aviation technology.
7. The science of children's development.
8. Criminal justice.
9. Health care services.
10. Sciences.
11. Business Management.
12. Nutritional Sciences.
13. Fashion design.
14. Photography.
15. Fire Science Technology.

In King Saud University, there are many programs that are mostly required for a community graduate to work through Riyadh and other Saudi cities. The programs are:

1. Web and Graphic Technologies.
2. Financial and banking management.
3. Human Resource Management.
4. Executive secretary.
5. Insurance business management.
6. Computer Science.
7. Marketing.

In King Saud University, students offered diploma (Associate Degree) and College students enroll in the first joint semester (general), after which the student specializes in one of the following specializations:

1. Human Resource Management.
2. Financial and banking management.
3. Marketing.

4. Insurance business management.
5. Computer Science.
6. Executive secretary.
7. Web and Graphic Technologies.

Jazan Community College, Saudi Arabia includes a number of technical specializations for applicants to join the college to obtain an intermediate university diploma in any of them, to help him obtain scientific and training experience to work in the field he graduated from. These specializations are:

1. Systems analysis and organization.
2. Office management.
3. Programming and computer operation.
4. Business Management.
5. Marketing.
6. Tourism and hotel.
7. Accounting.

The Community College in Abha is affiliated with King Khalid University. It was established in 1428 AH. Since its establishment, the college has been working on expanding the programs offered to female students who have obtained high school diplomas to help them obtain the perfect job opportunity for them in light of the developments in the labor market in the Kingdom, which made it need to specific specializations that were not required before, and the college has a number of departments and specializations, namely:

1. Intensive English Program.
2. Islamic culture.
3. language skills.
4. Kingdom history.
5. Business Management.
6. Office management.
7. Accounting.
8. Computer programming and operation.
9. Information Systems.

ROLE OF COMMUNITY COLLEGES IN RELATION TO JOB AND LABOR MARKET

After completing two years of study in community colleges, the student can join another university for the same major for another two years, after which he will obtain a bachelor's degree, or start working life immediately after the community college, including a number of academic disciplines:

- Medical analysis technician.
- Specialist Nursing.
- Nutrition specialist.
- Software and computer technician.
- Physiotherapy technician.
- Speech and behavior modification specialist.
- Administrative employee.
- Fashion designer.

At King Saud University, college of community in Riyadh through training and Employment Support Unit at the college, in cooperation with Leejam Company, conducted personal interviews at the college headquarters for the jobs offered by Leejam Company for the college graduates. She has an ambitious employee who has gained knowledge and skill, which qualifies him to be an ambassador for Al-Watan University, King Saud University (college of community, King Saud University https://news.ksu.edu.sa/ar/node/125338) .

Income level is another reason to consider a community college degree. According to the Bureau of Labor Statistics 2018, two-year college graduates earned an average of $42,952 in the third quarter of 2018, while high school graduates earned an average of just $2,727. One of the main reasons for the high growth rates and income levels of community college jobs is the specialized nature of for many degrees. Community college students often study majors in which they develop skills that are easily applied in the marketplace.

The vast majority of the highest-paid job vacancies (and those with the highest expected growth rates) fall into three categories: healthcare, engineering and other technology fields. All of these sectors are experiencing above-average expansion of job opportunities due to economic trends.

A committee in the Shura (Consultative) Council in Saudi Arabia agreed to include a community college diploma qualification for qualification requirements in the Civil Service Job Classification Manual. This recommendation came based on the committee's adoption of the content of what was submitted by Shura members to review the organizational structure of the job classification guide and include a community college diploma qualification for job qualification requirements on ranks of five and below for jobs that are suitable for graduates. The graduates of the Community College are a pressing force on the national economy and the social fabric, as they constitute a percentage of the graduates' unemployment and the rise in their continuous demands for employment, and that they do not ignore the offering of suitable jobs for them.

There are priorities for applying science of data and artificial inelegance through their strategy, which are:

- **Education sector:** Integrating data and artificial intelligence in the education sector in order to achieve compatibility between the education system and the needs of the labor market, and to develop the educational process of students.
- **Government sector:** Aligning the use of data and artificial intelligence in the government sector in order to reach a government sector based on the use of smart technologies, effective and productive
- **The medical sector:** Aligning the uses of data and artificial intelligence in health care systems with the aim of enhancing access to health services and preventive care along with accommodating the growing demand for health services/for health services
- **Energy sector:** Harmonizing data processing and artificial intelligence in the energy sector in order to increase the capacity, increase the efficiency of the energy sector, and develop the sectors supporting it
- **Transport and Communications sector:** Integrating data and artificial intelligence in the transportation sectors to build a world logistics center, establish systems based on the use of smart technologies in mobility, and enhance traffic safety in cities

TECHNOLOGY-BASED COMMUNITY COLLEGES IN ARAB WORLD

The results of several studies confirmed the importance of integrating technology into education, and qualifying teachers to carry out this task. Miqdad (2004) indicated that most university professors need training and educational preparation that enables them to acquire some skills necessary to raise their professional performance, such as skills of contemporary teaching methods, realistic assessment methods, classroom management methods, and methods of using and employing educational technology in university education. The study (Al-Maqtari, 2000) also concluded that university teachers should enroll in training programs dealing with the skills of operating devices. He was also assigned to prepare programs in the light of educational technology, and to identify realistic training bodies for university teachers, especially in the field of employing educational technology in university education. The study (Morriso, 2002) emphasized the importance of preparing and training teachers at the university or public level to face the digital age and providing them with a set of skills related to the use of modern technologies in education, such as: e-learning. This was confirmed by the study (Grabe and Grabe, 2004 p. 3333) regarding training teachers to design technology-based learning environments.

Younis (1997) lists the competencies of compulsory educational technology for a faculty member in teachers' colleges and their level in the Kingdom of Saudi Arabia in the areas: cognitive, performance and emotional. The study also found that there are no statistically significant differences between qualified and educationally unqualified faculty members in estimating the importance of competencies, and the degree of its suitability for teaching the subject matter of specialization. The study (Saraya, 2003) showed the importance of integrating technology into programs to train faculty members in higher education in this field. Norton and Sprague (2002) suggested some conditions to enable the learner to employ technology in education when joining the field education program, such as witnessing a live application of this technology by faculty members in colleges of education or teacher preparation, and training teacher teachers on the use of technology (Chuang, 2002, pp. 227-249). A study emphasized the importance of developing in-service training programs for teachers to meet their contemporary needs by including in these programs modern concepts and issues related to the field of education technologies, information and innovations Innovation Technological, such as activating distance learning through e-learning via the Internet and employing interactive multimedia educational programs Multi - Media Interactive, provided that these programs have a gradual level of training from general to specific. of information and communication technologies and their use in university education, identifying tribal training programs in this field, and trying to identify the difficulties that face It can hinder the employment of these technologies in university education. The study reached a set of results, including: There is a large discrepancy among faculty members in the degree of use of information and communication technologies in university education, which he attributed to the presence of a number of difficulties that prevent the use of these technologies in education.

COMMUNITY GRADUATE PRIVILEGE IN ARAB WORLD

As experts proposed that community Colleges" serve the needs of the labor market and save effort and expenses." The Ministry of Higher Education and Scientific

Research intends to launch a new educational system in universities, similar to the American system, which relies mainly on what is known as "community colleges."

The protocol stipulates the establishment of "community colleges" in several disciplines, with a 4-year system in two stages, and grants the student an initial certificate in his specialization, and allows him to join the labor market after only two years of study, provided that he returns to complete his studies at a later time for another two years, and no the student obtains a bachelor's degree only after completing the second stage.

Ain Shams University, announced that his university is the first university that has begun preparing and coordinating a number of study programs that provide its students with real opportunities in the labor market, pointing out that there is full coordination with the Ministries of Higher Education and International Cooperation and representatives of Egyptian universities for the success of the experiment. "Community Colleges".

The Director of King Fahd University of Petroleum and Minerals revealed that the university is heading to inaugurate a branch of the Community College in Jubail and Al-Ahsa, due to the success of the programs offered by the College of Community Service and keeping pace with the needs of the labor market today.

He also explained on the sidelines of the graduation of the sixth batch of community college students that the University of Petroleum is keen to carry out a process of continuous development of the college's programs as the university is, to bring the graduates to the level of service required by the current labor market.

The president of the university pointed out that the offers that are showered on college graduates in particular are the biggest evidence of the college's programs keeping pace with the needs of the external labor market, and that the graduates' salaries start from 9,000 riyals and they only spent 3 years in their studies. He explained that the University of Petroleum is pursuing an approach to modernizing its academic programs, whereby the college goes to the labor market and studies the need for it from graduates, and then new specializations are launched. As a result, the college inaugurated the Department of Non-Destructive Tests at the Community College in Hafr Al-Batin, specializing in chemistry laboratories, based on the requests of some companies.

He also pointed out that the Dammam Community College is looking for companies' need for specializations, and then they are included in the college to ensure the career future of students and to qualify them properly. A program from one to two years aimed at qualifying students in the English language and in specializations that serve the labor market, pointing out that this program is a joint program between the university represented by the Community College and the Human Resources Fund.

There are 5 advantages of community colleges could be summarized as:

1. *It is easier to be accepted than to other universities:* Admission to the Community College is open to all high school graduates of all scientific, literary and commercial types, in addition to graduates of scientific institutes and those of similar educational status. The community college - like any college of the university - requires that there be a proportionality between the student's specialization in the secondary level and the specialization he wishes to enroll in at the college. Any student of King Saud University can transfer to any of the college's majors in the qualifying programs, provided that his specialization in high school is suitable for the specialization he wishes to transfer to.
2. *It does not require high scores in language proficiency tests:* Admission to the college department is open to high school graduates. The admission requirements are summarized as follows:
   a. The student must obtain a high school diploma or its equivalent.
   b. The student is free to study according to the time periods required by the study, whether it is morning or evening. There is no obligation for English language proficiency required for student's enrolment and registration in community colleges. Later on, after two years graduation and being ready for upgrading his study level for college, English proficiency is required.
3. *It requires fewer years of study:* Study time required for community colleges is mostly two years and give students an exit point to enroll for another academic college within the university. That gives students another chance to enroll. The period of study in one major is four semesters as a minimum and seven semesters as a maximum, and summer semesters are not counted among them. Sometimes in few colleges it is available for students to have evening study system which give them opportunity to work and study at the same time.
4. *It enables its entrants to engage in practical life and be promoted in various jobs:* Many of students prefer to get two years study and being technician or specialist in information technology or business administration or other medical paramedics and start his career journey early as much as he can. There are many destinations available for community colleges graduate and mostly are private and entrepreneur pattern. Another good reason to consider community colleges is the expected rapid growth rate in professions that require an associate degree - the degree usually offered at community colleges. The Bureau of Labor Statistics projects that associate degree jobs will grow 17.6% by 2022, while bachelor's degree jobs will grow 12.1%.
5. *It accommodates the largest number of students at the lowest possible cost:* Why consider a community college instead of a four-year college? Certainly, cost is one factor. According to the College Board, the average tuition for in-state students at public two-year colleges (typical community college) for the

2018-2019 year is $3,660, while four-year public colleges charge $26,290 and private colleges have a tuition cost the median is $35,830. In addition, more than ten states now offer some form of free college education.

BENCHMARKING COMMUNITY COLLEGES IN ARABS VERSUS GLOBAL EXPERIENCE

The study of "Some requirements for implementing the system of community colleges in Egypt in the light of the experiences of some Arab and foreign countries" (Abdel-Hay, 2008) and determining the similarities and differences between community colleges and technical institutes in Egypt, then identifying the necessities calling for the application of the community college system in Egypt, and the requirements for converting intermediate technical institutes into community colleges, then developing a proposed vision for implementing the community college system in Egypt in light of the experiences of some Arab and foreign countries. The study concluded that community colleges are among the educational systems that are characterized by meeting the local needs of the labor market in various fields, in addition to the presence of some similarities between intermediate technical institutes and community colleges. The study also identified the requirements for converting intermediate institutes to community colleges.

The Higher Institute of Social Work in Cairo is a private higher institute affiliated to the Ministry of Higher Education, which awards a bachelor's degree in social work. The institute was established in 1937 as the first higher institute in the Middle East and Africa with the aim of preparing social workers. It was called the School of Social Work in Cairo at that time.

The first class of students graduated in 1940, and the number of its students was only 17; they represented the first building block of social work on which the Ministry of Social Affairs relied when it was established in 1939. In 1972, the name changed to the Higher Institute of Social Work in Cairo, which is the name of the current institute. The duration of study at the institute is 4 university years. The Higher Institute of Social Work contains a number of scientific departments such as:

1. Community Service Department.
2. Department of Community Organization.
3. Department of Foundation Sciences.
4. Department of Social Planning.
5. Individual service department.
6. Community Service Department.

Social work is an academic discipline and a practice-based profession that is concerned with individuals, families, and communities in an effort to enhance social functioning and general well-being. Social work is the way people perform their social roles, and the structural institutions that are provided to support them. Social work applies the social sciences, such as sociology, psychology, political science, public health, community development, law, and economics, to engage with client systems, conduct assessments, and develop interventions to solve social and personal problems; To bring about social change. Social work practices are often subdivided into microwork, which includes working directly with individuals or small groups; and holistic action, which involves working with communities, and within social policy, promoting change on a larger scale (Turner, 2005).

Social workers have a strong belief in working towards achieving equality, upholding social justice and redressing injustice against groups of society. The primary tasks of social workers include the provision of various services such as case management (linking clients with relevant agencies and programs that meet their psychosocial needs), medical social service, psychological counseling (psychotherapy), human services management, social care (care service) policy analysis, and community organization Providing support to others, defending their rights, and paying attention to education (in social work colleges and institutes), as well as focusing on research and study of social sciences (Rabah, 2017).

On October 21, 2020, the President of Alexandria University received the Director of International Relations at Ocean American Community University - New Jersey, to discuss the necessary executive steps to start implementing the community college program as of February 2021. The president of the university indicated that these programs aim to link graduates directly to the industry, To meet the needs of the jobs of the Fourth Industrial Revolution, he explained that the academic system in these programs depends on the student studying for two years at the University of Alexandria, after which the student obtains a certificate in the field of specialization from Ocean University, according to the joint protocol between the Supreme Council of Universities and the American "Ocean" University for the establishment of community colleges. He pointed out that this new model of education allows the student the freedom to choose the study system, develop his skills and acquire various experiences during and after university studies, which qualifies him to join the modern labor market (Alexandria University, 2018).

During the meeting, the President of Alexandria University stressed the importance of creating this unique model of community colleges, which allows the provision of modern study programs that are compatible with development plans and the needs of the evolving labor market, as well as providing training opportunities for students through working life, and obtaining the necessary experience. Dr. Larson at Ocean County University also expressed his happiness at implementing the idea of community colleges in Egypt, which is a starting point that allows students to study modern technical sciences in various fields. At the same time, there is a general agreement to establish community colleges in six Egyptian universities and in technical disciplines that the labor market needs.

The delay in the establishment of community colleges through Egyptian universities, where schools and technical institutes were established with the aim of preparing technicians in the fields of industry, agriculture, trade, administration and services. Students are accepted in the types of technical secondary education after completing the preparatory education stage in accordance with the conditions and rules issued by the Ministry of Education annually. At the end of the study, the student is granted a diploma according to the duration of his study and his specialization, either a "3 years" diploma, a technician category, or a "5 years" advanced diploma, a first technician category. Students of the 3-year system are allowed to complete their studies for an additional two years in advanced technical schools, the five-year system, considering the specializations and conditions Determined by the Technical Education Sector at the Ministry of Education (Ministerial Resolution No. 483 of 2014).

Globally there are many well narrated and prestigious colleges and universities of community studies, which allow their students to enroll in community colleges, and among the most famous and well-known of these universities are:

1. Cambridge University.
2. Bell University.
3. Shanghai Jiao tong University.
4. Oxford University.
5. University of Chicago.
6. University of Toronto.
7. Sydney University.

American community colleges are one of the most important and successful global formulas in linking the outcomes of university education with the requirements of the labor market, although it was not the main goal of the establishment of community colleges in the United States of America, but the primary goal was to rid the university of the burden of providing general education for young people, And to devote themselves to higher tasks such as specialized academic education and scientific research, and the evidence for this is the call of American educators - at that time - that the four-year colleges of poor quality should be transformed into intermediate colleges (Abel-Hay, 2008).

Community colleges were able to impose themselves on the arena of higher and university education in the United States of America through their role in eliminating the problems caused by the Great Depression that hit the United States of America in the thirties of the twentieth century; Where the mission of community colleges focused on vocational education to meet the demand side for employment by providing students with the skills required by the labor market (Dougherty & Townsend, 2006). Its programs that combine general academic education programs and technological education programs, which guarantees the graduate a job opportunity in the field of industry or to establish his own work that qualifies him to succeed as a businessman (Al-Bahnasawy, 2006).

Community colleges serve the local community, especially the rural community; The American countryside represents about 85% of the area of the United States of America, and despite this large percentage of the area, its population does not exceed 15% of the population. Community colleges serve the rural community through four: entertainment, education, cultural enrichment, economic development (Miller and Kissinger, 2007), and continuing education. Often, community colleges are the only way for the people of rural and remote areas to obtain a higher vocational and technical education, as these colleges work to provide the skills and experiences required by the local community in which these colleges are located. Thus, society benefits from all its children with their various competencies (Dembicki, 2006) as it is usual in rural areas is the transfer of qualified people to urban areas.

In Australia, community colleges continue to carry forward a tradition of adult education that was established in Australia around the mid-19th century when evening classes were held to help adults enhance their cognitive skills in writing, reading and arithmetic. Nowadays, courses are designed for the individual's personal development, career outcomes, or both. The educational program covers a variety of topics such as arts, languages, business and lifestyle; it is usually conducted in the evenings or on weekends, to suit people who work full hours. Community colleges may be funded by government grants or course fees; Most community colleges are not-for-profit organizations. There are community colleges located in urban, regional, and rural locations of Australia. The learning offered by community colleges has changed over the years. In the eighties, many colleges realized the community's need for computer training, and since that time thousands of people have been trained through courses in information technology. Before the late twentieth century, most colleges also became "registered with training organizations"; And realizing the need to educate individuals in a non-traditional educational environment to acquire skills that better prepare them for the workplace and available job opportunities. At community colleges, qualifications such as associate degrees and postgraduate degrees are not offered, although some community colleges offer certificate courses to four certificates (https://www.nsw.gov.au/education-and-training/adult-and-community-education) .

In Malaysia, community colleges in Malaysia are a network of educational institutions through which vocational and technical skills training is provided for all levels of school graduates before entering the labor market. Community colleges also provide the infrastructure for rural communities to acquire skills training through short training courses and opportunities for post-secondary education. At the moment, most community colleges award qualifications up to Level 3 under the Malaysian Qualifications Framework, (Certificate 3) in both the Skills Sector (Malaysia Kemahiran Seal or Malaysian Skills Certificate) and the Vocational and Training Sector, but few Community Colleges award Level 4 qualifications. (Diploma) began to increase. This is two levels lower than a Bachelor's degree (Level 6 on the Malaysian Qualifications Framework) and students intending to continue their studies to this level within the scheme usually seek entry to an advanced diploma program at public universities, polytechnics or accredited private providers (Malaysian Qualifications Register, 2008).

SUMMARY

Through the previous overview and clarification of community colleges in Arab region and ability to make it more advanced and cover all required abilities and competences required for labor market in Arab region country. Recently there are many colleges of community have been established in many Arab countries to facilitate professional and disciplined man power in different economical enterprises. There is great emphasize and focus on community college programs and how to be customized and oriented for each community aspects, features and requirements. Many of community colleges in Arab countries developed their academic program to be as professional and vocational education in very specific and required subjects fitting for community demand and inspiration. The overall numbers of community programs increased from just 6-8 professional programs or disciplines to be up to 60 programs and disseminated nationally based on the demand and requirements for each geographical community.

Based on the benchmarking comparison and considering the best practices all over the world, especially in United States and Europe countries, there are some recommendations and suggestion could be in mind for making breakthrough of community colleges in Arab lands, which include:

- Spreading community colleges within all Arab countries and focusing on required competencies especially for industry, agribusiness and information technology.

- Focusing on community colleges graduates as technicians and specialists for specific important domain more than as mediator point for university enrolment.
- Transfer all best practices for community colleges in US, Europe, Australia, United Kingdome and Canada to Arab world countries with giving specific patter for most domestic and required community and labor market needs.
- Emphasize on lifelong learning, exploration learning, artificial intelligence, and information technology through curricula structure and instruction.
- Initiation of Arab community colleges network that integrate facilities and expert exchange for best educational outcomes and covering labor demand.
- Encourage community colleges graduate with small projects loans and grants to facilitate entrepreneurship ability and economic prosperity for youth at early career ages.
- Study the domestic, regional and global demands of industrial and agricultural products and stuffs to facilitate exportation and encourage individual and family business enterprise.

TERMS AND KEY POINTS

Arab World

The Arab world refers to a large group of countries located primarily in West Asia and North Africa and sharing an Arab identity, linguistically or culturally. The majority of people in these countries are ethnically Arabic or Arabized and speak Arabic, which is used as a lingua franca throughout the Arab world. The Arab world is defined as the 18 countries where at least Arabic is spoken as a primary language.

Community College

Community colleges are considered one of the most prominent contributions and educational institutions that meet the needs of the labor market and mostly prolong for two years study period.

Transition Programs

These transitional programs are suitable for those students who did not have the opportunity to obtain university admission due to their low academic achievement at the secondary level.

Qualifying Programs

They are programs that qualify students in a number of specializations that are highly sought after in the labor market, and they are studied for two years, after which the student obtains a diploma qualification in intermediate colleges or an associate's degree in some cases, which qualifies the student or individual for career advancement.

Labor Market

The labor market is a theoretical hypothetical market and a type of economic market, in which there are job seekers and job offers from the owners of companies and others who create the workplace and search for their labor force.

Technology Based Community

Community Technology is the practice of bringing together the efforts of individuals, community technology centers, and national organizations using federal policy initiatives in broadband, information access, education, and economic development.

Artificial Intelligence (AI)

Artificial intelligence is the simulation of human intelligence processes by machines, especially computer systems. Specific applications of AI include expert systems, natural language processing, speech recognition, and computer vision.

Benchmarking

A benchmark is something that serves as a standard by which others can measure or judge. A reference point for making measurements.

REFERENCES

Abdel-Hay, A. A. I. (2008). *Some requirements for applying the system of community colleges in Egypt in the light of the experiences of some Arab and foreign countries*, [Unpublished master's thesis, College of Education, Mansoura University].

Al-Aqeel, A. (2005). Education Policy and System in the Kingdom of Saudi Arabia. Al-Rushd Library, Riyadh.

Al-Bahnasawy, F. A. (2006). *The Higher Education System in the United States of America*, 228. World of Books.

Al-Ghamdi, A. A. (2018). Problems facing students of transitional programs in the College of Applied Studies and Community Service at King Saud University. *International Journal of Educational and Psychological Studies, 4* (3), 416. www.refaad.com

Al Jazirah Newspaper. (2008). *G Issue,* 12981.

Al-Maqtari, Y. A. (2000). The needs of faculty members: Ibb University training on the use of educational aids and their attitudes towards them. *The Seventh Scientific Conference Learning Technology System in Schools and Universities: Reality 0/17-1000 and the Aspiration.* The Egyptian Association for Educational Technology.

Alexandria University. (2018). *Alexandria University receives a delegation from the American Ocean Society.* International Relations at Alexandria University. https://bit.ly/3J7EDH3

Alexandria University. (2018). *Meeting with a delegation from Alexandria University with the President of the American Ocean County University to establish a community college.* Alexandria University. https://bit.ly/34EJNLv

Almannie, M. (2015). Proposed Model for Innovation of Community colleges to Meet Labor Market Needs in Saudi Arabia. *Journal of Education and Practice, 6*(20).

Alsolaimi, K. S. (1996). *Quality Effectiveness of Technical Colleges as Perceived by Students, unpublished master theses, college of education.* Umulgra University.

Arab Encyclopedia. (2023). *Community college – Category, education, and psychology, 16.* Arab Encyclopedia. arab-ency.com.sy

Bubshait, G. I. (1997). *Establishing Community Colleges for Girls in the Kingdom of Saudi Arabia: Justifications, Objectives and Suggested Programs,* [Unpublished Ph.D. thesis, College of Education, Umm Al-Qura University, Makkah Al-Mukarramah].

Chuang, W. H. (2002). An Innovation Teacher Training Approach: Combine live instruction with a web-based Reflection system. *British Journal of Educational Technology*, *33*(2), 229–232. doi:10.1111/1467-8535.00256

Deanship of Community Service and Continuing Education. (2000). *The annual report.* King Saud University in Riyadh, Saudi Arabia.

Dembicki, M. (2006). Community colleges fill local economic needs. *Community college Times*. http://www.communitycollegetimes.com/article.cfm?ArticleId=75

Dougherty, K. J. & Townsend, B. K. (2006): Community College Missions: A Theoretical and Historical Perspective. *New Directions For Community Colleges*, (136), p. 8.

Fifi, N. (2013). *External Efficiency of the Community Colleges in Saudi Arabia*, [Unpublished PhD thesis, Faculty of Social Sciences, the Islamic University of Imam Muhammad bin Saud].

General Authority of Statistics. (2018). *Bureau of Labor Market Statistics, Third Quarter, 2018*. Saudi Arabia.

Grabe, M., & Grabe, C. (2004). *Integrating technology for meaningful learning* (4th ed.). Houghton Mifflin.

Hamad, T. A. (1984). *Graduates of Libraries and Documentation of Community Colleges, 1*(19), 21-26. Risalah-Jordan Library.

Hanafi, M. M. M. (2010). The role of American community colleges in meeting the requirements of the labor market and how to benefit from them in Egypt. *Journal of the College of Education in Port Said*, (7), 235 - 274 http://web.ebscohost.com/ehost/pdf?vid=5&hid=102&sid=52bdfe16-6a3a4799-922b-b43e71782327%40sessionmgr107

Introduction to Technology Park. (2019). *Vision 2030*. Kingdom of Saudi Arabia. https://www.vision2030.gov.sa/media/rc0b5oy1/saudi_vision203.pdf

Malaysian Qualifications Register. (2008). *List of Qualification – Community Colleges*. Wayback Machine.

Miller, M. T., & Kissinger, D. B. (2007, Spring). Connecting Rural Community Colleges to Their Communities. *New Directions for Community Colleges, 2007*(137), 27–34. doi:10.1002/cc.267

Ministerial Resolution No. 228. (2012). *Concerning the Establishment of an Integrated Educational Complex in Fayoum Governorate*.

Ministerial Resolution No. 483. (2014). *Admission Regulations for Young Education Three and Five Years*.

Ministry of Education Report. (2016). *Acceptance of students with technical certificates, the three-year system*. Ministry of Education.

Ministry of Higher Education. (2003). *The comprehensive national report on higher education in the Kingdom of Saudi Arabia.*

Miqdad, M. (2004). Educational Preparation for the University Professor, Seminar on Faculty Development in Higher Education Institutions, Challenges. King Saud University.

Morriso, G. R. (2002). *Integration Computer technology into the classroom*. Person Education.

Norton, P., & Sprague, D. (2002). Timber Lane technology tales: A design experiment in alternative field experiences for preservice candidates. *Journal of Computing in Teacher Education*, 40–60.

Rabah N. (2017). Introducing the profession of social work. *Social Science Magazine Forum – Palestine*. Wayback Machine website.

Saraya, A. A. M. (2003). Designing a training program in the field of employing technology in education for faculty members in teacher colleges in the Kingdom of Saudi Arabia. *The Fourteenth Scientific Conference - Education Curricula in the Light of the Concept of Performance,* (*Volume 1*, pp. 265 – 306). Ain Shams University - The Egyptian Association for Curricula and Teaching Methods.

Turner, F. J. (2005). Encyclopedia of Canadian Social Work. Wilfrid Laurier Univ. Press.

World Bank. (2010). *Study/Reviews of National Higher Education Policies - Higher Education in Egypt.* OECD.

Chapter 5
Workforce Skills and Agribusiness Needs for Development in Arab Countries:
Minding the Gap

R. Kirby Barrick
https://orcid.org/0000-0001-7827-0957
University of Florida, USA (retired)

ABSTRACT

The disconnect between the perceptions of the agriculture sector in terms of skills and competencies needed by workers and the abilities of agriculture graduates and others entering the workforce is existent throughout much of the world and especially in the Arab world. Economic development cannot be sustaining until this disconnect, a skills-gap, is resolved. This chapter provides an overview of the skills and competencies needed in agriculture and the perceptions of the extent graduates possess those skills and competencies. A system for identifying the workforce development priorities, namely high importance and low worker preparedness, is described. The last section of the chapter provides a mechanism to identify what skills and competencies agriculture teachers and faculty need to develop and enhance in order to teach students the skills they need to enter and be successful in the agriculture sector.

DOI: 10.4018/978-1-6684-4050-6.ch005

INTRODUCTION

At every station on the London Underground, the recorded message is the same: "Mind the Gap." Passengers are reminded that sometimes the floor of the train car and the level of the platform are not the same; sometimes the train car is several inches away from the platform. Be careful and mind the gap! For agricultural development successful in Arab countries and elsewhere, there is a need to "mind the gap," to be able to match the skills of potential workers in agriculture and competencies skills needed in the agricultural industry. This chapter is designed to provide leadership in addressing those issues and concerns. The objectives for this chapter are 1) to describe the need for a competent workforce for agricultural development Arab countries; 2) to describe the process of conducting a skill gap analysis; 3) To discuss methods for identifying agriculture industry workforce competence and skill needs; 4) To discuss methods for identifying the skills and competencies of potential agricultural workforce; and 5) To describe a process to institute agricultural competency and skill development into schools.

THE NEED FOR A COMPETENT WORKFORCE FOR AGRICULTURE DEVELOPMENT

In May 2017, the World Economic Forum issued an Executive Brief on the future of jobs and skills in the Middle East and North Africa. The report indicated several common themes regarding workforce issues and agricultural development. The labor markets show a low but increasing level of workforce participation by women and a reliance on foreign workers. Additionally, high rates of unemployment and under-employment especially among young workers and the relatively well-educated workers continues to be a detriment to an expanding workforce. Throughout the Middle East and North Africa region, it is estimated that only 62% of its full human capital potential is realized. And as the world economy continues to evolve, automation can become a concern when thinking in terms of welcoming new workers into the workforce. The Forum proposed that the current challenges will require broad reforms and public-private collaborative efforts.

A policy brief published by FAO in 2020 outlined various aspects of the need for skills development in the agriculture sector of Lebanon. In that country up to 25% of the Lebanese workforce draws its income from farming and related work. A similar situation exists in Syria. An issue raised in that policy brief was that the budget of the Ministry of Agriculture has remained well below 1% of the total expenditures of the government for most years over the past three decades. The overall lack of sufficient funding has led to challenges for the extension system throughout the country. These shortcomings disproportionately affect poor farmers and contribute to sustained inequalities in the agricultural sector. A similar situation exists for the technical and vocational education training programs. Enrollment in the agricultural technical schools is low, which leads to a lack of a trained workforce in agriculture.

Maiga et al. (2020) purported that there has been a declining trend of youth participation in agriculture since 2000 in low- and middle-income countries, mainly in favor of the service sector, which precipitates migration from rural to urban areas. Increased educational attainment for rural youth coupled with inability to rent or own land is a driver of urban migration. In addition, the increasing ageing farmer population in rural areas exacerbates the demographic pressure on land at the expense of the youth. A further constraint on youth engagement in agriculture is a lack of education in disciplines related to agriculture or skills training. Various studies of youth have shown that a large majority identified knowledge of farming practices as a pre-requisite to setting up a viable farm. Further, government initiatives to increase skills and productivity have generated interest among youth in joining the sector, and vocational training has been viewed as increasing the likelihood of a successful career in agriculture. Such findings challenge an assumption common in policy proposals that youth are not interested in agriculture. With the development of information and communication technology (ICT), young people have more opportunities to strengthen their skills and access relevant information and are therefore well positioned to understand market dynamics and institutional and financial systems, enabling them to initiate and capitalize on processes of change in the agricultural sector. Therefore, perhaps the youth themselves are not the central problem in agriculture workforce development.

There are many potential benefits to increased agricultural development in the Arab world. Obviously from an industry standpoint, investment in the agriculture sector can lead to increased profits. But from a societal and development perspective, increased success in the agriculture sector can also lead to improving the economic ecological social and security situation in the country where agribusinesses create jobs and provide education for the labor force. Improved economic conditions may also improve stability (The Hague Centre for Strategic Studies, 2015).

Figure 1. Benefits of Investment in Agricultural Education for Development
Source: Adapted from Gehem et al. (2015)

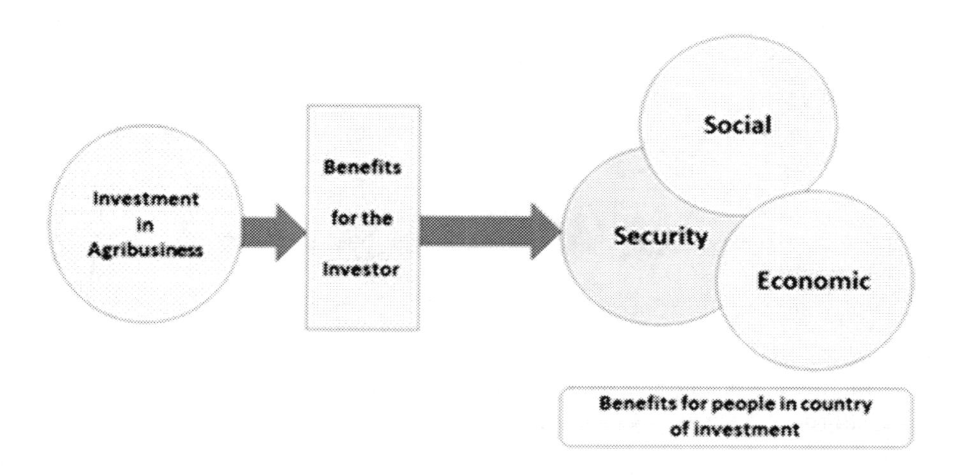

The importance of combining education and agriculture is evident when considering issues related to the agriculture sector (Wopertz, 2017). Agriculture continues to experience a decline in gross domestic product contribution and employment capacities when compared to other industries. The Middle East and North Africa region is the largest food importer in the world, making it highly vulnerable. Agriculture is the largest water consumer in the region and provides a substantial proportion of employment. In the past, agricultural expansion has been questionable because of its reliance on limited water resources and possibly producer subsidies. A well-developed agricultural education program would contribute to a greater understanding of the importance of efficient use of land and water resources rather than only concentrating on increasing agricultural production. Market access is an important consideration to improved agricultural economies.

In countries that are attempting to address the issues of the agriculture and food sector, skill development has become a high priority with the objective of creating programs for youth as well as for the youth to recognize the value of experience and knowledge. The focus of these programs has been on the quality of training, assessment, and certification, therefore helping to ensure high standards and greater market acceptability. Governments have been a force behind the needed change in designing, implementing, and financing such programs. The private sector, including potential employers as well as global partners, has been widely recognized for improving the scope, target, and outcomes of skill development programs. Increased technology creates demand for a better skilled workforce which should lead to better paying jobs and higher returns in farming and agribusiness. Innovative outreach models can also lead to substantial skill improvement for a low investment.

Several organizations and nonprofit agencies throughout the Arab region have also provided leadership for programs to enhance worker competence and therefore improve agricultural development. The Arab Organization for Agricultural Development (AOAD, n.d.) addresses a number of socioeconomic reasons for cooperation. Among them are:

- the desire to establish agricultural and economic sectors on solid foundations
- the recognition of the importance of the agricultural sector within the Arab economy
- the realization that Arab agricultural resources have not been exploited to their fullest nor in the most efficient and effective manner
- similarities clearly exist across the Arab world in environmental and socioeconomic conditions as well as their agricultural challenges
- an integrated Arab agricultural policy can be attained only through coordination among the countries.

In many countries it may take 8 to 12 years for basic skills attainment among children, which is extremely inefficient. Hungry, unhealthy children cannot learn; this is an immediate disadvantage in basic skills development. Students must also be able to use those basic skills. Additionally, skills in decision-making, problem-solving, and technical competence are essential for enhancing and expanding the agricultural workforce. High quality education at the primary school level generates higher returns, both at the primary level and later. Improvements in quality and efficiency of basic education is urgently needed (Fasih, 2008).

Workers with better skills become entrepreneurs rather than only laborers. Competence development leads to a need for and more interest in additional education. A productive, efficient workforce creates, sustains, and expands economic development. A productive, satisfied workforce leads to higher earnings, more civic engagement, and reduced crime. A productive workforce alleviates poverty. Most key factors for improving labor market outcomes are economic factors. But those economic factors are highly dependent upon education. Governments need to maintain economic stability in the region and private enterprise has to create and sustain labor opportunities.

A white paper by the Boston Consulting Group (BCG) (Puckett et al., 2020) posited that the problem, at least in part, is the result of ineffective or even nonexistent communication between the private sector and education. Without change, the system will continue to prepare people whose competencies and skills are outdated and no longer useful in the workforce. The authors identified several challenges to be addressed.

- A need to focus on jobs that do not yet exist
- Engaging the workforce in lifelong learning
- Workers lack motivation and accountability for personal development
- Access to labor market opportunities is limited, with nearly a third of the world population lacks internet access
- Job locations and not readily available geographically to potential worker locations

The BCG (2020) concluded:

The future economy demands a human-centric approach. This approach must be used to help workers acquire fundamental skill sets and create an enabling environment for lifelong employability, self-realization, and skills liquidity in a labor market that offers accessible opportunities and is inclusive and focused on people. At a time of high uncertainty, people need the cognitive and noncognitive skills and knowledge necessary for adapting to employers' changing requirements. They need skills that the labor market will continue to demand. At the same time, to be able to choose a career path, and in order to fully unlock their potential, people must take responsibility for their own professional development.

SKILL-GAP ANALYSIS

Armstrong's Deficiency Model (2002), as reported in Othayman, Mulyata, Meshari, and Debrah (2022), identified four components related to the potential gap between organizational needs and employee performance.

· Organizational analysis – short- and long-term goals, and factors affecting those goals
· Requirement analysis – requirements of the job
· Task and knowledge and ability analysis – specifying tasks required on the job; skills, knowledge, and attitudes needed to perform
· Person analysis – how well the employee performs

The first component, organizational analysis, includes establishing goals for the business or organization. In agricultural development, these goals may address productivity, efficiency, profit margin, personnel factors, and others. Short-term goals may focus on only one fiscal year, while long-term goals should probably capture a five-year plan. In both cases, the goals are not static but are dynamic, open to minor or even major updates as time elapses. For the goals to be manageable, industry leaders must be aware of the factors, both internal and external, that may affect the goals. Management itself is clearly an internal factor that affects growth and development. Externally, market demand, pricing, and regulations can play an important role in goal attainment.

Requirement analysis pertains to identifying the requirements of the specific job or jobs within the agricultural sector or business. Typically, those requirements would be enumerated in job descriptions and qualifications. For a skill-gap analysis to be helpful in decision-making, job requirements should include the specific skills and competencies that a contributing worker must possess.

The job requirement analysis leads to the task, knowledge, and ability analysis. At this point, each competency needed to complete the job is analyzed to identify the specific tasks and steps that are required to fulfill the competence or complete the required skills. Tasks and skills require specific knowledge (cognitive skill) and ability (psychomotor skill) to be completed. Further, and much more difficult to measure, attitude toward being engaged in the competency or skill can be important.

The last component of the deficiency model is person analysis. While the first three components focus on the required competency of the worker, person analysis addresses the actual worker – how well the employee performs the competencies and skills. Regardless of how well the goals, skills, and competencies are identified and clarified, well-prepared workers are the key to success. Information gained from such an analysis should lead to identifying the gap in skills and competencies among employees, leading to changes in training and development programs.

The Boston Consulting Group (2021) posited that skills-gap perhaps is more of a skills-mismatch. Their framework, labeled Future Skills Architect, allows governments, policy makers, and businesses to evaluate the labor situation in three constructs: capabilities, motivation, and opportunities. Capabilities include skill sets, motivation includes the opportunity for individuals to self-actualize and reach full potential, and opportunities involve equal access to jobs, irrespective of age or any other demographic or personal characteristic.

Skill-gap analysis is an outcomes assessment tool that is designed to measure skills and competencies that are more important to employers and in need of development by potential workforce participants (Vreyens & Shaker, 2007). The results of the skill-gap analysis provide guidance for curriculum development, utilizing the "backward design" approach of Wiggins and McTighe (1998). Using Armstrong's Deficiency Model (2002) components, the skill-gap analysis begins with specific tasks, skills, knowledge, and attitudes that are required to complete the job successfully. Concurrently, the ability of potential employees to perform is identified.

AGRICULTURE INDUSTRY NEEDS AND WORKFORCE COMPETENCE AND PREPAREDNESS

As noted in Chapter 1 of this text, a skill-gap analysis begins with an assessment by industry personnel to establish a comprehensive list of skills and competencies that are a part of a specific sector of agriculture development, such as field crop production. This process is developed over time, with input directly from employers, managers, and supervisors who are very familiar with the specific agriculture sector being studied. Industry and employer experts rate each skill and competency on the relative importance of each item on a 5-point Likert-type scale ranging from 5-Very Important to 1-Not Applicable for the Job. Additionally, the industry leaders and employers rate each item on the perceived level of competence of the graduates of the respective program, using the scale 5-Highly Competent to 1-Not Competent. This part of the skill-gap analysis forms the survey instrument used in the next phase of the process. Each skill or competency must be written so that people outside the current employment can respond adequately.

The second part of the skill-gap analysis process involves identifying recent graduates of related programs (following the example from above, field crop production). Program graduates are surveyed using the same list of skills and competencies on two scales. The first part of the instrument asks the graduates their perceptions of the importance of the skills and competencies on the scale 5-Very Important to 1-Not Applicable for the Job. The second part of the instrument identifies the perceptions of graduates regarding the level of preparedness to conduct or perform the skills and competencies on a scale of 5-Prepared to 1-Not Prepared. Table 1 provides a summary of the scales.

Table 1. Skill and competence importance, preparedness, and competence

Graduate Scale		Employer Scale	
Importance	**Level of Preparedness**	**Importance**	**Level of Competence**
1-Not applicable for the job	1-Not prepared	1-Not applicable for the job	1-Not competent
2-Not Important	2-Somewhat prepared	2-Not important	2-Somewhat competent
3-Somewhat important	3-Prepared	3-Somewhat important	3-Competent
4-Important	4-Very prepared	4-Important	4-Very competent
5-Very important	5-Highly prepared	5-Very important	5-Highly competent

Source: *Adapted from Vreyens and Shaker (2005)*

The results of the two surveys can be summarized into several categories.

- Industry and employer leaders rate a skill as Very Important and the graduates as Highly Competent, it can be concluded that there is no pressing need to adjust the curriculum of programs that are preparing graduates.
- The industry experts rate the skill as Very Important and graduates as Not Prepared, major curriculum and program enhancements are warranted.
- The industry experts rate the skill as Not Applicable for the Job, the level of competence of graduates is relatively unimportant. Perhaps the curriculum and program should consider removing that particular skill of competence from the preparation of graduates, providing time for including instruction for skills and competencies that need to be added.
- A major disconnect occurs when graduates perceive the skill to be Very Important, but they perceive that they are Not Prepared to perform the skill. Again, major considerations for program and curriculum enhancement are warranted.

AGRICULTURE WORKFORCE NEEDS AND EMPLOYEE COMPETENCY

A study conducted in Egypt (Vreyens & Shakar, 2005) provides several examples of the types of skills and competencies best utilized in a skill-gap analysis. If the requirements of the agriculture sector are not clearly stated as skills and competencies to be attained, the graduates of agriculture programs will not be able to identify their abilities.

The Egypt study grouped skills into three categories: Communication skills, Management skills, and Technical skills. These examples serve only as guidance for agriculture industry leaders in other Arab regions to identify skills in their sectors.

- *Communication skills*
    - Interpret written information
    - Write effectively
    - Analyze information
- *Management skills*
    - Think creatively
    - Apply problem-solving skills
    - Develop a basic budget
- *Technical skills*
    - Develop and irrigation management plan for multiple crops
    - Select and breed high quality fruits

  - Calculate costs of production for a specific crop (Adapted from Vreyens & Shakar, 2005)

In the Egypt study, graduates ranked the following skills as having *high importance* but ranked themselves *low on preparedness* to demonstrate these skills upon graduation:

- Analyze information
- Speak effectively with a target audience
- Think creatively
- Apply problem-solving skills
- Apply time management skills
- Develop a basic budget
- All computer skills including word processing, spreadsheets, presentations, e-mail for communication, access to the internet for resources and information
- Evaluate the competitive environment and identify opportunities
- Demonstrate knowledge of regulations impacting firm and industry
- Conduct situational analysis
- Manage and maintain financial records
- Use tissue culture to vegetatively propagate horticulture species
- Sort or grade fruits using international standards
- Operate milking machine equipment
- Evaluate and manage animal environmental
- Handle livestock
- Grade/evaluate wool
- Negotiate contract for buying or selling products
- Establish and develop contacts with suppliers and marketers
- Manage post-harvest handling of major fruit crops to ensure long shelf life and
- quality produce
- Calculate costs of production for a specific crop
- Design a year-round growing plan for a field/greenhouse

Graduates indicated a few areas where they ranked the importance of the skill to their current job essential as well as a high ranking for the level of preparedness in the following areas:

- Name scientific names of horticulture crops of Egypt
- Conduct, test and identify microbial analyses for pathogens
- Monitor sanitary milk delivery and transportation system

- Formulate complete balanced rations for livestock
- Identify diseases, assess risks and define mitigation strategy for crops
- Calculate application rates and apply herbicides

In contrast, employers listed the widest gap in importance of the skill on the job to the level of competence in the following areas:

- Analyze information
- Applying time management skills
- Develop a basic budget
- Access internet for resources/information
- Calculate credit options for investment or rates of return
- Analyze the chemical composition of a feedstuff
- Conduct, test and identify microbial pathogens
- Design appropriate packaging for processed foods
- Write a farm business plan
- Conduct cost benefit analysis of an agricultural project
- Evaluate competitive environment and identify opportunities
- Demonstrate knowledge of regulations impacting firm and industry
- Regulate fertilizer application, and make recommendations for soil amendment
- Develop an irrigation management plan for multiple crops
- Use tissue culture to vegetatively propagate horticulture species
- Select and breed high quality fruits and vegetables
- Sort or grade fruits using international standards

The data also revealed areas that were considered not important for graduates and for which they were not prepared as well. These skills appear to be of little importance to the employers as skills or qualifications of their new hires. Among these skills are: evaluating abnormal milk, command milk marketing skills, grade and evaluating wool, design a landscaping plan, identify key weed problems and define a mitigation strategy,calculate costs of production for a specific crop. While these results are specific to a study in Egypt in a carefully identified agriculture region, the concepts should provide excellent guidance to agriculture industry leaders and educators in other areas of the Middle East and North Africa. As Vreyens and Shakar (2005) summarized, the process of writing curriculum has been to convene a panel of experts in the field, select the topics and content to include in a course, and engage highly regarded experts to write a text for the course. The skill-gap analysis begins at the end; the skills and competencies needed and the competence and preparedness of the potential workforce.

In addition to the current assessment of job skills and graduate preparedness, the areas of potential growth in job opportunities should be considered. A study by Ebner, Ghmire, and Saleh (2020) reported the perceptions of employers and university faculty in agriculture regarding potential growth in various sectors of agriculture in Egypt. While there was not perfect agreement between the two respondent groups, the study did provide guidance that could be utilized in making curricular and training decisions. Examples of the higher growth sectors are listed below, only as an idea of how similar studies in other Arab nations may develop from the list of 24 agriculture sectors in the report.

- Animal production – poultry
- Food and beverage processing
- Horticulture – protected cultivation
- Water resource management
- Horticulture – vegetables
- Energy, biofuels, alternative energy
- Chemicals, pesticides, fertilizers

That same study investigated skills needed by new employees in agriculture. From an initial list of 35 skills, employer and faculty perceptions agreed on seven skill areas.

- Ability to function in a team
- Ability to access different resources for information
- Working with others from diverse backgrounds
- Self-motivation to learn new things and work
- Ability to make ethical decisions
- Knowledge of ethics and best practices in field
- Adaptability to changes in the fields or workplace

The important point to consider in future skill-gap studies is the need to include personal skills as well as technical skills in the analysis.

The final part of the Ebner et al. (2020) study was identifying the skills gaps in new university graduates as perceived and/or observed by potential employers. The complete list included 35 skills, with the following list only as example of those skills gaps appearing high on both lists.

- Time management
- Ability to plan and organize
- Knowledge of/ability to apply technologies specific to a job

- Adaptability to changes in the field or workplace
- Ability to function as a team
- Self-motivation to learn new things and work
- Knowledge of/ability to apply technical skills specific to a job

Employers want workers who possess communication, management, and technical skills required by the job. University and other school faculty want their graduates to be successful. It is therefore imperative that the two groups work closely together to identify the areas of need in agriculture graduates so that the agriculture sector employers will hire them, and the graduates will succeed.

PREPARING COMPETENT WORKERS FOR AGRICULTURAL DEVELOPMENT

The *Model of Agricultural Education for Development in Arab Countries* in Chapter 1 of this text includes three parts: Assessment, Content, and Process. The Assessment and Content sections, when utilized by employers and educators, lead to the final phase of the model: utilizing the information gained to develop and enhance education and training programs that result in a well-prepared workforce for agricultural development.

Othayman et al. (2022) raised several concerns regarding training needs assessment (TNA). First, the authors posited that there is typically a lack of any method for determining training needs, making it difficult to determine clear program goals. Training and development programs appear to be based only on the creator's own desires and beliefs. Oftentimes TNA is based on trial and error rather than on substantiated findings from inquiry.

A number of authors and reports have provided guidance for addressing the Process section of the model (Carnevale, Jayasundera, & Hanson, 2012; McKinsey & Co., 2011; Fasih, 2008; U.S Department of Health and Human Services, 2006; and Working Group for International Cooperation in Skills Development, 2001). While there are several economic factors involved in improving labor market outcomes, all are dependent upon education to be viable. Creating responsive education programs requires improved efficiency in education spending, curricular development and adjustments, supply vs. demand analyses, and enhanced teaching in the schools and industries.

Education in general has many challenges in the Arab region, but it is often technical and vocational education and training (TVET) that faces the greatest challenges simply because it serves the highest number of students and graduates compared to the academic stream (El Ashmawi, 2015). Various studies and reports have highlighted the lack of relevant technical and employability skills among the workforce as a major constraint to economic development and global competitiveness. There is a pressing need to change the current mind-set from one that views TVET as a marginal tool for easing the social impact of school dropouts and low performers to a means for aligning labor supply with industry demand. El Ashmawi further stated that most TVET systems lack the quality and relevance needed to create a competitive workforce. Stakeholders in the Arab countries stress the need to strengthen the link between education and training institutes, but the connect between both sides is weak and not institutionalized. Some countries have adapted different options, like expanding work-based learning with extensive involvement from employers at all stages, from planning to training schoolteachers on industry trends. Other countries have established groups that serve as the institutional link between TVET institutes and employers, specifying standards, developing qualifications, conducting research and developing needs assessments (El Ashmawi, 2015).

There is more than one route or pathway to entering and advancing in the agricultural workforce. Career and technical education in secondary and post-secondary schools (agriculture technical schools, community colleges), employer-based training, industry-based training, apprenticeships, and certificate programs are all viable methods to be used in workforce preparation. Early career exploration in primary/elementary schools can be a valuable part of the much-needed improvements in basic skills education.

The differences between what skills and competencies the agricultural industry needs and what potential employees possess could be referred to as the missing link. The question, then, becomes, what to do to span the link.

- Create industry/education partnerships to invest in training and education
- Upgrade school facilities to replicate industry standards
- Include personal development skills in the school curricula, including student organizations such as the Future Farmers of Korea and the Future Farmers of Japan
- Create additional education opportunities for technical school graduates (community college, etc.)
- Co-locate technical high schools and community colleges to share facilities and coordinate programs
- Devise career and technology courses for students in general high schools
- Expand instructor development programs

Agricultural development relies on economic strength of a country's rural region (Rasmussen, Pardello, Vreyens, Chazdon, Teng, & Liepold, 2017). While much of agricultural development has focused on the private sector and non-government organizations, leadership for change must also come from the public sector, including schools and universities as well as local organizations such as farmer associations. Identifying skills and competencies need by agricultural workers, then, must also incorporate areas such as leadership and strategic planning. A program developed and initiated in Morocco (Rasmussen et al., 2017) utilized a comprehensive inventory of skills and competencies to design and present a series of workshops for local farmer leaders. A brief list of some of the topics in the workshops provides guidance for others as similar programs are developed.

- Analyze the characteristics of leaders
- Identify action steps for change
- Identify skills needed for planning
- Differentiate among communication styles
- Analyze information or tasks needed to reach a goal
- Compare and contrast principles of leadership and management
- Plan an effective meeting
- Identify personal leadership strengths and opportunities for growth
- Determine the value of working with a diverse group of people
- Practice conflict management skills

Education, skills development and technical training are central to agricultural and rural employment. They prepare mostly young people for work in the formal and informal sector...and play an important role in poverty reduction. The better the training and the more refined the skills in terms of human capital, the higher the income and returns and the better livelihoods (Hartl, 2009).

If progress is to be made in developing and enhancing the agricultural workforce in the Arab region, it is imperative that leaders in public and private entities join together to establish base-line data for decision-making. The process includes a thorough and honest evaluation of the capabilities of potential workers upon entering the workforce from schools and colleges as well as a comprehensive study of the skills and competencies needed to in the agricultural sector. With that information, steps can be taken to establish, develop, and enhance training programs that will address the workforce issues.

DEVELOP TRAINING AND INSTRUCTIONAL PROGRAMS TO PREPARE THE AGRICULTURAL WORKFORCE

As noted earlier in this text, a culminating step in applying what was learned through the skill-gap analysis includes assessing the in-service needs of faculty and instructors who are responsible for curricular decisions and offering instruction in the skills and competencies needed for students to successfully enter and progress in the agriculture workforce (DiBenedetto, Willis, & Barrick, 2018). Consistent and timely in-service needs assessment can determine the topics, skills, and competencies to be included in professional development programs for faculty and instructors, part of the Process section of the *Model of Agricultural Education for Development in Arab Countries.*

Desimone (2009) proposed a conceptual framework for studying professional development, including five core features.

· Content focus
· Active learning
· Coherence
· Duration
· Collective participation

Context such as teacher and student characteristics, curriculum, school leadership, policy environment

Figure 2. Conceptual framework for studying the effects of professional development on teachers and students
Source: Adapted from Desimone (2009)

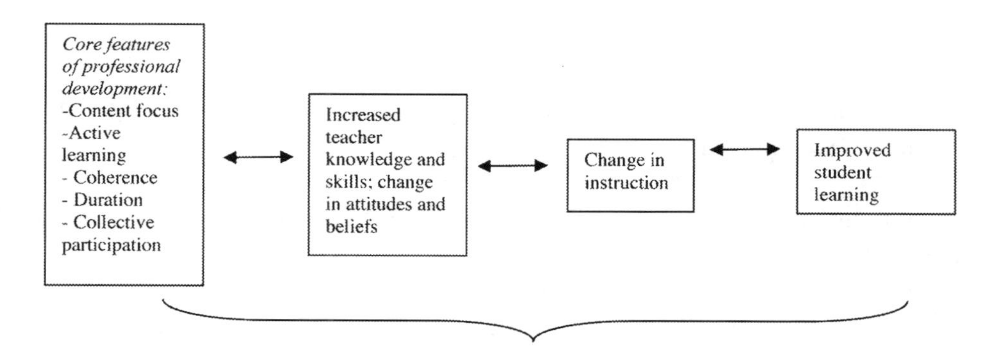

Desimone's framework indicates that the core features of successful professional development can lead to increased teacher knowledge and skills and a positive change in

attitudes and beliefs. Those improvements result in changes in instructional behaviors that lead to improved student learning. When students learn more and development new and enhanced skills and competencies in the agriculture sector, a well-educated workforce becomes available for the agriculture sector. Periodic changes in the core features and their adaptation to specific settings in Arab agriculture development ensures that new cadres of graduates will be relevant in the future.

A study by Swanson, Barrick, and Samy (2007) described the dilemma surrounding the instructional programs in Egyptian faculties of agriculture. Declining enrollments, of course, led to fewer well-prepared agriculture workers to enter the agriculture sector of employment. Limited funds prevented faculty from participating in activities to develop and enhance curricula and teaching or to engage in research. A lack of contact with the private sector added to the situation. Consequently, faculty members individually and collectively were unaware of the skills and knowledge needed by their graduates to be employed and be successful in in technical and management positions in the private sector. To address the issues, a series of workshops was designed and implemented over a period of several years, based on the identified needs of the faculty.

Observations and reflections of workshop presenters (Roberts et al., 2008) can provide guidance for similar endeavors. The following are among those observations.

- Knowing the background and situation where programs are conducted is essential. While similarities exist across the region, important differences must be considered.
- Clearly identified goals and objectives, shared with participants and leaders, must be used to guide the activities.
- Flexibility is essential; the program must be adjusted as the situation dictates.
- Participants must be carefully selected. Only those who are involved in curriculum and instruction in the agriculture sector, and who desire to improve programs, should be involved.
- Workshop leaders must know the potential audience, learn the culture, and understand the local situation.

If agricultural schools and colleges are designed to prepare society-ready graduates for the agriculture sector, then students must be prepared with cognitive skills that reflect real-world experiences. For that to occur, instructors in all levels of agricultural curricula must be adequately prepared to develop, coordinate, and supervise experiences that occur in classrooms, laboratories, school fields and facilities, and local agriculture industry (Thoron et al., 2010).

Active learning is engaging students in doing things and in thinking about what they are doing. Learning is an active process where new knowledge is acquired in relation to previous knowledge, and that information becomes meaningful when it is presented in a recognizable framework. Active learning involves providing opportunities for students to meaningfully reflect on the content, ideas, issues and concerns of the subject.

Many faculty believe that all learning is inherently active; therefore, students are actively involved when listening to a formal presentation. But research suggests that students must also read, write, discuss, and/or be engaged in problem solving activities to be actively involved. Active learning engages students in higher-order thinking (analysis, synthesis, evaluation) and helps students develop skills in handling concepts within the discipline, reflect upon ideas, and reflect on how to use those ideas.

Agriculture technical school teachers and university faculty typically are not formally prepared to teach, having attained an education in a technical field of agriculture. If the goal is to prepare work-force ready graduates, teachers must engage students in higher order thinking by utilizing active learning strategies in their classrooms and labs. Therefore, teachers need to learn new strategies and develop new skills in teaching.

Experiential learning activities such as internships are also relevant for teaching career skills to students. Experiential learning can help students learn to transfer knowledge, understand problems in agriculture, develop their self-confidence, connect practice and principle, improve psychomotor skills, develop problem solving skills, retain more knowledge, and become interested in learning. Therefore, instructors must be equipped to develop, conduct, and assess experience-based activities as part of the agriculture curriculum.

Needs assessments aid in identifying the competencies that should be emphasized in teacher development to better equip those teachers with adequate knowledge, experiences, resources, and materials (DiBenedetto et al., 2018). Teachers' needs also change over time, so it is important to consider the future as well as the present. Teachers often believe that their skills and knowledge are inadequate; those perceptions must be considered, in addition to industry needs, when planning teaching enhancement programs. Teachers generally have a desire and need for inservice training to ensure their skills will remain current and relevant (Barrick, Ladewig, & Hedges, 1983). The Barrick et al. study was among the first in agriculture teacher training to utilize the Borich Model for prioritizing topics for inservice education programs.

The Borich Model, as utilized by Barrick, et al., includes three factors: the *Importance* of the competency or skill in the industry, the *Knowledge* possessed by the instructor, and the *Ability* the instructor has to teach and perform the competence or skill. A basic premise is that, when asked what topics to include in an inservice program, many instructors will identify skill areas they enjoy teaching and have expertise in the area. To suggest other topics could be misconstrued as the instructor cannot teach or perform that task. The model includes the *Importance* rating to help overcome that concern.

Using the results of the skill-gap analysis, the list of skills and competencies that show disparity between industry needs and employee abilities is constructed. To be clear, these are the skills highly needed in industry that are perceived to be inadequate among graduates and workers. Those become the skills that need to be emphasized in training and development programs. Agriculture teachers (at any level of instruction) indicate their perceived level of *Importance*, their perceived level of *Knowledge*, and their perceived level of *Ability to perform* each skill or competency.

From the ratings, two weighted scores are derived.

Weighted Knowledge Score = (Importance Mean – Knowledge Mean) x Overall Importance Mean

Weighted Ability Score = (Importance Mean – Ability Mean) x Overall Importance Mean

The weighted scores are then ordered from highest to lowest. The higher the weighted score, the higher the priority in including that skill or competency in the agriculture instructor and/or faculty development program. Note that the actual values of the weighted scores are not a standard measurement but are only useful in relation to the other scores. The higher weighted knowledge scores typically indicate the need for instruction in the fundamentals of the content area; higher weighted ability scores indicate the need for "hands-on" instruction, allowing participants to apply what they know to carrying out activities.

An important aspect of the Barrick, et al. study was an analysis of the three rankings by the participants (secondary school agriculture teachers). The relative rankings of Importance, Knowledge, and Ability did not exceed a moderate correlation, indicating that using only one of the rankings may be misleading in identifying the most-needed topics for instructor development.

The preparation and inservice education for agriculture instructors is crucial to a successful implementation of the *Model of Agricultural Education for Development in Arab Countries.*

SUMMARY

This chapter provided an overview of several factors important in agricultural development in Arab countries. Attention was given to identifying the skills and competencies needed by workers as identified by agricultural employers and others.

Concurrently, the abilities of graduates from secondary schools and colleges to perform the needed skills and competencies was discussed. Using those two sources of information, the utilization of a skills-gap analysis was described. Finally, once the needs of potential workers are identified, a system of prioritizing inservice education programs for agriculture instructors and faculty was proposed.

REVIEW QUESTIONS

1. What are the most important skills and competencies needed by workers in a specific part of the agriculture sector?
2. What are the agricultural skills and competencies that agriculture graduates perceive they are ill-prepared to demonstrate in the sector addressed above?
3. Define skills-gap analysis.
4. What are the major factors involved in conducting a skills-gap analysis?
5. How can industry and education leaders utilize the results of a skills-gap analysis?
6. Describe the Borich Model.
7. How can the results of the Borich Model investigation be used to improve instruction in agricultural schools and faculties?

REFERENCES

AOAD. (n.d.). *Arab Organization and Development*. https://www.aoad.org/Eabout.htm

Armstrong, M. (2002). *Employee reward*. CIPD Publishing.

Barrick, R. K., Ladewig, H. W., & Hedges, L. E. (1983). Development of a systematic approach to identifying technical inservice needs of teachers. *The Journal of the American Association of Teacher Educators in Agriculture*, *24*(1), 13–19. doi:10.5030/jaatea.1983.01013

Bloom, B. S. (1956). *Taxonomy of Education Objectives: The Classification of Education Goals*. Edwards Brothers.

Boston Consulting Group. (2021). *Addressing the Middle East's growing skills mismatch*. Author.

Carnevale, A. P., Jayasundera, T., & Hanson, A. R. (2012). *Five ways that pay*. Georgetown Public Policy Institute.

Desimone, L. M. (2009). Improving impact studies of teachers' professional development: Toward better conceptualizations and measures. *Educational Researcher, 38*(3), 181–199. doi:10.3102/0013189X08331140

DiBenedetto, C. A., Willis, V. C., & Barrick, R. K. (2018). Needs assessments for school-based agricultural education teachers: A review of literature. *Journal of Agricultural Education, 59*(4), 52–71. doi:10.5032/jae.2018.04052

Ebner, P., Ghimire, R., Joshi, N., & Saleh, W. D. (2020). Employability of Egyptian agriculture university graduates: Skills gaps. *Journal of International Agricultural and Extension Education, 27*(4), 128–143. doi:10.5191//jiaee.2020.274128

El Ashmawi, A. (2015). *The skills mismatch in the Arab world: A critical view. British Council Cairo Symposium*.

Fasih, T. (2008). *Linking education policy to labor market outcomes*. The World Bank. doi:10.1596/978-0-8213-7509-9

Gehem, M., van Duijme, F., Ilko, I., Mukena, J., & Castelion, N. (2015). *Maooing agribusiness opportunities in the MENA: Exploring favorable conditions and challenges for agribusiness in the Middle East and North Africa*. The Hague Centre for Strategic Studies.

Hartl, M. (2009). Technical and vocational education and training (TVET) and skills development for poverty reduction – Do rural women benefit? Presented at the *FAO-IFAD-ILO Workshop on Gaps, Trends and Current Research in Gender Dimensions of Agricultural and Rural Employment: Differentiated Pathways Out of Poverty*, Rome, Italy.

ILO & FAO. (2020). *Skills development for inclusive growth in the Lebanese agriculture sector – Policy Brief*. FAO.

Maiga, W. H. E., Porgo, M., Zahonogo, P., Amengnalo, C. J., Coulibaly, D. A., Flynn, J., Seogo, W., Traore, S., Kelly, J. A., & Chimwaza, G. (2020, October). A systematic review of employment outcomes from youth skills training programmes in agriculture in low- and middle-income countries. *Nature Food, 1*(1), 605–619. doi:10.103843016-020-00172-x

McKinsey & Co. (2011). *Linking jobs and education in the Arab world*. Author.

Othayman, M. B., Mulyata, J., Meshari, A., & Debrah, Y. (2022). The challenges confronting the training needs assessment in Saudi Arabian higher education. *International Journal of Engineering Business Management, 14*, 1–13. doi:10.1177/18479790211049706

Puckett, J., Hoteit, L., Perapechka, S., Loshkareva, E., & Bikkulova, G. (2022, January). *Fixing the global skills mismatch.* Boston Consulting Group.

Rasmussen, C., Pardello, R. M., Vreyens, J. R., Chazdon, S., Teng, S., & Liepold, M. (2017). Building social capital and leadership skills for sustainable farmer associations in Morocco. *Journal of International Agricultural and Extension Education, 24*(2), 35–49. doi:10.5191/jiaee.2017.24203

Roberts, T. G., Thoron, A. C., Barrick, R. K., & Samy, M. M. (2008). Lessons learned from conducting workshops with university agricultural faculty and secondary school agricultural teachers in Egypt. *Journal of International Agricultural and Extension Education, 15*(1), 85–87. doi:10.5191/jiaee.2008.15108

Rosenshine, B., & Furst, N. (1971). Research on Teacher Performance Criteria. In B. Othanel Smith (Ed.), *Symposium on Research In Teacher Education* (pp. 37-72). Prentice-Hall, Inc.

Swanson, B. E., Barrick, R. K., & Samy, M. M. (2007). Transforming higher education in Egypt: Strategy, approach and results. *Proceedings of the 23rd Annual AIAEE Conference.*

Swanson, B. E., Cano, J., Samy, M. M., Hynes, J. W., & Swan, B. (2007). Introducing active teaching-learning methods into Egyptian agricultural technical secondary schools. *Proceedings of the 23rd Annual AIAEE Conference.*

Thoron, A. C., Barrick, R. K., Roberts, T. G., Gunderson, M. A., & Samy, M. M. (2010). Preparing for, conducting and evaluating workshops for agricultural technical school instructors in Egypt. *Journal of Agricultural Education, 51*(1), 75–87. doi:10.5032/jae.2010.01075

U. S. Department of Health and Human Services. (2006). *Linking education and employment for brighter futures.* Author.

United States Department of Education (2008). *Building an effective advisory committee. Mentoring Resource Center Fact Sheet. No. 21.* educationnorthwest.org

Vreyens, J. R., & Shaker, M. H. (2005). Preparing market-ready graduates: Adapting curriculum to meet the agriculture employment market in Egypt. *Proceedings of the 21st Annual AIAEE Conference.*

Wiggins, G., & McTighe, J. (1998). *Understanding by Design*. Association for Supervision and Curriculum Development.

Wopertz, E. (2017). *Agriculture and development in the wake of the Arab Spring*. International Development Policy, Revue international de politique de developpement. https://journals.openedition.org/poldev/2274

Working Group for International Cooperation in Skills Development. (2001). *Linking work, skills and knowledge*. Author.

World Economic Forum. (2017, May). *The future of jobs and skills in the Middle East and North Africa*. World Economic Forum.

ADDITIONAL READING

Barrick, R. K. (2012, October). *Missing links: Education and the job market*. Ministry of Education, Arab Republic of Egypt.

Barrick, R. K., Roberts, T. G., Samy, M. M., Thoron, A. C., & Easterly, R. G. III. (2011). A needs assessment to determine knowledge and ability of Egyptian agricultural technical school teachers related to supervised agricultural experience programs. *Journal of Agricultural Education*, *52*(2), 1–11. doi:10.5032/jae.2011.02001

Barrick, R. K., Samy, M. M., Gunderson, M. A., & Thoron, A. C. (2009). A model for developing a well-prepared agricultural workforce in an international setting. *Journal of International Agricultural and Extension Education*, *16*(3), 25–32. doi:10.5191/jiaee.2009.16303

Barrick, R. K., Samy, M. M., Roberts, T. G., Thoron, A. C., & Easterly, R. G. III. (2011). Assessment of Egyptian Agricultural Technical School instructors' ability to implement experiential learning activities. *Journal of Agricultural Education*, *52*(3), 6–15. doi:10.5032/jae.2011.03006

Chickering, A. W., & Gamson, Z. F. (1987, March). *Seven principles for good practice in undergraduate education*. AAHE Bulletin.

Dale, E. (1969). *Audiovisual methods in teaching* (3rd ed.). Dryden Press.

McKinley, B. G., Birkenholz, R. J., & Stewart, B. R. (1993). Characteristics and experiences related to the leadership skills of agriculture students in college. *Journal of Agricultural Education*, *34*(3), 76–83. doi:10.5032/jae.1993.03076

KEY TERMS AND DEFINITIONS

Agricultural Development: The process that creates the conditions for the fulfilment of agricultural potential including the accumulation of knowledge and availability of technology as well as the allocation of inputs and output.

Agricultural Sector: Farms and ranches, agribusiness sales and services that support farms, ranches and other entities, food manufacturing, processing and storing facilities, and restaurants.

Borich Model: A mechanism used to prioritize inservice education needs of instructors and faculty based on competency importance and instructors' knowledge and ability to perform.

Competency: Knowledge, behaviors, attitudes and even skills that lead to the ability to do something successfully or efficiently.

Skill-Gap Analysis: An outcomes assessment tool to measure skills and competencies that are important to employers and in need of development by potential workforce participants.

Chapter 6
Active Learning Strategies in Higher Education in the Arab World

Mohamed Hassan Abdelgwad
https://orcid.org/0000-0002-3500-6395
King Saud University, Saudi Arabia

ABSTRACT

Through this chapter there is reflection on using active learning strategies for agricultural education in the Arab world. There are difficulties with utilizing specific active learning strategies that encourage student learning and achieve important graduate attributes. Changing teacher and faculty perceptions about learning strategies and focusing on students' higher order thinking requires a lot of training, orientation, and mind shifting. Learning through specific active learning techniques, such as vocational, simulation, case studies, and using technology, supports student learning. It is particularly important to use instructional ways that encourage deep learning and analysis with agricultural problems and deficits. Professional development and hands-on learning changes perceptions of teachers through agricultural education in the Arab world. Many aspects relevant to using active learning in agricultural education perfectly in Arab world countries are clarified.

DOI: 10.4018/978-1-6684-4050-6.ch006

INTRODUCTION

This chapter will present and discuss many topics relevant to utilization of active learning concepts, rationality, importance, pillars and foundations, comparison with traditional learning, overview of agricultural education in Arab world, agricultural education development through MUCIA / USAID in Egypt, an overview of active learning utilization, and suggested strategies of active learning that could be applied for agricultural higher and vocational education in the Arab world. Vocational and higher agricultural education in the Arab world require new patterns for ways of instruction and giving more emphasis on cases, problem solving, and using technology through education. There is a high demand to transfer traditional education techniques in agricultural education based mainly on lecturing into more active strategies that inspire students and expose them to new technologies and recent advances in agricultural applications.

Establishing teaching excellence centers within schools and colleges will provide capacity building for teachers and faculty to be more effective and making breakthrough for graduates' capabilities and competencies. Through rural communities, closer links between higher educational institutions and agricultural companies will lead to increased awareness of farmers to accept the data and achievements of agricultural scientific research that are directly related to solving production problems.

According to Battisti and Passmore (2008), traditional experiential learning is viewed as a cycle of experimentation, generalization, reflection, and experience but does not always place much emphasis on the necessity of careful reflection or the requirement that the students' interests motivate the experience. Classic degree programs, in which students are typically expected to be information receptacles, and simulations are examples of traditional collegiate experiential learning pedagogies used in the classroom (in which students are exposed to the subject matter via a simulation). Capstone courses take place outside of the classroom but are still governed by the institution (in which students are invited to synthesize the coursework from their degree program into a final project, usually practical in nature). Additionally, as more active learning experiments are conducted in higher education, it is crucial that learning research be started as well. By doing so, our agricultural education will be able to better understand how active learning may change the way that individuals and groups think and behave.

Bransford et al. (1999) indicated that constructivist learning theory emphasizes that individuals learn by combining new ideas and experiences with existing knowledge and experiences to build their own knowledge and develop new and improved understandings. In addition, Freeman et al. (2014) revealed that students in traditional lectures are 1.5 times more likely to fail than students in active learning courses. In addition, incorporating active learning into the course design increased student performance on exams, concept inventory, or other assessments by an average of about half the standard deviation.

BACKGROUND

Agricultural Education is the preparation of agricultural human frameworks that are scientifically and technically qualified with information, experiences, and skills. These competencies support the effectiveness of their agricultural, plant, and animal performance, achieve increased production, improve quality, and reduce the cost of agricultural development projects.

Agricultural education is of great importance in the world, as 50-80% of the population of the continents of Asia, Africa and Latin America work in agriculture, which is a main source of income for citizens (Arab Encyclopedia, 2020). The countries of these regions are interested in increasing the areas planted with different crops in order to increase production to match the forced increase in the population and to get rid of the threat of famine. As for the industrially developed countries, the portion of workers in agriculture does not exceed 10% of their population. Since agricultural industries occupy an important place in the economies of these countries, and agricultural education helps to develop agriculture and increase the production of raw material needed for industry in these countries or abroad, agricultural education aims to rehabilitate technical frameworks by providing them with the necessary skills and competencies to positively interact with modern techniques, means, and machines while carrying out their agricultural tasks. This requires keeping pace with modern scientific progress, increasing the dissemination of agricultural scientific knowledge and values, and developing educational methods in addition to contributing to the organization of agricultural projects and activities proposed by the planning bodies in the concerned country, analyzing and evaluating them, preparing and developing educational literature, and working on establishing effective agencies and specific programs for agricultural extension.

The process of curriculum development has two correlative aspects that cannot be separated from the other, namely the content and the method of teaching. Therefore, any development in the content should be accompanied by a development in the method of teaching. Despite this, most of the processes of the so-called curriculum development in Egypt were not accompanied by any the development of teaching methods, which had a negative impact on the outcomes of the educational process. In active learning (Cohn et al., 1994), algorithms become active participants, and their goal is to learn the function in an economic manner (Settles, 2012). By carefully deciding which cases to explain, active learning algorithms can accurately learn a target function.

IActive learning algorithms are superior to random sampling and common baselines, which reduces the need for time, effort, and resource in the process of constructing a predictive model. For example, most active learning experiments in the literature employed only one classifier and one performance measure.

From all of this came the idea of using active learning in university classes to complement the efforts currently being made by the Ministry of Education to develop general education curricula in university curricula in particular.

AGRICULTURAL EDUCATION IN THE ARAB WORLD

The agricultural sector constitutes a cornerstone of the comprehensive economic and social development plans in most countries of the Arab world and is of particular importance as it represents a part of the value of the gross product in these countries. The importance of agricultural education institutions in the Arab region has emerged since the early nineteenth and twentieth centuries, as many agricultural schools were established, and a number of colleges of agriculture and veterinary medicine were established in the early twentieth century in some Arab countries such as Egypt, Sudan, Syria, and Tunisia. Other nations did not start building agricultural university institutions until the beginning of the 1950's. Agricultural education in the Arab region includes the following stages.

Agricultural Secondary School Education

Agricultural secondary schools are found throughout various Arab countries, with their number reaching about 200 secondary schools. The Syrian Arab Republic, the Hashemite Kingdom of Jordan, and the United Arab Emirates rank first among the Arab countries in the number of specialized agricultural secondary schools for the rural population. Libya, Iraq, and Tunisia come in second, Lebanon, Egypt, and the Sultanate of Oman come in third, and Algeria comes in fourth.

The overall features of the development of agricultural secondary school education can be identified as follows:

- Emphasizing the importance of preparing the agricultural teacher, and enrolling teachers in educational and scientific courses to increase their skills and provide them with the necessary suitable instruction methods and using active learning strategies and expertise, especially for the modern equipment and devices used in agriculture.

- Flexibility of study plans and curricula, provided that they achieve rural development goals, and develop them to contribute to meeting the needs of the agricultural labor market.
- Expanding the admission of rural students and providing the necessary funds to carry out scientific trips that help students develop their expertise and introduce them to agricultural activities in the various bodies and institutions in which they may work in the future.

Agricultural Intermediate Education

Interest in agricultural intermediate education has increased in recent years in most Arab countries, in order to prepare intermediate agricultural technical frameworks that combine specialization on the one hand and applied scientific capabilities on the other. There are more than 65 Arab agricultural intermediate institutes, distributed mainly in Syria, Iraq, and Morocco. For admission to these institutes, the student is required to obtain a high school diploma - the scientific branch in Syria, Iraq, Yemen - and some of them are accepted in addition to the general secondary school certificate, the agricultural secondary certificate. The duration of study is two years in Syria, Iraq, and Morocco, three years in Sudan, four years in Tunisia, and five years after the preparatory certificate in Egypt.

The student is granted a technical diploma or a technical bachelor's degree in various specializations of plant and animal production. It became necessary to develop and update requirements using modern educational means such as computers, video, light board, and others.

Higher Agricultural Education

The importance of higher agricultural education institutions in the Arab region has emerged since the early twentieth century, and the number of agricultural colleges has reached 57 colleges and higher agricultural institutes, and 15 colleges of veterinary medicine spread in the Arab region as follows: nine in Egypt, six in Libya, five each in Syria and Sudan, four in Iraq, three each in Morocco, Saudi Arabia, Algeria and Jordan, two in Yemen, and one college in the United Arab Emirates. The faculty members in these faculties of agriculture have completed their postgraduate studies in several universities around the world, which has given them a variety of experiences in the field of teaching, the quality of courses, teaching systems, examinations, and scientific research.

Colleges of Agriculture are affiliated with the Ministry of Higher Education in the Arab Mashreq countries of Egypt, Libya, and Sudan, while they are affiliated with the Ministry of Agriculture in Tunisia, Morocco, and Algeria. The duration of study varies according to the Arab country between 4-6 years, some of which follow the semester system and credit hours (Jordan, United Arab Emirates, Saudi Arabia, Lebanon), and others adopt the semester system and compulsory courses (Syria, Algeria, Sudan, Somalia, Iraq, Libya, Egypt, Yemen), while in Morocco the continuous annual system and compulsory courses extend for a period of six years. As for the period of study in veterinary colleges in the Arab world, it is five years.

CONCEPT AND RATIONALITY OF ACTIVE LEARNING

Saadeh et al. (2011) defined active learning as

a method of teaching and a learning method at the same time, where students participate in activities, exercises and projects with great effectiveness, through a rich and diverse educational environment, that allows them to listen positively, constructive dialogue, rich discussion, conscious thinking, and analysis sound, and deep contemplation of everything that is read, written, or presented from study material, issues, issues, or opinions, between each other, with the presence of a teacher who encourages them to take responsibility for their own education under his strict supervision, and pushes them to achieve the ambitious goals of the curriculum, which focuses on building the integrated and creative personality of the student.

Active learning is an educational philosophy that aims to activate the role of the learner and make the learner pivotal in the educational process. Active Learning Strategies target primarily higher-order thinking skills such as analysis, synthesis, and evaluation, depending on educational situations and various activities that require research, experimentation, work, and self- or group learning as well to acquire skills, obtain information, and form trends and values.

Active learning requires:

- Active and positive participation of the learner in various learning situations.
- Assigning responsibility for learning to the student or learner.
- Exiting the traditional education framework and opening up to the new.
- The learner's self-reliance in acquiring skills and obtaining information (and solving problems in general) without resorting to classical methods such as memorization.
- Develop the learners' thinking abilities so that they can solve his own problems.

IMPORTANCE OF USING ACTIVE LEARNING IN AGRICULTURAL EDUCATION

There are many benefits that active learning brings for students' deep learning. There are 10 benefits of using active learning in classroom (https://www.edapp.com/blog/benefits-of-active-learning/); in addition there are many other transferable skills and rational aspects for using active learning strategies in agricultural education in schools and university to develop student competencies in Arab world countries. The following advantages are expected as a result of applying active and interactive learning strategies.

1. **Develop teamwork skills:** Group work is one of the most effective active learning methods, as it helps develop students' communication skills.
2. **Encourage expression of opinion:** Active learning encourages the student to express and defend his or her ideas, rather than sitting down to take notes throughout the class.
3. **Encouraging the student to prepare for the subject:** Active learning applications require students to be prepared in each session to participate and interact. It differs from the traditional student attendance, which does not require any participation, and the participation of students improves the effectiveness of their thinking and formulation of ideas.
4. **Improving students' critical thinking:** Through active learning activities, students learn to share ideas positively and to make logical arguments and justifications.
5. **Memory stimulation:** Interactive active learning activities help learners understand and memorize ideas more than if they were just words, as Edgar Dale's cone theory of educational experiences. Dale's Cone of Experience indicated that students remember 10% of what they read, 20% of what they hear, and 90% of what they do with it.
6. **Technology activation:** Active learning applications contribute to activating the use of technology by the students themselves and not only by the teacher.
7. **Stimulating creative thinking:** Creativity is one of the most difficult skills that can be taught through traditional learning methods, while active learning provides a training environment that helps students to think and develop creativity.
8. **Enhance complex problem solving:** Active learning gives students the opportunity to find realistic solutions to complex problems.

Active learning strategies support students' thinking and lets them progress to higher-order thinking (synchronize, analysis, evaluation, and innovation). Flipped classroom is one of the best strategies to be used to make more time for students to ask and discuss more than articulation. Flipped learning applications are increasing at a fast pace, educators reported the deficits encountered in applying flipped learning strategy (Tsai et al., 2017). For example, students could not be able to recognize and understand the learning content when watching videos individually at home; teachers might have difficulties knowing students' individual readiness, which makes it difficult to conduct effective flipped classes (Hwang et al., 2015).

Active learning enables agricultural education students to follow up and keep pace with scientific innovations and provide them with various scientific, critical, and innovative thinking skills and scientific trends and values that help them understand the contemporary world, coexist with it, and confront its problems, in addition to providing them with self-learning and continuous learning skills and employing what they have acquired in knowledge, information, skills, trends, values, and methods. Thinking about solving the problems they face, that is, scientific education, can prepare these individuals for life by achieving comprehensive and integrated growth in all their aspects and help them to adapt to the current situations and prepare them to deal with the changes of the era and its contemporary and future developments (Al-Wasimi, 2004).

PRINCIPLES AND FOUNDATIONS OF ACTIVE LEARNING

Suggested by Douglas R. Barnes (1989) are Seven Principles for Successful Active Learning in the Classroom. These principles are essential to the full implementation of learning with students. These principles are the main principles of an active learning for the classroom.

- **Purposeful:** the task being taught is compatible with the needs of the students
- **Reflective:** students provide their own reflections after the content of the lesson is presented
- **Negotiable:** It is necessary to negotiate between the teacher and the student about learning objectives
- **Critical:** for students to recognize the diversity of content learning methods
- **Compound:** This will allow students to compare the content of the lesson in real life, and it will give them space to build a reflective analysis
- **Situation-oriented:** For the success of specific learning tasks, the situation should be clear
- **Participatory:** Teachers should make activities interactive and based on real life.

In addition, there are some foundations or pillars in active learning use that should be considered:

1. Employment of active teaching strategies centered around the learner.
2. Involve students in defining educational goals, work systems and rules.
3. Giving each student the opportunity to learn at their own pace
4. Provide an atmosphere of fun, fun and reassurance while learning.
5. Diversifying learning resources, tools, and means.
6. Relying primarily on self-evaluation and peer evaluation.
7. Familiarize students with self-management (self-reliance and decision-making).
8. Helping students when they need it (follow-up).
9. Helping the students discover the strengths in their personality and work.

In addition, Muhammad (2010) indicated that active learning depends on several foundations that helping students achieve their educational goals and objectives like:

- Availability of communication in all directions between the learners and the teacher.
- Helping the learners understand themselves and discover their strengths and weaknesses.
- Use of learner-centered teaching strategies that are compatible with students' abilities, interests, learning styles, and intelligences.

COMPARISON BETWEEN TRADITIONAL TEACHING AND ACTIVE LEARNING

In conventional instructing most of the course time is went through with the teacher addressing and the understudies observing and tuning in. The understudies work exclusively, and participation is debilitated. On the other hand, active learning changes the center of the action from the educator to the learners, in which understudies illuminate issues, reply to questions, define questions of them possess, talk about, clarify, talk about amid lesson; in addition, understudies work in groups on issues and ventures beneath conditions that guarantee positive interdependency and person responsibility. In spite of the fact that student-centered strategies have more than once been appeared to be prevalent to the conventional teacher-centered approach to instruction, the writing with respect to the adequacy of different educating strategies is uncertain. Educational comes about and writing demonstrated noteworthy contrast between active learning and conventional educating, where techniques advancing active learning to conventional addresses seem increment information and understand.

Hyun et al. (2017) investigated the effect of active learning education on student satisfaction with the learning process in ALC and traditional classrooms [TC]. Results show that Active Learning Classroom [ALC] helps increase student involvement and improve student performance. However, converting all traditional classrooms to ALC entails significant financial burden. Therefore, a compelling question for higher education institutions is whether active learning education can improve learning outcomes when classroom resources are limited. Not only are there numerous studies that show active learning methods work to enhance students` success in the classroom, but there are also numerous resources available to support faculty committed to applying active learning pedagogy to their courses. In addition, classroom space has become a focus of interest, in the light that changing traditional classrooms into spaces that more readily accommodate the active learning pedagogy would effectively promote learning outcomes (Hyun et al., 2017).

Numerous analysts feel that an active learning environment is a perfect way to induce data into the minds of understudies. Concurring to (Chickering and Ehrmann, 1996) an effective learning environment requires openings for interaction and input. Numerous learning speculations bolster the see that understudy learning is upgraded through openings to work collaboratively (Jecker, 1964). For illustration, cognitive scholars like Jecker9describehow learning takes put as understudies go up against and talk about conñicting opine-ions with peers. For understudies to be included in active learning, they must study, type in, talk about, or be locked in in fathoming issues. Bonwell and Eison (1991) reported understudies must lock-in in higher order considering assignments and examination, union, and assessment in arrange to be effectively included. In brief, it is proposed that procedures advancing active learning result in directions exercises including understudies in doing things other than basically tuning in and considering approximately.

The following table represents comparison between traditional and active learning strategies on many of educational criteria.

Table 1. Comparison criteria for traditional versus active learning strategies

Active Learning	Traditional Learning	Criteria
It is announced to the students, and they participate in its development and planning	Not announced to students	Objectives
Facilitate, manage, instruct, direct, regulate communication	Memorization	The role of the teacher
The students share the instructions with the teacher	Issued by the teacher	Instructions
Involve the teacher in choosing the work system	The teacher imposes it on the students	Working system
Enthusiasm - Fun - Collaboration	Rigor and firmness	Teacher character
Didactic	Educational	Instructional means
Diversity in sitting and freedom of movement	Fixed seats	Pupils sitting
Students are allowed to ask questions to the teacher and to their classmates	The teacher is the one who asks often	Questions
In all directions	In the direction of the teacher only	Communication
Each student learns at his/her own pace	One for all students	Fast learning
Understanding and solving problems, high and innovative levels, and skill and emotional aspects	Remember and save information	Outputs
Helping the student discover strengths and weaknesses and comparing the student with himself/herself	Making a judgment about success or failure and always comparing the student to others	Assessment

MUCIA, USAID PROJECT, AND AGRICULTURAL EDUCATION DEVELOPMENT IN EGYPT

Educational reform and professional development programs for agriculture faculties and Agricultural Technical Schools (ATS) instructors in Upper Egypt were offered through the Midwest Universities Consortium for International Activities (MUCIA) and funded by the United States Agency for International Development (USAID) in the early 2000's.

The objective of the projects was to provide workshops to educate ATS instructors and faculties of agriculture on how best to deliver the agricultural curriculum to students and inspire them in agriculture business, with the end result being a well–trained agricultural workforce in Egypt (Barrick et al., 2011). See Chapter 1 of this text.

Upper Egypt's Agricultural Technology Schools includes 130 secondary schools, with an average of 154 instructors and 2,750 students at each institution. USAID funded a value chain training program through MUCIA and provided in-service training to agricultural trainers at Agricultural Universities. The purpose of funding was to ensure that trainers carry out programs that meet the labor needs of the country's agribusiness (Thoron et al., 2008).

Developing better educators requires an understanding of psychology, principles, techniques, and a basic pedagogical understanding of the teaching and learning process (Abolaji & Reneau, 1988). Experiential learning has long been a central feature of American agricultural education programs (Roberts & Harlin, 2007). However, this teaching method had not been adopted by Egyptian ATS instructors.

If ATS instructors receives in-service training, can they carry out experiential learning activities? The basis of the work carried out in this project was the theory of Teacher Adaptive Expertise (Hammerness et al., 2005). This theory suggests that teacher expertise is developed along two aspects: efficiency and innovation. Efficiency expertise leads to the ability to get the job done with little attention, while innovation expertise leads to new attempts and changes to existing practices. Adaptability includes efficiency and innovation. Effective workshops for teachers are needed to strengthen their adaptive knowledge.

ATS teachers had the greatest discrepancies in knowledge of the following competencies: (a) explain the connection of internship and school room instruction, (b) identify the function of instructors in internships, and (c) explain the manner of studying with the aid of hands-on approaches. According to Roberts and Dyer (2004), agricultural instructors desire inservice training for their improvement to grow to be better educators.

Samy (2008) indicated that the MUCIA AERI Linkage Project designed an intervention strategy to promote needed curriculum changes and active learning techniques that would improve the employability of ATS graduates and integrate private sector needs into ATS educational programs.

OVERVIEW OF ACTIVE LEARNING UTILIZATION

Active learning is learning based on the various activities practiced by the learner and resulting in behaviors that depend on the active and positive participation of the learner in the teaching / learning situation. Active learning derives its philosophy from contemporary global and local changes, and after meeting these variables requires a review of the roles of the learner and the teacher. The focus of attention moves from the teacher to the learner and makes the learner the focus of the educational process (Ministry of Education, 2005, p. 22).

As DeCesare and McKinney (1998) explained, active learning is based on two assumptions, namely that learning is in its nature an active process carried out by the learner, and that learning reaches its maximum when the learner is included in the educational situation, and also emphasizes the possibility of using active learning in various educational stages. With small and large numbers, successful active learning depends on the extent of learners' readiness and the extent to which the educational environment is enriched with different activities and teaching methods (DeCesare and McKinney, 1998).

Active learning can be accomplished in or out of class in collegiate and secondary school settings but is not likely to occur in traditional settings since it requires participating in discourse with a group of peers, called an Action Learning 'set'. As expressed by McGill and Brockbank (2004) "action learning sets formalize gather reflective learning and legitimize the allocation of time and space to it, with consistent, voluntary membership over an extended period of time."

After the movement for elective and more experiential forms of college instruction began, Kolb (1984, p.3) remarked:

Programs of supported experiential learning are on the increment in higher instruction. Internships, field placements, work/study assignments, structured exercises, and role-plays, gaming reenactments and other shapes of experience-based instruction are playing a bigger part within the educational programs of undergraduate and professional programs. For numerous so-called nontraditional students – minorities, the destitute, and mature adults – experiential learning has ended up the method of choice for learning and individual development. Experience-based instruction has gotten to be widely accepted as a strategy of instruction in colleges and universities over the country.

In order to implement active learning, it is necessary to diversify the teaching methods, as well as to rely on the pillars of the educational process, with the need for good planning for the various activities. Among the active learning strategies that can be used in the classroom with any educational course, and with any number of students in different educational stages, are think - pair - share, learning cell, turn around and talk to those around you, whisper groups, write reports, write summaries, peer questions and answers, analyze scenarios or topics, build and rebuild concept maps around the lesson topic, round table, corners, education problem solving, or educational games (Razaitis, 2005).

Saadeh (2007) refers to active learning as a method of learning and teaching at the same time in which students participate in a variety of activities that allow them to listen positively, consciously think, and sound analysis of the subject matter (Badir, 2008, p. 35).

Many studies have focused on the issue of active learning, as the study of Hendy (2002, pp. 38-64) to identify the impact of the diversity of using active learning strategies on acquiring some biological concepts among first-year agricultural secondary students. The results of the study resulted in a high ability to acquire biological concepts and a tendency towards positive mutual dependence as a result of the diversity of active learning strategies, while it did not confirm the improvement of the learners' self-esteem level.

The study by Hamadeh (2005, pp. 225-265) aimed to identify the effectiveness of the strategy "think - pair – share" and the investigation based on the active learning method in school mathematics clubs in developing mathematical thinking skills and reducing mathematics anxiety among middle school students. The study by Hamadeh (2005, pp.101-141) also found the effect of using some active learning strategies in teaching social studies on academic achievement and developing problem-solving skills for preparatory stage students.

The study by Rowley and Jensen (2005) aimed to determine the extent to which home economics teachers achieve active learning through a longitudinal follow-up study of teachers in classrooms addressing social relations among learners. Also, a study by Mabrouk (2008) aimed to identify the impact of teaching some contemporary issues using active learning strategies through the home economics curriculum on the outcomes of the educational process.

As the vision of the Curriculum Development Research Division on "active learning and improving the educational process" said:

The teacher carries out the educational process as well as the teaching process, meaning that he transfers knowledge and facts and works to form certain concepts and generalizations among his students as he seeks to provide them with many tendencies, trends, values and aspects. Appreciation and appreciation, as well as helps them to acquire different forms of skills appropriate for them, and this means that the teacher seeks to bring about mental, emotional and behavioral changes in his students, the term education is an interactive activity between the teacher and his students, with the aim of helping students to make an intentional adjustment (cognitive, emotional, and skill). In their behavior (Muhammad, 2010, p. 15).

Stimulating and keeping college students' interest to efficaciously reap learning consequences is the primary objective of undergraduate education. The extent of interest in a designated discipline of study can have an effect on each academic performance and well-being (Schiefele, 2001). It has been proven that strong interest will increase college students' motivation who are looking for understanding and the variety of strategies they use for getting to know (Pressley et al., 1992). In higher education courses, the structure of sports and the gaining knowledge of surroundings wherein they are performed can significantly impact students' interest in a subject.

As a consequence, directors and instructors can attempt to enhance hobby inside the subject matter by restructuring the curriculum and the classroom surroundings.

School room interventions designed to boost pupil interest regularly target first-year foundation publications, with the aim of influencing students' whole undergraduate experience. One frequently used strategy is the addition of lively studying additives (Yuretich et al., 2001; Freeman et al., 2014).

Past inquiries hypothesized a positive criticism circle between topic-specific interest and active learning (Stahl & Feigenson, 2015). Actualizing active learning through social, bona fide, and problem-based exercises creates students who are intrigued and inspired by supporting independence and connectedness (Deci, 1992). Active learning can be actualized through numerous methodologies, each with distinctive impacts on students' learning and interface.

Generally, little is known about the sorts of exercises and highlights of learning situations that best back the improvement of understudy (Rotgans & Schmidt, 2011). Active learning and understanding are not broadly considered in horticulture courses. Nevertheless, there is potential to make students intrigued by including active learning in these courses. The Garton Consider of Learning Inclinations showed that agribusiness understudies lean toward collaborative and intuitive learning situations (Garton, 1997).

Agribusiness undergraduate students at King Saud University portrayed an "enthusiastic and curiously instructing style" and an "interactive classroom environment" as the characteristics of the classroom that propelled them the foremost to memorize, and chose "a long, boring lecture" as an essential calculate in diminishing inspiration (Mankin et al., 2004). Constructivist learning hypothesis is the key to learning (Palmer, 2005). Consideration as a component of inspiration has been recognized by a few consideration of inspiration and scholarly execution (Rotgans & Schmidt, 2011; Wagner and Stefan, 2012).

Recent work has shown that attention can lead to higher task engagement, improved learning, and increased academic achievement (Harackiewicz et al., 2016). Attention theories generally characterize attention as having two forms: situational interest and individual interest (Hidi, 1990). Situational interest refers to a transitional psychological state caused by external factors, while individual interest is a more permanent tendency shaped by personal values and experience (Zhu et al., 2009). Increased situational interest in tasks precedes development of individual interest in topics.

ACTIVE LEARNING STRATEGIES

A study was conducted to evaluate the implementation of active learning strategies in the classroom by teachers in the Egyptian Agricultural Technology Schools (ATS). The Stages of Concern Survey was conducted on 230 participants in the Active Learning

Workshop. After eliminating principals, supervisors, and those who no longer teach, the population became 160 teachers and there was a viable response from all of them. ATS teachers who attended the Active Learning Workshop expressed concern about their readiness to adopt and use innovation (Myers, Barrick, & Samy, 2012). Overall results indicated that participants categorized themselves as very interested in animal science research and found that they were even more interested in all active learning strategies implemented. Among the evaluated learning strategies, students jointly preferred problem-oriented experimental station activities. The results of these findings can inform the development of educational strategies that support the development of individual interests of students. In an active learning environment, individual interests can lead to greater cognitive involvement and sustained learning.

Erickson et al. (2019) suggested a link between active and essentially motivated learning and subject-specific curiosity that teachers can use to support academic performance. Case studies, think-pair-shares, exam review sessions, lab stations, lab handouts, clicker questions, important lab considerations were included in the instruction. The majority of students stated that their interest in animal science was very high or extremely high before and after the course. These results indicate that active learning strategies have generated more interest among students curious about animal science, and interactive group-based teaching methods such as case studies and laboratory stations are for students in this group.

An overview of US universities' contributions to the education of foreign agronomists, published in The Agricultural Scientific Enterprise, provided an overview of agricultural education (secondary school and higher education), including formal and informal agricultural training programs and institutions that are relevant to Egypt. The report states that (1) supporting only one faculty of agriculture, (2) developing faculties in new locations that enable the development of appropriate educational and research facilities, and (3) new faculties were encouraged to emphasize practical skills as well as training and theoretical knowledge necessary for graduates to fully contribute to the development of the agricultural sector. This final recommendation means that the institution's curriculum should be directed towards practical problem-solving (Swanson, 1986).

Agricultural faculties in many developing countries face the serious challenge of keeping their academic programs up-to-date and related to the talent needs of the agricultural sector. In many cases, faculty members have limited opportunities to conduct research, have weak ties to the private sector, and academic programs do not effectively prepare students for their technical and administrative role in agribusiness. The report described the strategies being implemented in collaboration with five universities in Upper Egypt. These universities will bring about certain institutional changes that will transform academic programs and strengthen collaboration between these universities and private enterprises and commercial farms (Swanson, Barrick, & Samy, 2007).

In addition, as demonstrated in the AERI Linkage Project with a relatively modest USAID investment, faculty at all five partner universities are more hands-on and employment-oriented to improve the employability of graduates. Two independent reviews confirm that these activities have successfully transformed these academic programs, with most faculty renewing their courses and now using active teaching and learning methods. Course content has been successfully modified and implemented changes that provides subject skills for degrees in agriculture (Swanson, Barrick, & Samy, 2007).

The twenty-first century has created many challenges facing all societies of the world without exception with the rapid change at all levels of knowledge and technology. The rapid revolution of technology is only one of the many changes that the twenty-first century has brought to our lives. The world that is more competitive and people are dependent on each other, currently facing many diverse problems that require creative solutions. These great changes impose a new form of learning and new skills that students in the twenty-first century must master to enable them to interact with the data of the times, and education in its current state has failed to prepare students for the requirements of the twenty-first century. Attention is needed on a set of life and employability skills that help students to live and learn in this century. These skills include applying knowledge and problem-solving skills, transforming knowledge for meaningful purposes (knowledge economy) and cooperating with others (Fahnoe & Terry, 2013). Learning for the twenty-first century requires a teacher of a special style, who is educated, creative, and contemplative to provide students with these skills. Therefore, there is an urgent need to prepare distinguished teachers and curricula that achieve twenty-first century skills through active learning methods and techniques (Hefny, 2015).

Many Arab countries have reconsidered their academic programs, preparing and training teachers by providing them with distinguished educational knowledge and experiences, and improving and developing teaching methods to support students' learning and provide them with professional skills (Almady, 2013). Also, in the new and developed educational system, it was necessary to train teachers in the Arab world to use many competencies of information and communication technology and educational technology in order to provide them with a technological culture that greatly helps to transfer many experiences and skills and enable continuous communication with students (Noby, 2014).

In the Arab context, the program for preparing Arab youth for the labor market, entitled "Strategy for the inclusion of entrepreneurship and twenty-first century skills in the Arab education sector", which is sponsored by the INJAZ Al-Arab Foundation in cooperation with ALECSO and the Arab Bank, confirmed the integration of twenty-first century skills and active learning skills in general education and technical education. To help students keep pace with and develop future work environments (Al-Khuzaim & Al-Ghamdi, 2016), the teacher's skills for the twenty-first century are the skills that students need to succeed in study, work and life, and consist of learning and creativity skills (critical thinking and problem solving), digital culture (information culture and ICT) and career and life skills (flexibility, adaptation, initiative, self-direction and social interaction, leadership, and responsibility).

Several educational conferences also emphasized the need to pay attention to teacher training in order to bring about comprehensive professional development in the field of specialization and the need to keep pace with these trainings to the variables of the era of knowledge, and the requirement of technological progress.

Interactive teaching techniques (Yee, 2020) indicated 226 active learning techniques that could be used for instruction. These techniques have multiple benefits: the instructor can easily and quickly assess if students have really mastered the material (and plan to dedicate more time to it, if necessary), and the process of measuring student understanding in many cases is also practice for the material—often students do not actually learn the material until asked to make use of it in assessments such as these. Finally, the very nature of these assessments drives interactivity and brings several benefits. Students are revived from their passivity of merely listening to a lecture and instead become attentive and engaged, two prerequisites for effective learning. These techniques are often perceived as "fun," yet they are frequently more effective than lectures at enabling student learning. The following table briefly and categorizes all 226 active learning strategies based on instructor and students sharing and activities and engagement. It was classified based on frequency of use.

Table 2. Different techniques utilized within active learning strategies.

Active Learning Technique / Pattern	Number of Strategies
Instructor Action: Lecture	33
Instructor Action: Lecture (Small Class Size)	9
Student Action: Individual	49
Student Action: Pairs	12
Student Action: Groups	31
Second Chance Testing	7
YouTube	7
Mobile and Tablet Devices	1
Audience Response Tools	6
Creating Groups	8
Icebreakers	18
Games (Useful for Review)	11
Interaction Through Homework	7
Student Questions	5
Role-Play	6
Student Presentations	5
Brainstorming	6
Online Interaction	5
Total	**226**

Derivative of Interactive Activities by Kevin Yee 2020.

Not all of the 226 techniques listed above have universal appeal, and factors such as teaching style and personality influence the choices that may be appropriate. The following active learning strategies could be used with agricultural education classes or lectures in the Arab world, which empower their deep learning and intellectual output, and mostly require more detailed information and faculty capacity building.

§ Concept Mapping
§ Mind Mapping
§ Problem based learning (PBL)
§ Team based learning (TBL)
§ What If?
§ Muddiest Point
§ Audio and Videotaped Protocols
§ Interactive Video Quizzes

§ Think-Pair-Share
§ Discovery learning
§ One-Minute Papers
§ Student Learning Communities
§ Student Questions (Index Cards)
§ Minute Paper Shuffle

There are a few active learning techniques that should be applied for agricultural education and making better learning and wildly acceptable and recommended in many different educational associations for technical and vocational learning in agricultural education.

SPECIFIC ACTIVE LEARNING STRATEGIES FOR AGRICULTURAL EDUCATION

The remainder of this chapter describes several important active learning techniques widely utilized within agricultural education and mostly recommended in Arab World countries even for secondary or university agricultural education. Strategies include, but not at the least: experiential learning, filed experience, project methodology, simulation, agricultural laboratories and case studies for problem solving.

Experiential Learning

Each individual and every setting has a different definition of experiential learning. Learning activities including direct student participation in the phenomenon under study are referred to as experiential education. The involvement is direct and intentional, tackling a real-world issue in a natural environment. For experiential learning to be successful, the teacher is a crucial component. To enhance learning and comprehension, agricultural education must take on additional responsibilities in the classroom and promote student engagement with their surroundings. The primary goal of learning should be to apply what is learned in the classroom to "real world" circumstances.

Experiential learning (Battisti and Passmore, 2008) requires the learner to start the learning prepared, the encounter to be veritable, and the reflection to lead to modern thoughts that can be attempted in an unused circumstance. Genuine collegiate experiential learning pedagogies within the classroom incorporate: project-based learning (in which understudies are displayed with a real-world issue and challenged to work as a group to come up with an arrangement), facilitated considerations (in which interdisciplinarity is emphasized), and free majors (in which understudies define, propose and seek after a special major that fits their special instructive interests). True collegiate experiential learning pedagogies are exterior the classroom, but still beneath the support of the college, incorporate field ponders (in which understudies pick to induce out of the classroom and into the field to induce firsthand experience with their subject matter), and agreeable instruction (in which understudies are combined with commerce, government, or non-profit organizations to pick up encounter working in).

Field Experiences

Field experiences give understudies an opportunity for experiential learning within the setting of their choice. There are a few diverse opportunities inside the Division of Agrarian Instruction and Ponders for understudies to have these non-classroom encounters. It is imperative to know that understudies win scholarly credit by completing self-evaluations, creating competencies, making introductions and other scholarly work in association with their encounter. Credit is not earned basically by having a work encounter (*Field Experiences,* n.d.).

By utilizing modern techniques and instructing applications (Shannon et al., 2006), agrarian teachers can continually improve their programs and emphatically affect understudy learning and development. During application, teachers must allow important encounters and dynamic understanding using Kolb's cycle to help within the procurement and absorption of subject matter. Teachers should implement different shapes of experiential learning into their courses, such as internships, field placement, work/study assignments, and organized work, to extend understudy learning.

In the University of Florida College of Agricultural and Life Sciences is a course centered on instructing strategies in agrarian instruction. One way to make strides in education is to watch others and take courses of action to watch the collaborating instructor (or internship location) centering on the instructing viewpoint. Reflections will be done through journaling to record perceptions.

Project Methodology

The project methodology is an integral part of the triad model of agricultural education programs. However, it is unclear how sound the modern practice for implementing the project method is in theory. The purpose of this philosophical treatise, therefore, was to summarize the theoretical and historical underpinnings of the project methodology and compare it with contemporary best practices. This will provide insight and direction to advance practice and research (Roberts and Harlin, 2007) .

As the scope and mission of agricultural education expands, educators must be cognizant of that awareness and mission. Implement effective and meaningful teaching and learning strategies in the classroom. Research and empirical evidence identify this transition to a new age and strongly support the benefits of experiential learning in agricultural education (Cheek et al., 1990).

Simulation Techniques

A crop simulation model is a computer program that uses weather information, crop conditions, and soil conditions to depict the growth and developmental stages of a crop in order to solve real-world problems. Crop simulation models are crucial in the decision-making process since they can help you save time and money. One of the most important factors in the decision-making process is the simulation models' ability to forecast outcomes accurately. As a result, the information gleaned from qualitative analysis of traditional research has numerous limits in light of the necessity for decision support systems that can last for more than 10-15 years and changing climatic and meteorological variables, computerized statistical tools are needed. This is why crop simulation models might be a crucial instrument in future agricultural research and measures to combat climate change (Asma et al., 2021).

Agnew and Shinn (1990) indicated that teachers must continuously decide how to create lessons that best fit the needs of the students. These choices encompass picking strategies and tactics. In addition to knowing how to employ different strategies, the selection process demands that the teacher also define what levels or types of knowledge may be taught with various procedures, what types of pupils learn best using various approaches, and under what circumstances should these techniques be employed. In the absence of real equipment or if safety is a concern for basic cognitive instruction in direct current electricity and hydraulics, the simulation technique can be utilized to teach students in specialist agricultural mechanics courses in secondary schools.

Agricultural Laboratories

Students need to be able to handle problems related to scientific material in order to keep up with trends in the agriculture industry. Agricultural laboratories are regarded as a key element of secondary agricultural education and are ideal for giving students the chance to practice problem-solving techniques. The purpose a study by Shoulders and Myers (2012) was to investigate the current state of agricultural laboratories in secondary agriculture education, their utilization, preparatory needs, and impediments to their use, as well as their link to teacher perceptions of student learning. Although there are numerous facilities that are readily accessible and often used during instruction, the results show that instructor perceptions of student learning, preparation needs, and barriers vary by facility (Shoulders & Myers, 2012).

Currently, it is believed that agricultural laboratory techniques, which can include mechanical laboratories, greenhouses, livestock facilities, land laboratories, and aquaculture laboratories, among others, are a way to provide students practice in applying the theories taught in the classroom. However, the development of scientific agricultural education may present opportunities for these laboratories (McCormick, 1994). Agricultural laboratories tools and techniques become a cornerstone in the instruction of scientific knowledge and problem-solving techniques. By organizing laboratory education to focus on scientific problem solving, teachers can increase student experiences to prepare them more effectively for agricultural vocations based on science (Parr & Edwards, 2004).

Case Studies for Problem Solving

In the Leite and Marks (2005) essay, the case study design is discussed as a research methodology for agricultural and extension education. Three intrinsic characteristics—specificity, intensity, and multiplicity of sources of evidence—are used to describe case study throughout the paper. For case studies, a typology is recommended considering the goal of the study, the number of examples, and the researcher's area of interest. In order to assess the methodological consistency of case studies with regard to their applicability, subjectivity in the data gathering and analysis processes, dependability, and credibility, the study examines various elements.

More than 100 case studies that Farming First (a global coalition for sustainable agricultural development, https://farmingfirst.org/case-studies) has gathered are displayed. Climate Change, Food & Nutrition Security, Gender, Sustainable Agriculture, and Water are the five categories into which they are divided. The farmers featured in In

The NACTA Journal teaching tips / notes (https://bit.ly/3B17Lxm) mentioned that case studies could be used as teaching strategy for promoting critical thinking in the 21st century for agricultural education students to improve critical thinking through six specific stages. Case studies can be used in a variety of fields and on a variety of subjects.

Students should be given the following instructions once teachers have introduced the topic and the problem or issue to be solved to the class:

1. Tell us what you know about the problem or concern being raised.
2. Clearly describe and discuss the issue or situation.
3. Describe the steps needed to resolve the issue or problem.
4. Analyze the steps needed to resolve the issue or problem.

5. Discuss the advantages and disadvantages of resolving the issue or problem.
6. Write a summary of the findings after the problem or issue has been resolved.

In every setting, there are examples that require consideration in relation to agricultural education. Finding a real problem and thoroughly expressing it, usually in writing, is all that is required (but a video explanation could work, too). The stages above are simply completed by students using the information that has been provided to them.

For many ages, Egypt has been renowned for its successful and influential agriculture. Its agricultural sector has seen numerous limitations and problems over the past 20 years, reduced yields and reduced economic returns. The Agricultural Extension and Advisory Service has made a significant contribution to improving crop yields, which has helped rural livelihoods and the national economy. The results of case studies and tests showed that strong ties between the National Agricultural Extension Service and Egyptian Agricultural Departments (Agricultural Research and Development) can help realize sustainable agriculture and attain food security (NAES) in order to increase both departments' productivity and meet national needs (Mirza, et. al, 2018). In addition, improving farmer methods for sustainable agriculture to attain food security. Farmers, planners, and policymakers may find the material offered useful in their efforts to build the agricultural sector on good foundations through the application of sustainable agriculture methods and scientific concepts.

The Midwest Universities Consortium for International Activities (MUCIA) is a USAID-funded network of American universities and upper-Egypt agricultural schools working to educate small-scale farmers and connect them to export markets In Egypt. Specific case study groups worked through 10 months with agricultural family business established at the beginning of last century in upper Egypt and focusing on ordinary agricultural business system with no more profit and return. The case focused on different production styles or scenarios through advantages of new reclamation desert areas or using different production systems. The case study teams are different agricultural education disciplines and tried to use different agribusiness norms to reach three different production scenarios. some sort of interprofessional education followed this case and introduced for college student's instruction, especially for agribusiness and animal science collective cohort of students and instructors. This interprofessional education class energized students and faculty learning and added more values through agribusiness and animal science backgrounds and concepts.

CONCLUSION

Conclusively, there are many methods and strategies of active learning should be deployed for agricultural education in Arab world countries to maximize students deep learning ad using the cognitive higher thinking order and possess all required skills and competencies from agricultural education and being ready for labor market demand. Having good learning context, well designed agricultural curricula and train faculty and teacher for inspiring students and deep learning are obligatory by educational leaders and decision makers to develop agricultural education in Arab world.

Traditional methods and techniques of teaching have endured throughout the world, including within Arab nations. However, those methods tend to focus only on the instructor and the topic being taught. Engaging students in their own learning can bring about better and longer-lasting learning. The background and concepts of active learning presented in this chapter should be incorporated into agricultural education instruction in secondary schools and colleges throughout the Arab world.

REFERENCES

Abolaji, G., & Reneau, F. W. (1988). Inservice needs and problems of agricultural science teachers in Krara State, Nigeria. *The Journal of the American Association of Teacher Educators in Agriculture*, *29*(3), 43–49. doi:10.5032/jaatea.1988.0343

Agnew, D. M., & Shinn, G. C. (1990). *Effects of simulation on cognitive achievement in agriculture mechanics*. Faculty Publications: Agricultural Leadership, Education & Communication Department. https://digitalcommons.unl.edu/aglecfacpub/4

Al-Khuzaim, K. M., & Al-Ghamdi, M. F. (2016). *An analysis of the content of mathematics books for the upper grades of the primary stage in the Kingdom of Saudi Arabia in the light of the skills of the twenty-first century*. The Mission of the College of Education and Psychology.

Al-Wasimi E. A. A. (2004). *Preparing the science teacher in the light of scientific variables*. Reference research submitted to the Permanent Scientific Committee for Education to promote assistant professors.

Almady, S. M. (2013). *A proposed conception for developing the basic education teacher through educational system in the Arab world in light of professional standards and requirements*. Culture and Development Magazine.

Arab Center for Education and Development. (2010, July). *The Fifth International Conference on the Future of Arab Education Reform for the Knowledge Society - Experiences, Standards and Visions*. Arab Center for Education and Development, Arab Open University.

Arab Encyclopedia. (2020). *Agricultural education - Enseignement Agricole*. http://www.arab-ency.com.sy/ency/details/2915

Asma, F. Y., Kumar, R., Bilal, A. L., Sandeep, K., Dar, Z. A., Faisal, R., Abidi, I., Nisar, F., & Kumar, A. (2021). Crop Simulation Models: A Tool for Future Agricultural Research and Climate Change. *Asian Journal of Agricultural Extension, Economics & Sociology, 39*(6), 146-154.

Badir, K. (2008). *Active Learning*. Dar Al-Maysara for Publishing, Distribution and Printing.

Barrick, R. K., Samy, M. M., Roberts, T. G., & Easterly, R. G. (2011). Assessment of Egyptian Agricultural Technical School Instructors' Ability to Implement Experiential Learning Activities. *Journal of Agricultural Education*, *52*(3), 6–15. doi:10.5032/jae.2011.03006

Battisti, B. T., & Passmore, C. (2008). *Action Learning for Sustainable Agriculture*. Academic Press.

Bonwell C. & Eison J. (1991). Active Learning: Creating Excitement in the Classroom, in 1Ashe-Eric *Higher Education 1*(3), 1-121.

Bransford, J. D., Brown, A. L., & Cocking, R. R. (Eds.). (1999). *How people learn: Brain, mind, experience, and school*. National Academy Press. doi:10.17226/9853

Bu Shehidim, S., Awwad K., Khalila, S., Al-Amad, A., & Shadeed, N. (2020). *The Effectiveness of E-Learning in the Light of the Spread of the Corona Virus from the Point of View of Teachers in Palestine*. https://st. najah.edu /media/publishedresearch

Cairo University. (2003, April). *The Sixth Scientific Conference of the Faculty of Education in Fayoum, Teacher's Sustainable Professional Development*. Cairo University, Fayoum Branch.

Cheek, J. G., Arrington, L. R., Carter, S., & Randell, R. S. (1990). Relationship of supervised agricultural experience program participation and student achievement in agricultural education. *Journal of Agricultural Education*, *35*(2), 1–5. doi:10.5032/jae.1994.02001

Chickering, A. & Ehrmann, S. (1996). Implementing the Seven Principles: Technology as Lever. *AaheBulletin*, *1*(4.8).

Cohn, D., Atlas, L., & Ladner, R. (1994). Improving generalization with active learning. *Machine Learning, 15*(2), 201–221. doi:10.1007/BF00993277

DeCesare, M., McKinney, K., Beck, F. D., & Heyl, B. S. (1998). Sociology through Active Learning: Student Exercises. *Teaching Sociology, 29*(3), 376–377. doi:10.2307/1319200

Deci, E. L. (1992). The relation of interest to the motivation of behavior: A self-determination theory perspective. In K. A. Renninger, S. Hidi, & A. Krapp (Eds.), *The role of interest in learning and development* (pp. 43–70). Lawrence Erlbaum Associates.

Egyptian Association for Curricula and Teaching Methods. (2004, July). *16th Scientific Conference for Egyptian Association for Curricula and Teaching Methods, Teacher Formation*. Ain Shams Hosting House.

Erickson, M. G., Guberman, D., Zhu, H., & Karche, E. (2019). Interest and active learning techniques in an introductory Animal Science course. *NACTA Journal, 63*, 293–298.

Fahnoe, C., & Terry, L. (2013). What Knowledge Is of Most Worth: Teacher Knowledge for 21st Century Learning. *Journal of Digital Learning in Teacher Education, 29*(4). https://files.eric.ed.gov/fulltext/EJ1010753

Field Experiences. (n.d.). Iowa State University. Retrieved from https://www.ageds.iastate.edu/undergraduate/field-experience
s

Freeman, S., Eddy, S. L., McDonough, M. M., & Smith, N. (2014). Active learning increases student performance in science, engineering, and mathematics. *Proceedings of the National Academy of Sciences, 111*(23), 8410-8415. 10.1073/pnas.1319030111

Garton, B. (1997). Agriculture teachers and students: In concert or conflict? *Journal of Agricultural Education, 38*(1), 38–45. doi:10.5032/jae.1997.01038

Hamadeh, M. M. (2005). The effectiveness of the two strategies (think - pair - share) and the survey based on the active learning method in school mathematics clubs in developing mathematical thinking skills and reducing mathematics anxiety among middle school students. *Journal of Educational and Social Studies, 113*, 183 – 237.

Hammerness, K., Darling–Hammond, L., Bransford, J., Berliner, D., Cochran–Smith, M., McDonald, M., & Zeichner, K. (2005). How teachers learn and develop. In L. Darling–Hammond & J. Bransford (Eds.), *Preparing teachers for a changing world* (pp. 40–87). Jossey–Bass.

Harackiewicz, J. M., Smith, J. L., & Priniski, S. J. (2016). Interest matters. *Policy Insights from the Behavioral and Brain Sciences, 3*(2), 220–227. doi:10.1177/2372732216655542 PMID:29520371

Hefny, M. K. (2015, August). Teacher Skills in 21st Century. *The 24th Conference for Egyptian Association for curricula and instruction. Programs of teacher's preparation at universities for Excellence,* 288 – 311.

Hendy, M.H. (2002). The effect of the diversity of using some active learning strategies in teaching a unit in the biology course on the acquisition of some biological concepts. Self-esteem and the tendency towards mutual dependence among first-year agricultural secondary students. *Egyptian Association for Curricula and Teaching Methods, Studies in Curricula and Teaching Methods, 79*, 1-59.

Hidi, S. (1990). Interest and its contribution as a mental resource for learning. *Review of Educational Research, 60*(4), 549–571. doi:10.3102/00346543060004549

Hwang, G. J., Lai, C. L., & Wang, S. Y. (2015). Seamless flipped learning: A mobile technology-enhanced flipped classroom with effective learning strategies. *Journal of Computers in Education, 2*(4), 449–473. doi:10.100740692-015-0043-0

Hyun, J., Ediger, R., & Lee, D. (2017). Students' Satisfaction on Their Learning Process in Active Learning and Traditional Classrooms. *International Journal of Teaching and Learning in Higher Education, 29*(1), 108-118. https://www.isetl.org/ijtlhe/

Jecker, J. D. (1964). The Cognitive Effects of Conñict and Dissonance. In L. Festinger (Ed.), *Conñict, Decision, and Dissonance*. Stanford University Press Stanford.

Kolb, D. A. (1984). *Experiential learning: Experience as the source of learning and development*. Prentice-Hall.

Leite, F. C. T., & Marks, A. (2005). Case study research in agricultural and extension education: Strengthening the methodology. *Journal of International Agricultural and Extension Education, 12*(1), 55–64. doi:10.5191/jiaee.2005.12106

Mabrouk, A. A. A. (2008). *The effect of teaching some contemporary issues using active learning strategies through the home economics curriculum on the outputs of the educational process* [Doctoral dissertation]. Faculty of Home Economics, Helwan University.

Mankin, K., Boone, K., Flores, S., & Willyard, M. (2004). What agriculture students say motivates them to learn. *NACTA Journal, 48*(4), 6–11. https://www.jstor.org/stable/43765890

McCormick, F. G. (1994). The power of positive teaching. Malabar, FL: Kreiger Publishing Company.

McGill, I., & Brockbank, A. (2004). *The action learning handbook: Powerful techniques for education, professional development and training*. Routledge Falmer.

Ministry of Education. (2005). *Active learning guide.* Curriculum and Materials Development Center.

Mirza, B. B., Straquadine, G. S., Qureshi, A., & Asaf, H. (2018). *Sustainable Agriculture and Food Security in Egypt: Implications for Innovations in Agricultural Extension: Regional Case Studies from Three Continents*. Climate Change, Food Security and Natural Resource Management. doi:10.1007/978-3-319-97091-2_5

Myers, B. E., Barrick, R. K., & Samy, M. M. (2012). Stages of Concern Profiles for Active Learning Strategies of Agricultural Technical School Teachers in Egypt. *Journal of Agricultural Education and Extension*, *18*(2), 161–174. doi:10.1080/1389224X.2012.655968

Noby, A. M. (2014). Designing electronic content in light of the principles of active learning and its impact on improving learners' learning methods and motivation towards training among learners. *Journal of Educational Technology*, *24*(4), 95–124.

Palmer, D. (2005). A motivational view of constructivist-informed teaching. *International Journal of Science Education*, *27*(15), 1881. doi:10.1080/09500690500339654

Parr, B., & Edwards, M. C. (2004). Inquiry–based instruction in secondary agriculture: Problem solving – an old friend revisited. *Journal of Agricultural Education*, *45*(4), 106–117. doi:10.5032/jae.2004.04106

Pressley, M., El-Dinary, P. B., Marks, M. B., Brown, R., & Stein, S. (1992). Good strategy instruction is motivating and interesting. In K. A. Renninger, S. Hidi, & A. Krapp (Eds.), *The role of interest in learning and development*. Lawrence Erlbaum Associates, Inc.

Razaitis, B. (2005). *Some Basic Active Learning strategies*. Center for Teaching and Services.

Roberts, T. G., & Dyer, J. E. (2004). Inservice needs of traditionally and alternatively certified agriculture teachers. *Journal of Agricultural Education*, *45*(4), 57–70. doi:10.5032/jae.2004.04057

Roberts, T. G., & Harlin, J. F. (2007). The project method in agricultural education: Then and now. *Journal of Agricultural Education*, *48*(3), 46–56. doi:10.5032/jae.2007.03046

Rotgans, J. I., & Schmidt, H. G. (2011). Situational interest and academic achievement in the active-learning classroom. *Learning and Instruction, 21*(1), 58–67. doi:10.1016/j.learninstruc.2009.11.001

Rowley, M., & Jensen, J. (2005). A longitudinal study of active learning in Family and Consumer Sciences. *Journal of Family and Consumer Sciences Education, 23*(1), 37–46.

Saadeh, J. A., Akl, F., Zamil, M., Arqoub, H. A., & Shtayyeh, J. (2011). Active learning between theory and practice (2nd ed.). Dar Al-Shorouk for Publishing and Distribution.

Schiefele, U. (2001). The role of interest in motivation and learning. In J. M. Collis & S. Messick (Eds.), *Intelligence and Personality: Bridging the gap in theory and measurement*. Lawrence Erlbaum Associates.

Settles, B. (2012, June). Active learning. Synthesis lectures on artificial intelligence and machine learning. Morgan & Claypool Publishers. doi:10.2200/S00429ED1V01Y201207AIM018

Shannon, A., Warner, W. J., & Osborne, E. W. (2006). Experiential learning in secondary agricultural education classrooms. *Journal of Southern Agricultural Education Research, 56*(1), 30–39.

Shoulders, C. W., & Myers, B. E. (2012). Teachers' use of agricultural laboratories in secondary Agricultural Education. *Journal of Agricultural Education, 53*(2), 124–138. doi:10.5032/jae.2012.02124

Stahl, A. E., & Feigenson, L. (2015). Observing the unexpected enhances infants' learning and exploration. *Science, 348*(6230), 91–94. doi:10.1126cience.aaa3799 PMID:25838378

Swanson, B. E. (1986). The Contribution of U.S. Universities to Training Foreign Agricultural Scientists. In L. Busch & W. B. Lacy (Eds.), *The Agricultural Scientific Enterprise: A System in Transition*. Westview Press.

Swanson, B. E., Barrick, R. K., & Samy, M. M. (2007). Transforming higher agricultural education in Egypt: Strategy, approach and results. *AIAEE Proceedings of the 23rd Annual Meeting*, 332 – 342.

Thoron, A. C., Barrick, R. K., Roberts, T. G., & Samy, M. M. (2008, March). Establishing technical internship programs for agricultural technical school students Egypt. *Annual conference, Association for International Agricultural and Extension Education. AIAEE Proceedings of the 24th Annual Meeting*, 468 – 475.

Tsai, C. W., Shen, P. D., Chiang, Y. C., & Lin, C. H. (2017). How to solve students' problems in a flipped classroom: A quasi-experimental approach. *Universal Access in the Information Society*, *16*(1), 225–233. doi:10.100710209-016-0453-4

Wagner, E., & Stefan, S. (2012, March). Effect of direct academic motivation-enhancing intervention programs: A metal analysis. *Journal of Cognitive and Behavioral Psychotherapies*, *12*(1), 85-98.

Yee, K. (2020). *Interactive techniques*. https://www.usf.edu/atle/documents/handout-interactive-techn iques.pdf

Yuretich, R. F., Khan, S. A., Leckie, R. M., & Clement, J. J. (2001). Active-learning methods to improve student performance and scientific interest in a large introductory oceanography course. *Journal of Geoscience Education*, *49*(2), 111–119. doi:10.5408/1089-9995-49.2.111

Zhu, X., Chen, A., Ennis, C., Sun, H., Hopple, C., Bonello, M., Bae, M., & Kim, S. (2009). Situational interest, cognitive engagement, and achievement in physical education. *Contemporary Educational Psychology*, *34*(3), 221–229. doi:10.1016/j.cedpsych.2009.05.002 PMID:26269662

ADDITIONAL READING

Youssef, M. M. A. (2021). Attitudes of university youth towards entrepreneurship: A field study at the Faculty of Agriculture, Damanhour University. *Alexandria Journal of Academic Exchange*, *42*, 1–31.

KEY TERMS AND DEFINITIONS

Agricultural Simulation: It is a computer simulation created to facilitate learning on the part of students or trainees. Educational simulations are abstracted or simplified representations of a target system, which are neither as complex nor as realistic as the relevant scientific simulations.

ALECSO: The Arab Organization for Education, Culture and Science is one of the organizations of the League of Arab States, which as a body concerned with the preservation of Arab culture.

Arab Encyclopedia: *Arab Encyclopedia* is an encyclopedia in 24 volumes in the Arabic language in 1953.

Arab World Countries: Consists of the 22 Arab countries which are members of the Arab League. A majority of these countries are located in Western Asia, Northern Africa, Western Africa, and Eastern Africa

Capstone Course: Senior seminar (in the U.S.) or final year project (more common in the U.K.) that make peers work together on specific project and represent results.

Case Study: A case study is an in-depth, detailed examination of a particular case (or cases) within a real-world context.

Collegiate/Non-Collegiate Settings: Collegiate means affiliated to a college or university or related to college student or life depends on the situation we use it, while non-collegiate means not regularly affiliated to the college or university.

Curriculum Reform: Curriculum reform can be seen as a process that aims to change the objectives of learning and the way learning takes place.

Deep Learning: Deep learning (deep structured learning) is part of a broader family of machine learning methods based on artificial neural networks with representation learning.

Experiential Learning: The experiential learning model allows youth to participate in engaging, stimulating activities that have a real-world basis.

INJAZ Al-Arab Foundation: Non-profit organization for education and training in workforce readiness, financial literacy, and entrepreneurship across the Arab World.

Interactive Teaching: Interactive teaching is a means of instructing whereby the teachers actively involve the students in their learning process by way of regular teacher-student interaction.

Intermediate Education: Vocational agricultural training is considered as an essential part of the educational system at secondary education institutions by all national partners in Egypt.

MUCIA/AERI Project: Midwest Universities Consortium for International Activities (MUCIA)/The Agriculture Exports and Rural Incomes (AERI) Project funded by USAID and concentrates much of its financial and human resources on organizing low-income farmers from the eight governorates of Upper Egypt into effective nonprofit business associations.

Teacher Character: Some qualities of a good teacher include skills in communication, listening, collaboration, adaptability, empathy, and patience. Other characteristics of effective teaching include an engaging classroom presence, value in real-world learning, exchange of best practices and a lifelong love of learning.

Teaching Excellence Centers: The Centre for Teaching Excellence collaborates with individuals, academic departments, and academic support units to foster capacity of effective teaching and meaningful learning.

Traditional Learning: The teacher directs students to learn through memorization and recitation techniques thereby not developing their critical thinking, problem solving, and decision-making skills.

Vocational Education: Vocational education is education that prepares people to work as a technician or to take up employment in a skilled craft or trade.

REVIEW QUESTIONS

1. What is meant by active learning strategies in agricultural education?
2. Which active learning strategies deepen student learning?
3. What are the differences between traditional learning and active learning?
4. What is the role of ICT within agricultural modern enterprises in the Arab world?
5. Describe the linkage between students' instructional techniques and their career development.
6. Why are case studies, simulation, and problem solving recommended for agricultural education?
7. Discuss the skills gap analysis pattern of agricultural education graduates in Arab countries.
8. Suggest specific active learning strategies that encourage student learning within your agricultural subject.
9. As a decision maker, what plan do you suggest improving agricultural vocational learning in Arab world countries?
10. Through Interactive teaching techniques given by Yee (2020), which are the most applicable for agricultural education?

Chapter 7
Education, Community, Industry:
Partnerships for Agricultural Development

Andrew Charles Thoron
https://orcid.org/0000-0002-9905-3692
Abraham Baldwin Agricultural College, USA

R. Kirby Barrick
https://orcid.org/0000-0001-7827-0957
University of Florida, USA (retired)

ABSTRACT

Agricultural and economic development in Arab countries is dependent upon having a well-prepared agricultural workforce. Among the components that are central to that mission are the utilization of an active and supportive external advisory committee and the incorporation of real-life, hands-on experiences for students through internships. This chapter provides background information to assist school and college administrators in establishing and maintaining an advisory committee as well as guidance for school and college personnel to incorporate student internships into the curricula.

DOI: 10.4018/978-1-6684-4050-6.ch007

INTRODUCTION

In most countries in the Middle East and North Africa, schools at all levels are strictly overseen by the Ministry of Education. Curricula for general and technical high schools as well as for colleges and universities are highly prescribed with little flexibility for adaptation to local needs. The concept of a "community" surrounding a secondary school or technical school is mostly non-existent; community members have no chance for input regarding the operation of the school or an assessment of how well the school meets its goals. Similarly, for technical schools, the area agriculture industry is not included in any decision-making processes and therefore have limited ability to determine the extent to which h technical school graduates are being prepared for industry.

Successful school and college programs rely on the involvement of all: the community, including families and agricultural businesses, the school or college instructors and administrators, and the students enrolled in the programs and activities. Various studies and reports (O'Brien, 2012; Epstein, 2008) indicate that teachers, parents and students indicate that parent engagement has increased considerably over the last 25 years. Linking schools with the communities they serve helps ensure that programs are relevant to the needs of the community and business/industry and are providing opportunities for students to develop and enhance competence and skills for entering and succeeding in the workforce. Parents want and need more and better information to guide their students, students benefit from family and community involvement, and educators must develop partnership programs that reach families and help students succeed. Schools should take the lead in creating those partnerships, including involvement of the agriculture and agribusiness industry.

A longitudinal study by Sheldon and Epstein (2004) indicated that school, community, and family partnerships can significantly reduce chronic absenteeism. Communicating with families about the importance of attendance and celebrating good attendance may reduce the percentage of students who miss twenty or more days of school each year. Students two are not in school obviously cannot learn as much or gain as many skills as those who are present.

For more than two decades, many schools have developed and implemented professional learning communities (PLCs). A PLC, as defined by Burnette (2002), is a school where "people are united by a common purpose, shared vision, collective commitments, and specific, measurable goals" (p.52) designed to enhance student learning. Al-Mahdy and Sywelem (2016, October) used the Professional Learning Communities Assessment—Revised to assess teachers' perspectives of PLCs in three Arab countries: Egypt, Saudi Arabia, and Oman. The Saudi and Omani teachers viewed PLCs in their schools positively, while Egyptian teachers showed negative perceptions. Further, female teachers viewed the PLC significantly more positive than their male counterparts.

Technical education development, including agricultural education, has resulted from attempts to achieve economic development through the expansion of arable lands, establishment of new industries and factories, the exploitation of mineral resources, and the shift to a modern economy. While progress has been made over the past 50+ years, the need still exists to improve and expand technical education in agriculture to achieve national goals. Technical education in Arab countries is typically a responsibility of the state, provided for and funded by national budgets (Harby, 1965). As in much of the populace throughout the world, the emphasis in public, government-funded education has continued to be on theoretical and academic education. Additionally, technical education is considerably more expensive that "academic" education, presenting an obstacle for underfunded schools and colleges.

In most Arab countries, technical schools are supervised and administered by the Ministry of Education. However, agriculture technical schools are often under the administration of the Ministry of Agriculture. At the college level, programs fall under the responsibility of the Ministry of Higher Education. The concern, then, lies with the issue of how the various ministries include business and industry, community groups, parents, school and college personnel, and students in designing and conducting programs for students at all levels to develop skills and competencies to be successful in the agriculture workforce and contribute to agricultural development. Utilizing an external advisory committee or council and providing educational internship experiences can assist in addressing that concern. As noted in Chapter 1, an advisory committee is an integral part of the instructional model for successful agricultural education programs that lead to a well-prepared agricultural workforce in Arab countries.

The institution of an external advisory committee and the initiation of field-based internship programs can help build the partnerships that are vital to school and college success and industry development. The objectives of this chapter are: 1.) to provide the foundations and procedures needed to successfully utilize an external advisory committee, and 2.) to provide the foundations and procedures needed to successfully implement field-based internship programs.

ESTABLISHING AND UTILIZING AN EXTERNAL ADVISORY COMMITTEE

What Is an External Advisory Committee?

In educational settings, advisory groups may be referred to as external advisory committees or councils, industry advisory committees or councils, or career and technical advisory committees or councils. For ease of reference, the term External Advisory Committee (EAC) will be used throughout this section.

Advisory committees are designed to increase the participation of industry and the community in career and technical education programs, including agriculture, for secondary school and college-level programs. An active EAC can provide valuable perspectives on policy, articulation, and educational programs from a variety of relevant industries and professions. Career and technical education programs should be tailored to meet the workforce development needs of the community and the needs and interests of the individual students. EACs strengthen collaboration between those responsible for career and technical education programs and the communities they serve. The dialogue between advisory committee members and career and technical educators promotes a shared responsibility for preparing students for a place in the agricultural workforce and in society (Ohio Department of Education, n.d.). The EAC can advise educators and administrators on the development, implementation, and evaluation of technical education programs to ensure that the programs meet workforce development needs of the community (McHenry County College, n.d.). The overall function of the EAC is to advise, assist and advocate career and technical education programs. EACs support and strengthen the partnerships among business, labor, the community, and education under the purview of state or federal guidelines and/or mandates (Skyline College, n.d.).

Roles of the External Advisory Committee

Advisory committees have three major roles:

- To advise — assess specific areas of the career and technical education programs and make suggestions and recommendations designed to improve that specific area. Such recommendations could include the modification of curriculum, purchase of new instructional materials and equipment, or adoption of a new safety policy.
- To assist — help instructors and administrators carry out specific activities. These activities could include judging competitive skill events, setting up a scholarship program, or obtaining media coverage for special events.

- To advocate — promote the career and technical education program throughout the community, offering internships, tours, job shadowing, and more work-based learning opportunities. Promotion or marketing could include talking to industry leaders, speaking at community functions, writing articles for local newspapers, or arranging publicity for successes. (Ohio Department of Education, n.d.)

Specific opportunities for EACs to carry out those three major roles may include some of the following (Skyline College, n.d.).

- Job opportunities:
  - Assist in surveying workforce needs and new and emerging occupations
  - Monitor the changing nature of the competencies in industry and career fields
  - Assist in gathering information for job placement of graduates
  - Identify opportunities to place students in full-or part-time jobs
- Facilities and program equipment:
  - Consult on existing equipment, facilities, and resources and compare with industry norms
  - Consult on lab equipment and compare with the current and future technology and industry/profession standards
  - Consult on lab safety program
- Course content:
  - Consult on the development of educational objectives
  - Review the program's sequence and scheduling of courses and delivery options
  - Review and suggest content for courses of study and standards of proficiency in areas that are essential to becoming successfully employed in a career path
- Instructional and learning experiences:
  - Identify or suggest resource personnel to enrich the instructional content
  - Assist in locating and obtaining equipment and supplies on loan, as gifts, or at special prices
  - Suggest qualified persons for needed instructor/personnel vacancies
  - Participate as a resource person to enhance the instructional process
- Promoting education:
  - Participate in programs designed to promote career and technical education
  - Testify at governing agencies in support of career and technical education

- Encourage other businesses to stimulate development of work experience programs
- Build interest and understanding between the school/college and community and professional organizations

External Advisory Committee Membership and Selection

Effective advisory committees should be large enough to reflect the diversity of the community, yet small enough to be managed effectively. Committees with fewer than five members may have a limited perspective, inadequate information on the career fields, and too little diversity. Committees with more than 15-20 members can become unmanageable.

EAC members should be persons whose opinions are respected. The value of any recommendation of the committee will be essentially equal to the collective respect that the community, the agriculture industry, and the school and administration have for the members of the committee. Consider having the members of the committee formally appointed by the appropriate administrator or school personnel. Appointments to advisory committees should be made for definite periods of time. Procedures should be established to address members' appointment dismissals or resignations and the possible use of alternates.

To allow for both continuity and change, it may prove beneficial to incorporate a rotational three-year term of service. To establish this rotation with a new committee, the members draw lots for one-, two-, or three-year terms, with one-third of the committee in each category. New members are appointed as terms expire. See Appendix 1 for additional information regarding the operation of the External Advisory Committee.

Here are some additional points to consider in selecting potential members of the EAC (Frala, 2022).

1. Get people involved who will understand the program, such as:
    a. A graduate of the program who is working in the local area.
    b. The program administrator (dean, department chair, etc.).
    c. Program administrator from a neighboring college, with a program similar to yours.
    d. Local program instructor.
    e. Business and industry representatives.
2. Invite the community businesses to participate.
3. Hold face-to-face contact with the community.
4. Involve government offices entities that have an interest in the particular programs.

5. Invite local political leaders who share a voice in the community.
6. Involve your school/college counselors who can help with job placement.

Once appointed, advisory committee members should attend meetings regularly, participate in discussions, and respect the rights of fellow members. Systematic controls should be established to deal with committee operations. Meetings should be held as often as the committee has important business. Officers should include a chair, a vice chair and a secretary. The business of the committee should be conducted with:

- A working agenda
- Use of parliamentary procedures with recording of minutes
- Appointment of subcommittees as necessary to carry out work of the committee
- Decision by a quorum vote of a simple majority.

Advisory Committee Responsibilities

The External Advisory Committee should assume responsibility for providing advice to and support of the instructional program (McHenry County College, n.d.).

- Curriculum and Instruction
    - Identify and expand the use of new technologies in the industry and review technology standards in the curriculum
    - Compare course content with occupational competencies and tasks
    - Analyze course content and sequence for relevance
    - Assist in developing and validating skills tests
    - Advise on local labor market needs and trends
    - Recommend safety policies and procedures
    - Promote and assist in maintaining quality programs
    - Review curriculum to ensure that it meets business needs and industry standards
    - Assist with incorporating employability skills in the curriculum
    - Assist in student competency assessment
    - Provide hands-on work-based learning and internship opportunities.
- Program Review
    - Review student performance standards
    - Assess, recommend, and/or provide equipment and facilities to replicate industry
    - Conduct community and occupational surveys

  - Identify new and emerging occupations
- Recruitment and Job Placement
  - Notify program instructors of entry-level job openings for students
  - Provide work experiences, internships/externships, apprenticeships, work/study, or work-based learning opportunities for students
  - Assist in developing strategies to recruit students into program
  - Assist in identifying additional work-based learning experiences
- Community/Public Relations
  - Promote the program to employers, communities, and the media
  - Assist in recognizing outstanding students, teachers, and community leaders
  - Conduct workplace tours.

Committee officers have additional responsibilities in providing leadership for the External Advisory Committee (Ohio Department of Education, n.d.)

Chair. The chair's leadership is key to the success of the advisory committee. It is suggested that a member other than a school representative assume this role. The chair should possess skills and characteristic such as:

- Experience in business/industry in the community served by the program
- Ability to manage meetings, plan and adhere to schedules, involve members in ongoing activities and reach closure or consensus on issues
- Skill in oral and written communications as well as willingness to make appearances before school and community representatives
- Experience as a committee member
- Ability to delegate responsibility as well as willingness to accept responsibility for the committee's actions
- Personal characteristics such as empathy, fairness, tolerance and sound judgment

The responsibilities of the chair include:

- Work with school and community representatives to plan and carry out the committee's program of work
- Prepare agendas and assist the instructor in handling details regarding meetings
- Preside at meetings
- Keep group efforts focused and all members involved in tasks
- Delegate tasks and follow-up work
- Arrange for presenting background information and reports to the committee

- Represent the committee at official meetings and functions
- Submit recommendations of the committee to appropriate administrators and groups
- Follow-up on committee recommendations or actions

Vice Chair

The skills and responsibilities of the vice chair are identical to those of the chair. The vice chair takes charge when the chair is absent or cannot serve.

Secretary

The secretary records meeting minutes and performs other clerical duties.

Committees may use a school representative in this position because of their access to computers and reproduction facilities.

The responsibilities of the secretary include:

- Take minutes at meetings; prepare and distribute minutes
- Mail agenda, announcements, minutes and other information to members
- Help assemble and distribute necessary background information to members
- Correspond with representatives of the school and community as needed.

Once the committee has identified its priorities, the discussion will become more specific as the committee determines exactly what it wants to accomplish. Possible activities in each priority area include:

- Community relations activities
    - Present programs to community organizations and groups
    - Establish ways to recognize outstanding students, teachers and community leaders
    - Obtain financial contributions to promote programs
    - Participate in and promote special school/college events
    - Provide information at administrator meetings
    - Set up and support a student scholarship program
    - Promote secondary/postsecondary connections
- Curriculum activities
    - Review instructional materials for technical accuracy
    - Assist in obtaining instructional materials
    - Recommend equipment and supplies
    - Recommend core curriculum content

  - Recommend safety policies
  - Provide equipment and facilities for specialized training needs
  - Encourage/promote secondary/postsecondary connections
- Community resource activities
  - Identify community resource people
  - Provide tours and field trip experiences
  - Provide speakers
  - Promote awareness of career opportunities
- Student organization activities
  - Assist in developing competitive skills events
  - Judge competitive skills events
  - Sponsor student activities
  - Collect skill events contributions of equipment and supplies
- Job placement activities
  - Organize employer/student conferences
  - Notify teachers of job openings for students
  - Provide training sites for students
  - Encourage other employers to provide training sites
  - Assist students to develop interviewing skills
  - Recommend employability skills
  - Hire career and technical education graduates
- Program review activities
  - Review program goals and objectives
  - Participate on program evaluation teams
  - Compare student performance standards to business/industry standards
  - Review adequacy of the facility
  - Make recommendations for program improvement
  - Review secondary/postsecondary possibilities
  - Support dual credit opportunities for students (secondary/postsecondary)
- Staff development activities
  - Provide in-service activities on new and current business/industry methods and processes for instructors
  - Provide skilled technicians to supplement instructors' experience
  - Provide summer/part-time employment to instructors for technical upgrading
  - Support instructor participation in professional development activities
  - Recruitment activities
  - Assist in recruiting new staff
  - Assist in recruiting potential students

This list of activities is not all inclusive but should give the advisory committee some ideas to consider. The functions and activities chosen for the program of work should match the needs of the program and community.

External Advisory Committee members serve at the pleasure of the school or college administration. They give freely of their time and talent in support of the agricultural education program and to promote agricultural and economic development for the local community and the country. Efficiency is of paramount importance; not wasting members' time is essential. EAC meetings should focus on two major components: providing clarifying information to the members and listening to the committee's recommendations and feedback. Meetings should be facilitated by the administration in conjunction with the committee chair, keeping in mind the following tips for effective meetings (McHenry County College, n.d.).

1. Put times for each agenda item on the agenda to keep the committee on task.
2. Make sure that any equipment/technology to conduct the meeting is ready to go in advance.
3. Start and end the meeting on time.
4. Try to identify "theoretical discussions" and make sure those items are saved for the end of the agenda.
5. Make sure all "action" items occur toward the first half of the meeting.
6. If possible, assign any individual tasks or information gathering prior to the meeting. This will eliminate unnecessary meetings.
7. Give an end of meeting evaluation.

Benefits of an External Advisory Committee

An effective External Advisory Committee can be of benefit to the school, administrators, and students beyond the expected duties and responsibilities of the committee.

The advisory committee may benefit administrators by:

- Advising the school authorities on the development of its long-range and annual plans for career and technical education
- Advising on policy matters rising out of the administration of the plan
- Interpreting to the school authorities the career and technical education needs of individuals and the community
- Assisting the school authorities in identifying the career and technical education needs of individuals and the community
- Assisting in assessing present and foreseeable needs in the area labor market

- Evaluating the impact of career and technical education programs on the people and community
- Assisting school personnel in understanding the postsecondary training needs for meeting the labor market's needs

The advisory committee may benefit the school and students by:

- Assisting in conducting surveys of local workforce needs
- Reviewing long-range and annual plans for a career and technical education program
- Advising on the establishment and maintenance of quality and relevant career and technical programs
- Reviewing career and technical education budget requests for instructional materials, labs, equipment and supplies
- Evaluating the adequacy of CTE facilities and equipment
- Reviewing course content to ensure relevancy
- Assisting in identifying skills/competencies needed in specific occupations
- Assisting in development of work-based learning opportunities
- Assisting in development of postsecondary training opportunities
- Promote career guidance activities in school counseling
- Serve as mentors for students seeking career guidance
- Reviewing child labor laws and local student employment
- Arranging plant/field trip visits for teachers, students and counselors
- Assisting in developing/maintaining library of visual aids, magazines, and books concerning the industry
- Assisting in developing and securing samples of industry products/materials for exhibit and instructional purposes
- Providing scholarships and other financial assistance for outstanding graduates desiring to continue their education and training
- Providing recognition such as awards to outstanding students
- Supporting and becoming involved with career and technical education student organizations
- Assisting in conducting clinics and in-service training for teachers
- Providing resource persons from industry to assist teachers
- Arranging industry experiences for teachers
- Providing recognition such as awards to outstanding teachers
- Paying industry organization membership dues for teachers
- Helping teachers, financially, to attend out-of-town industry and teachers' organization meetings

The advisory committee may benefit the community by:

- Providing speakers to address trade or civic groups or appear on radio and television concerning career and technical education programs in the schools
- Providing news stories on career and technical education programs to local news media and to industry trends magazines
- Attending meetings in support of career and technical education programs that may be called by local or state schools officials, boards and legislative groups
- Contributing funds to promote specific programs by way of advertisement in newspapers or through other media

To help ensure that the EAC remains focused and on-task, a periodic review of the committee can help in determining:

- The extent to which it is accomplishing its goals
- The extent to which the recommendations and actions have strengthened the career and technical education program
- Future direction and activities for the committee

The review can be either formal or informal, with the goal of helping the committee determine its overall effectiveness. Table 1 identifies some "Do and Don't" suggestions for EAC members to keep in mind.

Table 1. The do's and don'ts for EAC member involvement

Do	Don't
Make a commitment	"Wait and see"
Attend meetings	Have a record of absenteeism
Stick to the agenda – meet for a purpose	Waste others' time
Align recommendations with available resources	Impose pre-established, personal opinions
Remember advisory role	Add to the "wish list"
Advise when improvement is desired	Fail to suggest alternative solutions
Be a fact finder	Be a fault finder
Make commendations for a job well done	Be afraid to give recognition
Invite all opinions	Avoid those who disagree
Develop good rapport	Remain in isolation
Consider a variety of subjects	Usurp school administrators' authority

Source: *Developing a Local Advisory Committee*, Ohio Department of Education, n.d.

Factors that Contribute to Advisory Committee Success

External advisory councils can provide specialized expertise and be ambassadors to your community and connect to a greater constituency. They can perform important duties, such as fundraising, advocacy, and program evaluation. Well-conceived and well-executed advisory councils can be greatly beneficial. Unfortunately, some may not be well conceived nor well executed. Here are nine keys to a successful advisory council (Board Source, 2021).

1. **Be prepared to give it time and resources:** Advisory councils require care and feeding to be effective. School staff need to have the time and resources to effectively manage and support the committee.
2. **Know exactly what it is the council is to accomplish:** A lack of clarity in purpose, role, or scope is a common problem with advisory councils. It is important to have a written statement of purpose that addresses the following topics:
   a. Reason for the advisory council to exist: its goals.
   b. Relationship of the advisory council to the school or college administration.
   c. Relationship of the staff to the advisory council.
   d. Criteria for membership.
   e. Description of the selection process and to whom the advisory council reports.
   f. Length of term of service and duration of the group.
   g. Job descriptions that identify the specific responsibilities or expectations of individual members.
   h. Titles and duties of officers.
   i. Number and frequency of meetings.
3. **Find the right advisory council members; take matchmaking seriously:** The council's purpose and goals should determine its size, its meeting frequency, and its credentials for membership. Form must follow function, not the other way around. Seek out the skill sets, expertise, and insights that are required to meet the council's goals. The quality and commitment of those selected to serve on the advisory council will impact its effectiveness. Explain to prospective members why they were chosen and what will be expected of them.
4. **Prepare external advisory council members for service through an orientation program:** Set the stage for success by teaching the council members about the organization's mission, vision, values, and strategic plan. Explain the role of the council and expectations for their involvement.
5. **Support your council with well-developed meeting agendas, adequate supporting material sent in advance of the meeting, and skilled meeting**

facilitation: in this aspect, effective advisory councils follow the example of well-planned administrative meetings.

6. **Create mechanisms for communication and opportunities for dialogue between the school/college administration and the advisory council:** To facilitate ongoing communication and to provide the board with the council's expertise and advice, consider the following:
   a. Invite the chair of the advisory council to serve as a nonvoting member of the school/college governing board or on a board committee related to the council's purpose.
   b. Invite the chair of the advisory council to make regular reports at administrative board meetings or include a report in the consent agenda with other committee reports.
7. **Periodically assess the council's performance to identify its strengths and weaknesses or determine if it has outlived its initial purpose:** This can be done through self-assessment, internal review by an internal ad-hoc committee, or third-party reviews by an external consultant. The school/college and council should define the criteria by which the council's work can be judged. Possible questions to ask council members include the following: What would this school/college be missing if this council was not in existence? In what ways could we add greater value?
8. **Keep the group fresh and informed through continuing education:**
   a. Invite school/college administrative board members to make presentations at advisory council meetings.
   b. Enable council members to observe the way the school/college works by experiencing its programs and operations firsthand. Maintain a reasonable flow of information between meetings that keeps the council members informed of key school/college activities.
   c. Plan a retreat or special council meeting to give the members the opportunity to review their responsibilities, identify priorities to strengthen their performance, and get to know each other.
   d. Invite council members to organizational social events.
9. **Disband the advisory council when…**

…there is a lack of funds or staff time to support the group, when the council is not meeting its goals, or simply because the group has achieved its goals and its work is done. At that point, it is time to re-think the need and use of an external advisory committee and, literally, start the process anew.

In all cases, the school/college should show its gratitude to the group for the time and effort each member contributed.

Recognizing the Service of External Advisory

When individuals volunteer their time, appropriate recognition can let advisory committee members know that their investment of knowledge and time is worthwhile and appreciated. Their recommendations are given careful consideration. Knowing they make a difference can inspire advisory committee members. Therefore, inform the advisory committee when recommendations are implemented.

Rewards encourage attendance and involvement. Recognition activities also attract the attention and interest of other qualified people who may someday serve on the advisory committee. They also bring public attention and goodwill to the organization because they demonstrate that the organization appreciates the efforts of its members. Advisory committee members are not paid for their efforts; therefore, rewards and recognition are especially important to advisory committees. Rewards should not be given indiscriminately but should be based on actual contribution to the committee's activity.

The best types of rewards or recognition are those that can stimulate productivity, improve committee interaction and increase member satisfaction. Most members are willing to attend regularly and work hard as long as their expertise and talents are used, their recommendations are seriously considered, and they are given feedback concerning their efforts. The following are some ways to recognize committee members:

- Issue press releases announcing member appointments
- Report periodically at meetings and in the media on the results of committee
- recommendations and the ways the committee has been of service
- Invite members to visit programs to see the results of their recommendations
- Invite members to attend special career and technical education events
- Introduce advisory committee members at program or CSO meetings or events
- Hold a banquet in honor of the committee and present certificates of service
- Schedule a meeting whereby administrators of the school or institution can attend
- Place members' names on a display board or plaque at the school or institution
- Include members' names on program information disseminated to the public
- At the end of the year, send each member a letter of thanks and appreciation, signed by the appropriate school official
- Send a letter of appreciation to the committee member's supervisor and/or company, explaining the work being done by the member and committee and thanking the company for its support

- Provide a certificate that the member can display at work, identifying him or her as a current advisory committee member (Ohio Department of Education, n.d.).

FIELD-BASED INTERNSHIP PROGRAMS

Internships bring together skills and knowledge development that are connected to the classroom curricula. The field-based aspect to internships involves students working for employers for a specified time to learn about the industry or an occupation and extend the formal classroom learning into the community and agricultural industry. Students that participate in an internship put skills, knowledge, and abilities into practice through a supportive learning environment that not only focuses on the development of skills, but also student learning and development. Internships provide an outlet for a student or trainee to work for an organization to gain work experience or skills for qualifications. Internships involve communities, students, parents, teachers, headmasters, and industry/employers (Shoulders, Barrick, & Myers, 2011).

Field-based internships can be *career discovery*, *skill acquisition*, or a *capstone experience* for students. All three could co-exist in a curriculum but should be defined and be purposeful in the objectives achieved. A *career discovery* internship would focus on a student gaining career knowledge and the student would focus on what skills they need to obtain in the industry or occupation. This internship may be shorter in length. A career discovery internship may be as little as a week to three weeks. A *skill acquisition* internship would be central to gaining hands-on experiences in a field-based experience. Skill-focused internships would be best suited for agricultural industry occupations that are specific to the scope in which the school setting can provide. This type of internship may last during a planting or harvest season up to eight weeks. Finally, the *capstone experience* internship (an advanced internship) would be utilized near the end of a student's formal educational time and the internship would put all formal learning and classroom-based skills learning into practice in an industry setting. Typically, this internship would last a considerable length of time ranging from six to sixteen weeks or longer. Students may earn credits toward their education by completing an internship. This is possible at all three levels described above.

Internships may be paid or unpaid. Internships differ from a part-time job or work when a student's formal school is not in session in two aspects. First, internships are planned and have learning outcomes that are decided by the school system, agricultural industry employer (partner), and the student. Second, students (interns) are supervised by school or teaching staff and industry supervisors.

Students should seek an internship based on their area of interest and focus with the assistance and advisement from their school internship coordinator or teaching faculty advisor. The goal of a younger individual may be for career exploration, while the goal of an advanced individual should be for career development and experiential-based skills acquisition (Barrick, Samy, Roberts, Thoron, & Easterly, 2011). This should be decided by the student and teacher. Once decided, the student can either be placed into an internship or seek a collaborator with whom they conduct their internship. A discussion between the agricultural employer, the student, and the teacher should be conducted to ensure learning goals, skills practice, and expectations while on-the-job. The focus is upon learning. Individuals who may be completing their first internship may experience general aspects of the agricultural business and assist with job performance tasks. For more advanced internships, trainees should be utilizing skills, developing new skills, and begin developing management decision-making competencies.

Examples of Field-Based Internships

The headmaster, departmental coordinator, and school faculty should work to make connections with the agricultural industry to set up internship possibilities for students. Agreements should be signed agreements with the headmaster or departmental leadership/administration.

Table 2. Examples of field-based internships

Agricultural Safety	Grape and Vegetable Operations
Beekeeping and Honey Production	Wheat Production and Agronomy
Communication Skill Development at Agriculture Commodity Companies	Horticultural Value Chain
Marketing and Transportation of Commodities	Soil Science and nutrients
Post-harvest Handling	Fertilizers and Herbicide Industry
Packaging	Planting and Harvesting of Commodities
Input and Supply Chain Areas	Irrigation Instillation
Inventory and Facility Management	Drip and Spray Irrigation
Animal Nutrition and Feeding	Landscaping at Hotels
Dairy Handling	Turfgrass
Livestock Value Chain	Integrate Pest Management
Veterinary Technician	Greenhous Management
Large and Small Animal Care/Husbandry	Landscape Technician
Food Science and Food Development	Tractor Operation and Maintenance
Food Processing	Small Engine Repair and Maintenance
Food Safety	Equipment Management

Schools that have developed internships exclaim that before working with students to participate in field-based internships they only used traditional teaching methods and had few teaching aids, other than simple textbooks and a chalkboard in each classroom. This left students with little or no access to practical training in or out of school. As a result, student interest in the school was low and the dropout rate was high (about 20%). At the same time, the agribusiness community had urgent need of skilled technicians to work in areas of horticulture and livestock value chains and in food processing factories.

Individuals Involved in the Field-Based Internship Process

Development of successful internships for students begins with identifying the type of internship (career discovery, skills-based, or capstone), learning goals/objectives, and involvement of community, parents, industry partners (employers), and the student:

- Type of internship - should be determined by the teaching professional in collaboration with the student and parent. Factors include the level of education the student has achieved.
- Learning goals/objectives - clearly defined learning goals and completion of tasks by the student that helps direct their learning and final assignments as a participant of an internship.
- Involvement of the community – to create awareness of educational goals, build relationships with the local community, and add value to the community through an educated workforce that reflects agriculture within the community or region.
- Parents should be involved in the decision-making process to create parental support for their student's learning and creation of understanding of the commitment from the student intern.
- Industry partners (employers) – relationships with employers that are educated on the value and buy-in to the concepts that they are professionally developing individuals for a better prepared workforce and not just providing a younger individual with a job. Industry partners must recognize the value of learning that is involved in an internship.
- Student – should be a motivated learner, conduct themselves professionally in a workplace environment, maintain workplace standards (showing up on time, completing tasks, and having supervision).

Best Practices for Internship Programs

The National Association of Colleges and Employers (NACE, 2022) outlined 15 best practices for internship programs.

1. Provide real work assignments: provide interns with real work experiences that benefit their learning and objectives that are valuable to the business. The student is learning because of their internship and the employer is also receiving a benefit to their business or organization while developing a young professional.
2. Conducting orientations: development of a meeting or orientation so that the employer, teacher, student, and parents (if appropriate) are all in agreement with the learning goals and expectations that will result from the internship. This will ensure clarity from the beginning, and all involved will have time to ask questions for clarification and have agreed upon learning outcomes and work expectations.
3. Development and use of a handbook: this provides a copy of agreed upon goals/learning objectives, procedures, rules, assignments, and frequently asked questions that can provide clarity and documented expectations.
4. Plan for housing or living arrangements: if a student will not live at home or a school setting, housing is often a limiting factor for internships that require relocation for a shorter period. Cost of housing should be considered. Students and parents need to know well in advance who is required to pay for housing if the student is not going to live in their normal living arrangement. Some employers may assist or pay for housing.
5. Scholarships or additional assistance: are their supplemental or can supplemental funding be developed to support students that are conducting their internship?
6. Planned work arrangements: developing work hour expectations and a normal working routine will often supply students with the ability to plan and know their commitment for their internship. Outline procedures for who they contact if the student is sick and cannot be at their internship on a working day.
7. Having an intern manager: outline who the student intern reports. A dedicated manger/supervisor at the school and at the business who is hosting the intern is important for student accountability and provides the intern with one person (at the school and employer) who they may ask questions to and who is supervising.
8. Involvement of a team: the student, school supervisor, and employment supervisor are a team that is working to develop the student individual for a well-prepared agricultural workforce.

9. On-site visits: the school's internship manager or faculty member should visit the intern at the internship site during their working hours. An on-site visit is strongly encouraged at least one time. This helps strengthen school and industry relationships. This also allows school employees to understand the context of the student's internship. Visits will also help promote that the internship is a learning experience.
10. Hold new-hire panels: having former students that conducted an internship and are now employed will create interest and buy-in among younger students. This will strengthen student motivation and interest in previous internship sites.
11. Invite guest speakers from industry into the school: hosting guest speakers to talk about the agriculture industry sector and their company to speak to younger individuals strengthens relationship between the school and community for future internships. This also will promote businesses to plan for future or expanded internship opportunities through the interaction with students. This will benefit teachers as they will hear first-hand what is important within the industry directly from potential employers.
12. Offer out-of-class training: this training may focus on employability skills, speaking skills when meeting with potential employers, or planning for living arrangements during an internship.
13. Conduct focus groups or surveys: survey students that have participated in an internship and also survey employers as to the benefits or changes needed for successful internships. This will show employers that their input is valued and will maintain a positive working relationship. Surveying students that have completed an internship will allow for students to state positive or negative experiences with employers.
14. Showcase internships: displaying placement of student internship using posters or computer technology in school hallways or showcases will show that the school is proud of student accomplishments, that internships and learning outside of the formal school setting is valued and will provide student motivation by younger students to aspire to participate in an internship.
15. Conduct exit interviews: ask students to complete an exit survey (as described above) and then follow up with a face-to-face conversation is an excellent way to gather feedback and gauge their experience during their internship experience.

Development of an Internship

During the development of an internship process, criteria should be established to help define and differentiate an internship from work. The National Association of College and Employers (NACE, 2011) suggested the following criteria:

1. The experience should be related to and an extension of the classroom. This addresses that an internship should be focused in the agricultural sector and within a curricula that is related to what the student is learning or an application of what has been learned in the formal class setting.
2. The internship should not simply only advance the operations of the employer. The employer must allow for the intern to learn skills and be able to apply a new skill set. Not to simply use the student intern as an additional labor force.
3. Skills and knowledge that the intern learns should engage in transferable skills. While aspects of the internship may be specific to a segment of the industry, interns should not solely learn skills that are unique to only that business.
4. There should be defined learning objectives and professional goals achieved by the student through the internship experience.
5. Set dates of beginning and end of the internship, a job description with duties and qualifications.
6. Supervision of the intern by the employer and the school or educational entity. Supervisors should be professionally qualified. Thoron, Barrick, Roberts, and Samy (2008) noted that instructors need to engage in practical skills and experience as part of their professional development and maintain up-to-date knowledge and skills with industry experiences.
7. The intern should receive feedback from the on-site and educational supervisor in a planned manner on a routine basis.
8. The employer should supply resources, equipment, and materials needed for the intern to complete their internship objectives.

During the development stage of internship by an educational organization faculty must recognize the importance of their students' learning through experience. Industry partnerships must be developed and maintained through agreements and recognition that the purpose of an internship is for students to gain practical experiences and that employers will benefit through a well-educated and prepared workforce. Teachers, Headmasters, and school administration should work with businesses and the agricultural industry through the skill gap analysis (noted in chapter 1). The completion of the skill gap analysis will identify potential employers for internship placement.

Next, schools will develop a handbook and procedures. This includes internship definitions, requirements, eligibility, guidelines, an internship agreement (between the employer, school, student, and in some cases parents),

Example of Internship Requirements

Definitions

- Advisor – the student's assigned academic advisor
- Cooperating Supervisor – industry or government, agency, business or organization that has agreed to participate in the internship program and whose participation has been agreed to by the school.
- Departmental Coordinator – department head or person designated by same, who coordinates activities of the internship.
- Faculty Supervisor – faculty member who supervises the student's internship experience. Such person should be knowledgeable in the area of work in which the student is gaining experience and may or may not be the student's advisor or departmental coordinator.

General Information

- To be eligible for the internship, students must have met a standard of academic performance. Criteria may include grade point average or successful completion of coursework at an acceptable level, class rank, among other applicable academic and/or work ethic and ethical factors. It may also outline the year in school they need to be to enter an internship.
- It should be noted if the internship is for academic credit and if the internship will be graded by the school or academic institution. There should be an internship application with noted deadlines for students to apply to an internship.
- Course prerequisites – schools may establish prerequisites which are appropriate to the experience. If the school chooses to utilize the three types of internships (career discovery, skills-based, or capstone) each should be explained when a student would complete the type during their academic standing. Clear and specific outlines for each should be explained.
- Academic credits – schools should establish if any academic credit will be granted for internships and if there are any limits to internship credit. If credit is granted, schools should establish a formula to determine the number of credits awarded for a given experience. Credit should be granted based on student's accomplishments and internship objectives and not merely length of time. It should be determined if internship credits can be repeated.
- Grading – the school should determine the grading system to be used, which could be either a satisfactory/unsatisfactory or letter grades. The final grade is

determined from three sources: a) internship site supervisor recommendation, b) university supervisor recommendation, and c) student portfolio/evaluative report. Students will be evaluated by internship site supervisors, based upon how well they accomplish the tasks designated on the Training Plan. Final responsibility for assigning the grade rests with the faculty supervisor.

- Exit Interview - The internship employer supervisor will conduct an exit interview at the end of the internship to evaluate the progress made over the course of the semester. Students should treat the exit interview/evaluation as if students were "really" employed.

Student Responsibility

- Initiate Participation – students should discuss their intention or interest in completing an internship with their advisor. If a school required an internship, the school should determine at what stages during a student's academic program they would complete their internship.
- Consideration of Employment Opportunities – students should obtain information concerning potential opportunities in their area of interest from the coordinator. Internship opportunities should be presented to students or be placed in areas in the school so that students can learn about potential internship placements. Schools should hold internship showcases as described earlier in this chapter.
- Prepare a Resume – students should prepare a resume and/or personal interest sheet that can be used in seeking an internship. Assistance in preparing such materials should be provided through classroom instruction or out-of-class training as described earlier in this chapter.
- Determination of Specific Objectives – students should develop ideas/learning goals they wish to achieve by completing an internship in consultation with their advisor or future supervisor. If done well, this will lead to easier placement by the school and matching student interest with an internship employer.
- Assignments – students should complete assignments and communicate with their supervisors (faculty and employer supervisor).
- Students' responsibilities in the internship also include academic work, in addition to regular on-the-job requirements. Specific on-the-job expectations will be outlined in the *Training Plan*. This plan lists both the core requirements, as well as other requirements outlined by the internship site supervisor in conjunction with the faculty supervisor.

Departmental/Program Coordinator Responsibility

- Implementation of policies – assume primary responsibility for implementing those standards established by the faculty (course requirements, grades, credit-awarding etc.)
- Selection of students – discusses internship concepts with prospective participants and works with academic advisors in selection of students for participation. Collects applications from students seeking internship experience and checks that appropriate prerequisites have been completed.
- Creation of memorandum of agreements – seeks and maintains agreements with internship employers and works to develop new agreements and seeks to strengthen relationships with the industry partners.

Faculty Supervisor Responsibility

- Identification of objectives and activities – assist the student in identifying internship objectives that are consistent with the intention of the program and the student's interests and capabilities. Identifies activities that are consistent with the objectives and are feasible at the selected work site. Indicates support of objectives and activities with signature on the agreement.
- Collaboration with cooperating employer – should work to develop objectives and complete student internship learning objectives collaboratively with the employer.
- Supervision of Internship – provide indirect supervision by reviewing progress reports that are submitted by the intern and any additional assignments. Serves as the contact person when cooperating employer finds it desirable to discuss aspects of intern's program. On-site visit by the faculty supervisor is desirable whenever practical.

Cooperating Employer Responsibility

- Approval of objectives and activities – the employer should have input and final approval of the student learning objectives. Sign agreement that signifies that stated objectives can be achieved and stated activities are feasible at the proposed work site.
- Provide a professional educational experience – understand that the internship opportunity is for the educational goals of the student intern. Agree to work with the intern and supply a professional educational experience.

- Evaluation – may assist the faculty supervisor in the final evaluation of the student's performance. Tasks under advisement suggestions made by intern in final evaluative report as to how future internship at the site could be improved.
- School visits and career showcases – when possible, attend school career showcases and meet with students that would be seeking internships.

Overall, the goal of an internship is that the intern benefits from the experience and the host organization should benefits from the intern's work. This mutually beneficial experience is organized, designed, and delivered through the formal education organization. The most positive outcome resulting from an internship is employment for the intern. The benefit from an employer of an intern is to have a well-prepared employee that already knows the business, work requirements, and has an established set of skills. This will lead to a well-prepared agricultural workforce.

During the creation of internships teachers will need to address opportunities with potential industry partners and students. Teachers and coordinators will have the responsibility of securing agricultural partnerships. To best accomplish this, teachers need to develop clear objectives for the internship and note benefits to host organizations. In-person visits to agricultural companies and sharing curricular goals are vital in communicating the focus of internship work. Asking industry partners their desires and goals as well as addressing a skill gap (presented in Chapter 1) will assist. The teacher will need to develop responsibilities of the employer and selection criteria for placement of the student that will serve in that internship. It is suggested to reference Chapter 1 of this text for additional responsibilities that are addressed in that chapter. Appendix 2 supplies an outline and example for a handbook.

CONCLUSION

Establishing and maintaining effective educational programs that contribute to agricultural and economic development goes beyond the development of curricula. Faculties of agriculture in colleges and universities and instructors in agricultural technical schools must broaden involvement in their programs by utilizing a well-structured external advisory committee that advises as well as supports the programs. A well-developed workforce is incumbent upon school and college graduates having knowledge but also experience in their chosen career path in agriculture and agribusiness obtained through internship and field experiences.

Review Questions

1. What is an external advisory committee and why is it important in a college or school environment?
2. What are the key roles and responsibilities of EAC members, individually and as a group?
3. Who should serve on the EAC, and how are they selected?
4. What are some perceived benefits of the EAC for the school or college, the students, and the community?
5. What factors may contribute to the success of the EAC?
6. What is the main purpose of a field-based internship?
7. How can field-based internships benefit the school, student, and agriculture industry?
8. What are the three types of internships and how do their goals differ?
9. What are three examples of field-based agriculture internships?
10. Which of the fifteen best practices would apply best for implementation in your context?

REFERENCES

Al-Mahdy, Y. F. H., & Sywelem, M. M. G. (2016, October). Teachers' perspectives on professional learning communities in some Arab countries. *International Journal of Research Studies in Education*, *5*(4), 45–57. doi:10.5861/ijrse.2016.1349

Barrick, R. K., Samy, M. M., Roberts, T. G., Thoron, A. C., & Easterly, R. G. (2011). Assessment of Egyptian agricultural technical school instructor's ability to implement experiential learning activities. *Journal of Agricultural Education*, *52*(3), 6–15. doi:10.5032/jae.2011.03006

Board Source. (2021). *Advisory councils nine keys to success*. Author.

Burnette, B. (2002). How we formed our community. *National Staff Development Council*, *23*(1), 51–54.

Epstein, J. L. (2008, February). *Improving family and community involvement in secondary schools*. Condensed from *Principal Leadership, 2007*. National Association of Secondary School Principals.

Frala, J. (2022). *How important is an advisory committee to your vocational program success?* Academic Senate for California Community Colleges.

Harby, M. K. (1965). Technical education in the Arab States. UNESCO.

McHenry County College. (n.d.). *Career and technical advisory committee manual*. Author.

National Association of Colleges and Employers (NACE). (2011, July). *Positions Statement on United States Internships*. Retrieved from https://www.naceweb.org/about-us/advocacy/position-statement s/position-statement-us-internships/

National Association of Colleges and Employers (NACE). (2022). *15 Best Practices for Internship Programs*. Retrieved from https:// www.naceweb.org/talent-acquisition/internships/15-be st-practices-for-internship-programs/

O'Brien, A. (2012). *The importance of community involvement in schools*. Edutopia.

Ohio Department of Education. (n.d.). *Developing a local advisory committee*. Author.

Sheldon, S. B., & Epstein, J. L. (2004, October). Getting students to school: Using family and community involvement to reduce chronic absenteeism. *School Community Journal*, *14*, 39–56.

Shoulders, C. W., Barrick, R. K., & Myers, B. E. (2011). An assessment of the impact of internship programs in the Agricultural Technical Schools of Egypt as perceived by participant groups. *Journal of International Agricultural and Extension Education*, *18*(2), 18–29. doi:10.5191/jiaee.2011.18202

Skyline College. (n.d.). *CTE program advisory committees*. Author.

Thoron, A. C., Barrick, R. K., Roberts, T. G., & Samy, M. M. (2008). *Establishing technical internship programs for Agricultural Technical School students in Egypt*. Paper presented at the Association for International Agricultural and Extension Education Conference, Costa Rica. Retrieved from https://www.academia.edu/51021584/Establishing_technical_int ernship_programs_for_agricultural_technical_school_students_ in_Egypt

KEY TERMS AND DEFINITIONS

Community: Parents, agriculture industry leaders, and other non-school personnel who have a special interest in the operation of a college or school and are willing to lend support to programs and students.

External Advisory Committee: An organization consisting of school or college personnel, parents, faculty and instructors, and community and agricultural industry leaders formed for the purpose of providing guidance to and support of the agriculture programs in the school or college.

Field-Based Experience: Activities that connect knowledge and skills learned in classrooms and labs with applying that knowledge in an actual agricultural setting.

Internships: Activities designed for students enrolled in school or college programs of agriculture to provide real-life and hands-on experiences related to the knowledge and skills needed in a chosen field of the agricultural industry.

APPENDIX A

Sample External Advisory Council Bylaws

Article I: Name

The name of this organization is the ______________________ External Advisory Council.

Article II: Purpose

1. Functions of External Advisory Council

    a. Assist in the preparation and evaluation of the school improvement plan
    b. Ensure that the funds provided in the annual budget for use by external advisory councils are used for implementing the school improvement plan.
    c. Assist in the preparation of the school's annual budget with technical assistance from the Ministry of Education.

Article III: Membership

1. The dean/chair/headmaster (administrator) is responsible for ensuring the membership of the council is representative of the ethnic, racial, and economic community served by the school.
2. This council shall include parents, teachers, education support employees, community members, students, and the (administrator). A majority of the membership shall be non-district employees.
3. For purposes of EAC membership, parents shall be defined as anyone who has a student currently enrolled at the school/college.
4. Council members representing teachers, education support employees, students, and parents shall be selected by their respective peer group in the school/college in a fair and equitable manner as follows:
    a. Teachers shall be elected by teachers.
    b. Education support employees shall be elected by education support employees.
    c. Students shall be elected by students.
    d. Parents shall be elected by parents.
5. Elections shall be held at the beginning of each school year.
    a. The (administrator) will inform the school/college and community that nominations to the external advisory council for the upcoming school

year are now being accepted. The invitation for nominations must clearly indicate a deadline for nominations, membership categories, and the time, date, and event by which the voting will be taken.
   b. Written ballots will be provided for each peer group. The (administrator) will organize and hold formal voting for each membership category, collect ballots, and report results to the school/college and community.
6. Community members shall be appointed by the principal.
7. Term of office for newly elected members shall be for a period of three years.
8. Members not attending for two consecutive meetings, without an excused absence, will be replaced. Vacancies for parents, teachers, educational support personnel and students will be filled by peer election. Elections to fill vacant positions may be held as necessary throughout the year.
9. Membership in the EAC shall be limited to (8-15) voting members. Non-members are encouraged to attend meetings but may not vote. Voting in EAC meetings shall be limited to duly elected/appointed school board approved members.

Article IV: Responsibilities

1. The External Advisory Council is a resource to the school/college and the (administrator). The term "advisory" is intended to mean 1) inquiring, 2) informing, 3) suggesting, 4) recommending, and 5) evaluating.
2. The External Advisory Council, since it is advisory only, has some limitations.
   a. It may not dictate school/college policy.
   b. It must deal with issues rather than a particular person, whether they are administrators, teachers, students, citizens, or parents.

Article V: Officers

1. Officers and their election

   a. The officers of the External Advisory Council shall be a chairperson, a vice chairperson, and a secretary. All officers must be members of the EAC.
   b. These officers shall be elected, by written ballot, at the first meeting of EAC at the beginning of the school year, provided notice of election has been served.

Article VI: Roles of the Administrator and EAC Members

1. The Administrator

- Serves as a resource, providing information regarding the local school education program.
- May appoint community EAC members.
- Maintains the appropriate statutory composition of the EAC.
- Acts as an active resource.
- Encourages leadership from within the council.
- Assists in training members in leadership skills.
    - Arranges for presentations of interest to the council.
    - Informs the EAC of policies, curriculum, etc.
    - Establishes, maintains, and consults with the EAC on a regular basis involving it in decisions in accordance with stated purposes and policies.
    - Develops, through positive actions, feelings of trust and understanding among EAC, community, and staff.
    - Serves as the administrator in charge of the school/college with total responsibility to arrange all affairs including general control and supervision of its employees.

2. The EAC Chairperson
    - Works closely with the principal and the council to plan each meeting and establish an agenda in time to notify the community of the purpose of each meeting.
    - Calls the meetings to order, maintains order, and sees that the meeting is properly adjourned.
    - Instructs the secretary and other officers in their duties.
    - Sees that minutes are taken, prepared, read, approved, and properly filed.
    - Sees that business is ordered, considered, and disposed of properly.
    - Is an impartial, conscientious arbiter of discussion and debate, and insists on fairness in the actions and debate of the members.
3. The Secretary
    - Keeps accurate and complete minutes and files them for inspection. A copy of the minutes should be provided to the (administrator).
    - Keeps accurate records of council membership, attendance, duties and special assignments.
4. Parents and other community representatives
    - Act as council members according to established procedures by making suggestions and recommendations representative of the views of parents, citizens, and community organizations of the school/college community.
    - Participate regularly in EAC meetings and carry out council assignments

- Become knowledgeable about personnel and material resources of the school/college and community and the school's education program.
- Act as resource persons for the EAC, especially in the solutions of community-related problems which affect the school/college and its students
- Assist in obtaining community resources to aid the school's education program
- Serve as a communication link between EAC, the community, and the school/college
- Participate in activities aimed at obtaining parent and community support and assistance for school/college programs

5. Faculty and school/college staff representatives
    - Act as members of the council to represent the views and interest of the school/college staff.
    - Participate regularly in EAC meetings and carry out council assignments.
    - Act a resource person for the EAC by making available specialized information about the educational programs, innovative ideas, and available resources.
    - Assist in identifying community resources that can aid in the educational programs
    - Serve as a communication link between the EAC and the school/college staff, and keep the staff informed of actions can activities of the council.
    - Participate in efforts to encourage school/college staff support for goals and activities of the External Advisory Council.

Article VII: Committees

The EAC may create such committees as necessary to carry on the work of the council.

Article VIII: Meetings

1. The EAC shall meet as often as necessary to perform its duties, but no less than four times per year. The first meeting shall be held in September.
2. Meetings must be scheduled when parents, students, teachers, education support personnel and community members can attend.
3. All meetings must be open to the public.
4. A quorum must be present to conduct business. A majority of the membership shall constitute a quorum.
5. Three days written notice will be given to members concerning any item that will be voted on at the upcoming meeting.

Article IX: Amendments to Bylaws

1. No amendment can be made to these bylaws that is incompatible with current laws. If any amendment inadvertently violates that requirement, corrective action must be taken as soon as that oversight is discovered and should proceed as if the amendment had never been approved.
2. These bylaws shall be amended at any regular meeting of the EAC by a majority of the membership.
3. The amendment shall become effective immediately upon passage.

Article X: Ratification of Bylaws

1. The bylaws of the EAC will be ratified by a majority of if its members.
2. The school/college may review all proposed bylaws of the advisory council and shall maintain a record of minutes of council meetings.

Article XI: Parliamentary Authority

1. The rules contained in the current edition of "Robert's Rules of Order, Newly Revised" shall govern the council in all cases in which they are applicable and in which they do not conflict with these bylaws.
2. The chairperson has the option to appoint a member of EAC as a parliamentarian at any meeting deemed necessary.

APPENDIX B

Example Handbook

Agriculture
Internship Handbook
Course Name and Number
School Logo or Graphic
Revised:
Date & Year
Table of Contents

Introduction to Internship

The agriculture internship is a cooperative effort between the SCHOOL and an entity that has a goal of educating or working with the SCHOOL through educational efforts that are agriculturally based. This internship can serve **one** of three purposes for the student: 1) career discovery where a student will gain career knowledge and focus on skills needed to obtain within the agriculture occupation, 2) skills focused to gain further foundational knowledge mid-way through their school experience (where the focus will be on expanding agriculture content knowledge and skills), or 3) to become the capstone experience in the teacher education specialization.

The support network for the internship is made up of a faculty supervisor and a cooperating internship supervisor/employer. This may be a paid or unpaid internship. The cooperating internship supervisor will provide ongoing support and feedback on the intern's daily progress and should set aside time (on a weekly basis) to mentor the intern concerning the intern's duties and implementation of job duties. The faculty supervisor is involved in defining and communicating the purpose and expectations to be fulfilled by the intern and the cooperating internship supervisor. An important role for the faculty supervisor is to assist in keeping channels of communication open between the internship supervisor and the student intern.

The purpose of the *Agriculture Internship Handbook* is to assist interns, cooperating intern supervisors, and faculty supervisors in understanding their responsibilities related to the internship (COURSE). This handbook contains the information that the intern will need to complete their internship.

This handbook contains descriptions of the internship and requirements to intern, information about the core experiences that the intern should strive to gain during the internship, and descriptions of the internship assignments. **Students are required to provide a copy of these guidelines to their internship employer supervisor.**

The internship is designed to provide students with the opportunity to professionally engage through supervised practical preparation – in conjunction with academic assignments. Places suitable for this internship include, but are not limited to: LIST ACCEPTED TYPES OF PLACES STUDENTS WOULD HAVE INTERNSHIPS HERE. The goal is for interns, in an earlier experience, to emerge from this internship with greater knowledge, skills, confidence, and maturity. The goal for the capstone internship is to exhibit in-depth knowledge and ability to put agricultural knowledge, skills, communication, confidence, and maturity into practice in a professional setting.

HEADMASTER NAME AND TITLE

DEPARTMENTAL COORDINATOR NAME

NAME OF COORDINATOR/FACULTY SUPERVISOR

NAME OF ADDITIONAL FACULTY SUPERVISORS

Internship Requirements

Prerequisites:

To be eligible for the internship, students must have:

1. A 2.5 or higher overall GPA

Additional for the Capstone experience:

2. Completed at least 15 hours of agriculture courses, **with no grade lower than a C**.

Students may take an internship **for academic credit** any semester **after** the above requirements are met.

Internship Applications:

Internship applications must be submitted to the departmental coordinator no later than **four weeks immediately prior** to the start of an internship.

Internship Approval:

Students are to work with their academic advisor and departmental coordinator secure their internships. Students will be given several sources and internship site possibilities. The internship **must** strengthen agricultural knowledge and the capstone experience must be educationally focused or contain training and development.

Students will submit their application form, with the contact information of the internship site supervisor. The departmental coordinator will determine, based on the completed application form, if the internship is acceptable. A student's application submission constitutes an agreement to accept assignment to a site where it is determined that the objectives of the internship program can best be achieved.

Credit Hours:

Students receive variable credit, based on the number of work hours they perform. Students enrolled in the **internship** for six credit hours are expected to perform approximately **320 hours** (8 weeks at 40 hours a week) of job-related work – as designed and approved by the internship site supervisor and university supervisor.

In the capstone experience there may additional hours – for a 9-hour experience; it is required 12 weeks at 40 hours/week.

Grading:

The final grade is determined from **three** sources: a) internship site supervisor recommendation, b) university supervisor recommendation, and c) student portfolio/evaluative report. Students will be evaluated by internship site supervisors, based upon how well they accomplish the tasks designated on the *Training Plan*. Final responsibility for assigning the grade rests with the university supervisor.

Exit Interview:

The internship site supervisor will conduct an exit interview at the end of the internship to evaluate the progress made over the course of the semester. Students should treat the exit interview/evaluation as if students were "really" employed.

Forms to Submit:

The forms found in this packet should be completed by the person described below at the following times during the internship.

Weekly Journal Electronic Form – Completed by the intern (student) and submitted electronically to the university supervisor each week.

Internship Training Agreement – Completed by the intern (student), internship site supervisor, and university supervisor and submitted to the university supervisor **no later than the completion of the first week of the internship.**

Internship Training Plan – Completed by the intern and the internship site supervisor. The *Training Plan* should be submitted within the first two weeks of the internship.

Intern Rating Sheet – Completed by the internship site supervisor at the **end** of the internship. The internship site supervisor submits the *Rating Sheet* to the university supervisor at the completion of the internship.

Internship Roles

Student/Intern

- Initiate Participation – students should discuss their intention or interest in completing an internship with their advisor. If a school required an internship, the school should determine at what stages during a student's academic program they would complete their internship.
- Consideration of Employment Opportunities – students should obtain information concerning potential opportunities in their area of interest from the coordinator. Internship opportunities should be presented to students or be placed in areas in the school so that students can learn about potential

internship placements. Schools should hold internship showcases as described earlier in this chapter.

- Prepare a Resume – students should prepare a resume and/or personal interest sheet that can be used in seeking an internship. Assistance in preparing such materials should be provided through classroom instruction or out-of-class training as described earlier in this chapter.
- Determination of Specific Objectives – students should develop ideas/learning goals they wish to achieve by completing an internship in consultation with their advisor or future supervisor. If done well, this will lead to easier placement by the school and matching student interest with an internship employer.
- Assignments – students should complete assignments and communicate with their supervisors (faculty and employer supervisor).
- Students' responsibilities in the internship also include academic work, in addition to regular on-the-job requirements. Specific on-the-job expectations will be outlined in the *Training Plan*. This plan lists both the core requirements, as well as other requirements outlined by the internship site supervisor in conjunction with the faculty supervisor.
- Students' responsibilities in the internship also include academic work, in addition to regular on-the-job requirements. Specific on-the-job expectations will be outlined in the *Training Plan*. This plan lists both the core requirements, as well as other requirements outlined by the internship site supervisor in conjunction with the university supervisor.

Departmental Coordinator

- Implementation of policies – assume primary responsibility for implementing those standards established by the faculty (course requirements, grades, credit-awarding etc.)
- Selection of students – discusses internship concepts with prospective participants and works with academic advisors in selection of students for participation. Collects applications from students seeking internship experience and checks that appropriate prerequisites have been completed.
- Creation of memorandum of agreements – seeks and maintains agreements with internship employers and works to develop new agreements and seeks to strengthen relationships with the industry partners.

Faculty Supervisor

- Identification of objectives and activities – assist the student in identifying internship objectives that are consistent with the intention of the program and the student's interests and capabilities. Identifies activities that are consistent with the objectives and are feasible at the selected work site. Indicates support of objectives and activities with signature on the agreement.
- Collaboration with cooperating employer – should work to develop objectives and complete student internship learning objectives collaboratively with the employer.
- Supervision of Internship – provide indirect supervision by reviewing progress reports that are submitted by the intern and any additional assignments. Serves as the contact person when cooperating employer finds it desirable to discuss aspects of intern's program. On-site visit by the faculty supervisor is desirable whenever practical.

Cooperating Employer

- Approval of objectives and activities – the employer should have input and final approval of the student learning objectives. Sign agreement that signifies that stated objectives can be achieved and stated activities are feasible at the proposed work site.
- Provide a professional educational experience – understand that the internship opportunity is for the educational goals of the student intern. Agree to work with the intern and supply a professional educational experience.
- Evaluation – may assist the faculty supervisor in the final evaluation of the student's performance. Tasks under advisement suggestions made by intern in final evaluative report as to how future internship at the site could be improved.
- School visits and career showcases – when possible, attend school career showcases and meet with students that would be seeking internships.

Internship Site Information

Students should develop an understanding of the organizational structure and personnel involved in communication activities at their internship site. To enable students to develop an understanding of workplace procedures, students should discuss the following items with their internship site supervisor soon after starting the internship:

- office hours
- parking arrangements

- location of assigned workstation
- security
- safety
- secretarial support available
- telephone procedures
- interoffice communication
- facility management and maintenance
- office privileges and/or policy
- professional dress/working attire
- travel policies
- other "survival" issues

Experiences During Internship

Each intern's specific on-the-job experiences will vary. Below is a list of experiences that the intern may perform during their internship. Not every intern will perform each experience, but it is expected that the intern be exposed to several of the following:

Education & Communication Experiences:

Writing and reports. Materials, handouts, worksheets, and planning materials.

Oral communication. Organizes ideas and communicates orally.

Small group and nonverbal communication. Effective interaction in a professional environment through participation in meetings, strategy sessions, interviews, and other interpersonal communication.

Special event coordination/publicity planning. Participation in the development and design of special events or publicity-generating events.

Leadership Experiences:

Professional development. Participation in activities that are a part of the life of the organization such as attendance, as appropriate, at staff meetings, administrative conferences, and professional organization meetings.

Problem-solving. Participation in problem-solving or program/campaign development sessions to learn about strategy selection, matching strategies and plans of action.

Development of objectives and evaluation criteria. Learn to establish objectives and evaluative criteria for judging the success of programs.

Creative thinking. Demonstrate creativity in generating new ideas.

Leadership. Communicates ideas to justify position, persuades and convinces others, responsibly challenges existing procedures and policies.

Technical Experiences:

Agricultural technical development. Outline of the technical skills that should be developed.

Internship Assignments

Students enrolled in COURSE must produce formal **internship portfolios** before receiving a grade. To assist students in this process, students must maintain a **daily journal** and a **portfolio.** Following are the assignments required during the internship:

- **Weekly journal** – Students are to reflect on the happenings of each week and commit to writing their perceptions of these incidents. (Students may wish to record journal entries daily, instead of weekly. An *Example Weekly Journal Electronic Form* is provided on page 9.) Students are encouraged to be open with their thoughts. The aim of the journal is self-discovery. Students **are required** to e-mail the weekly journal entry to the university supervisor **no later than noon on the Monday following the completed week.** The complete, weekly journal will be submitted as part of the student's portfolio. The student should keep all weekly reports electronically. Students should provide – but are not limited to – the following information in their journal entries:
    - What did you do that week?
    - What skills did you learn?
    - Was the job the same type of work as you had done previously?
    - Did you do anything unusual? Travel? Meet anyone?
    - How much time was spent on various projects and activities?
    - What did you learn through your activity/activities?
    - What can you do to improve?
- **Portfolio** – Interns will maintain a **portfolio** – a record of all exhibits of work and other job-related materials, as appropriate. Interns must describe the contribution they had in the materials they include in their portfolio. Some materials provided in the portfolio may not be as easy to "see." For example, interns may be part of planning for a conference. The intern should provide a detailed narrative of what the intern did to help plan and carry out the conference.
    - The **portfolio** should be **typewritten, edited, and packaged** in a professional manner. Just as the internship showed the student's abilities, so should the portfolio.
    - Interns should consider the **portfolio** as a presentation of their best professional work to the world.
- At the end of the internship, an **evaluative report** (three to five pages) must be submitted to the university supervisor as part of the portfolio. This paper describes the internship experience, comments on the strong and weak points of the internship, evaluates the intern's level of preparation for the internship,

evaluates the worth of the internship, suggests what could be done by both the employer and the student to make the experience better, and gives advice for future interns. The **evaluative report** also serves as a self-critique of the intern's abilities and learning experiences over the course of the internship.

Example of Weekly Journal Electronic Form

Interns are required to submit this form to the university supervisor each Monday. Failure to do so will result in a lower grade for the internship. The university supervisor will provide the intern with a blank, electronic copy of this form. (can be more than one page, if needed)

Table 3.

	Agriculture Development Internships NAME OF SCHOOL		**Internship Journal** **Weekly Report of Activities**
Intern Name			**Ag Business, Organization, Agency**
Report for Week: ______________			
Goals Planned for the Week			Hours Worked
		Monday	0
		Tuesday	0
		Wednesday	0
		Thursday	0
		Friday	0
		Saturday	0
		Total	0
Accomplishments – Knowledge and Skills Developed (Refer to "Experiences During Internship")			

Monday –
Tuesday –
Wednesday –
Thursday –

Friday –

SCHOOL NAME HERE

Internship Training Agreement

To provide a basis of understanding and to promote sound business relationships, this agreement is established on

(date) . Placement will begin on and will end

on or about unless the arrangement becomes unsatisfactory to either party.

Intern: ____.

Intern's e-mail address: ____.

Internship site supervisor: ____.

Internship site supervisor's title: ____.

Name of company/organization: ____.

Company/organization's address: ____.

____.

Internship site supervisor's phone: ____.

Internship site supervisor's e-mail address: ____.

Usual working hours will be: ____

The internship site agrees to:

- Provide the student with opportunities to learn how to do well as many jobs as possible, with particular reference to those contained in the ***Training Plan***.
- Instruct the student in ways of doing his/her work and handling his/her management problems.
- Help the university make an honest appraisal of the student's performance.
- Avoid subjecting the student to unnecessary hazards. The internship site's general liability insurance will provide coverage for injuries to an intern occurring at the workplace.
- Conform to all federal, state and local laws and regulations regarding employment and worker's compensation.
- Notify the university immediately in case of accident or sickness and if any serious problem arises.
- Assign the student new responsibilities in keeping with his/her progress.

- Provide reimbursement for job-related travel and special expenses.
- Reserve the right to discharge the intern for just cause from the cooperating site.

The student agrees to:

- Act professional; be punctual, dependable, loyal, courteous, and considerate of the employer and other employees.
- Follow instructions, avoid unsafe acts and be alert to unsafe conditions.
- Dress properly for work and conform to the rules and regulations of the agricultural business, organization, or government agency.
- Be courteous and considerate of the employer and his/her employees.
- Keep records of experiences and make required reports.
- Achieve competencies indicated in ***Training Plan***.
- Notify the university immediately in case of accident or sickness or if any serious problem arises.

The university internship supervisor agrees to:

- Provide a copy of the agreement to the internship site.

- Maintain contact with the intern and internship site throughout the internship experience.

STUDENT ______________________________ Phone Number ________________________

(Signature)
INTERNSHIP SITE Phone Number ________________________
SUPERVISOR (Signature)
FACULTY Phone Number ________________________
SUPERVISOR (Signature)
SCHOOL NAME HERE

Internship Training Plan

To be completed by Intern & Internship Site Supervisor so both parties will know the internship's expectations. Please use the *Training Plan* as the basis for your exit interview/ evaluation of the intern's progress during the semester.

STUDENT: ____
INTERNSHIP SITE SUPERVISOR: ____________________________________

INTERNSHIP BEGINNING DATE: ____________________________

____

INTERNSHIPENDINGDATE:____________________________

INTERNSHIP SITE SUPERVISOR'S EXPECTATIONS (To be completed by **intern and internship employer supervisor**.) Please list specific activities the intern will be expected to do in the areas shown. Not every intern will have activities in each of the areas below. Please refer to the "*Experiences During Internship*" section for descriptions of each content area below.

Technical skill development

Oral communication

Visual communication

Small group and nonverbal communication

Application of communication technology

(pg. 1 of 2)

SCHOOL NAME HERE

Internship Training Plan

INTERNSHIP SITE SUPERVISOR'S EXPECTATIONS (To be completed by **intern and internship site supervisor**.)

Professional development

Problem-solving

Development of objectives and evaluation criteria

Creative thinking

Research (or research partnerships if applicable)

(pg. 2 of 2)

SCHOOL NAME HERE

Intern Rating Sheet

To be completed by Internship Employer Supervisor at the completion of the internship. Please use the *Training Plan* as the basis for your exit interview/ evaluation of the intern's progress during the semester.

Intern: ____________________________.

Internship supervisor: ____________________________.

Internship site supervisor's title ____________________________

Name of company/organization: ____________________________

Company/organization's address: ____________________________

____________________________.

Internship site supervisor's phone ____________________________.

E-mail address: ____________________________.

Description of major duties performed by the intern this semester:

Your evaluation of the performance of the intern. Please indicate, on a scale of 1-5, your assessment of the student, with 5 indicating superior performance. If not applicable, please write NA.

writing focus appearance speaking

flexibility potential enthusiasm cooperation

judgment creativity punctuality preparation

Other comments about the intern:

Recommended Grade (circle one): A B+ B C+ C D+ D E

____

(employer supervisor's signature)

(pg. 1 of 2)

SCHOOL NAME HERE

Intern Rating Sheet

Intern site supervisor's evaluation of intern's progression in core experience areas. Not every intern will have activities in each of the areas below. Please refer to the "*Experiences During Internship*" section for descriptions of each content area below.

Technical skill development

Oral communication

Small group and nonverbal communication

Special event coordination/publicity planning

Professional development

Problem-solving

Development of objectives and evaluation criteria

Creative thinking

Research (if applicable)

Important Dates

Interns:

Submit Internship Application to university supervisor 12/21/21

Report to Internship Site DATE

Submit Internship Training Agreement to university supervisor first week

Submit Internship Training Plan no later than: second week

Submit Weekly Journal (weekly goals, hours, accomplishments) continuous

Last Day of Internship: 04/26/22

Submit Portfolio (including evaluative report) to faculty supervisor 04/30/22

Internship Site Supervisors:

Intern Rating Sheet – Completed by the internship employer supervisor at the **end** of the internship. The internship site supervisor submits the *Rating Sheet* to the university supervisor at the completion of the internship.

Please submit the intern rating sheet to faculty supervisor 04/29/22

Please submit raw score out of 100 points to faculty supervisor 04/29/22

Departmental Coordinator Contact information
Faculty member name
School name
Email address
Phone number

Chapter 8
Establishing Career Advancement Centers in Egypt's Faculties of Agriculture for Human Resources Career Development

Emad El-Shafie
Faculty of Agriculture, Cairo University, Egypt

ABSTRACT

The agriculture sector is extremely important for Egypt's economy. For achieving food security objectives, within the SDGs 2030, the Agricultural Human Resources (AHRs) need systematic and regular training and lifelong learning. Faculties of agriculture (FAs) prepare and qualify researchers, technicians, and specialists in all areas of agriculture. Career advancement centers (CACs) need to be established in all FAs to provide lifelong opportunities to all AHRs' categories that need to cope with the ever advancing agricultural developments. To secure effective functioning of CACs, the different units needed, within their structures, are suggested. As a conclusion, establishing a CAC in each FA, is necessary and urgent for activating and maintaining its extension education, out-reach function of development, and serving the surrounding communities, especially rural communities. The currently working CACs need all types of support from the government, agri-business, and civil society organizations.

DOI: 10.4018/978-1-6684-4050-6.ch008

INTRODUCTION

Egypt's agriculture is one of the oldest types of human economic handling of living nature and was the origin of the very old history of the agriculture-based civilization around the river Nile for thousands of years. However, and due to the fact that the majority of Egypt's area is desert land, as indicated by FAO (2016), most of the cultivated land is located close to the banks of the Nile River, its main branches, and canals, as well as in the Nile Delta. Rangeland is restricted to a narrow strip, only a few kilometers wide, along the Mediterranean coast, and its bearing capacity is considering the importance of agriculture in the Egyptian economy and based on the recognition that an efficient, functioning economy is a precondition for addressing the environmental and social pillars of sustainability, the green economy is seen as a key implementation tool for sustainable development. The green economy, thus, is that economy that results in "improved human well-being and social equity while significantly reducing environmental and ecological scarcities." As a result, "Greening the Economy with Agriculture" (GEA) refers to ensuring the right to adequate food as well as food and nutrition security—in terms of food availability, access, stability, and utilization—as well as contributing to the quality of rural livelihoods while efficiently managing natural resources and improving resilience and equity throughout the food supply chain (FAO, 2012).

Agriculture is also a vital economic sector, since an estimated 55% of the labor force is engaged in agricultural activities, which consume about 80% of the freshwater resources and contribute about 13.5% of GDP (World Bank Group, 2021). Moreover, the authorities of Egypt have consistently underlined the sector's important role in reducing food imports and increasing self-sufficiency, both of which are necessary to assure food security. These targets have become particularly important in the current economic circumstances, as the 2016 devaluation of the pound made imports more expensive and the COVID-19 pandemic in 2020 disrupted global supply chains (Oxford Business Group, 2022).

Agriculture is a profession that is constantly evolving and progressing around the world. It is becoming a science- and research-based, highly sophisticated, and knowledge-intensive business or industry. Agricultural advancement and progress stem from several essential pillars, among which the three most important and indispensable are:

1. Heavy reliance on the latest developments in agriculture and life sciences and applied scientific research results, through applying the scientific method for diagnosing location-specific agricultural production and marketing problems, finding the best-fit applicable resolutions or practices for handling and resolving

these problems, and facilitating the diffusion and application of these practices by their end beneficiaries.

In Egypt, this pillar represents an important concern of the 2030 vision, as indicated by the National Strategy for Science, Technology, and Innovation 2030 (2019, p. 6–8). This vision is: An Egyptian scientific society that, in construction and development, depends on a perpetually learning generation that generates and uses the knowledge to provide scientific practical solutions to society's problems and exports the knowledge within a system that supports innovation and stimulates a knowledge-based economy.

2. Focusing on innovation in agricultural production and marketing. Specifically, FAO (2019, p. 93) defines agricultural innovation as "the process by which individuals or organizations introduce new or existing products, processes, or organizational methods for the first time in a specific context in order to increase effectiveness, competitiveness, resilience to shocks, or environmental sustainability and thus contribute to food security and nutrition, economic development, or sustainable natural resource management." This process is not only restricted to the use of technologies but also includes social, organizational, institutional, or marketing processes or arrangements.

As reviewed by Matthew et al. (2020), emerging technologies in agriculture are producing new machinery and process techniques for production, postharvest handling, and agribusiness. They are also resulting in innovations in agriculture that make agriculture a sustainable, profitable, and competitive enterprise. Examples of these emerging technologies include sensor technology, vertical farming, organic farming, smart farming, precision agriculture, robots, drones, automation, machine learning, big data, block chain, Radio Frequency Identification (RFID), and the internet of things.

3. Providing adequate, easily accessible, and high-quality education and lifelong learning opportunities, activities, and programs for all agricultural human resources (AHRs). Adult education and lifelong learning are essential to securing an individual's competencies to positively respond to and manage the ever-changing work needs, challenges, and environment.

As concluded by Shubenkova et al. (2017), adult education and lifelong learning are the basis of the social and employment paths of the modern man. Demir (2020) also concluded that today, the speed of change and increased knowledge in almost every field means that lifelong learning is now a necessity. No matter how qualified an individual's education was in school, after entering the business world, if he or she does not follow the developments and renew himself or herself, the increment gap between life and this individual will always appear as negativity and failure.

These AHRs include all relevant individuals, groups, organizations, and communities involved in all agricultural production and marketing stages, processes, and activities. The educational content and materials needed for these programs should be based on local-specific, applicable research results and best-fit practices for achieving sustainable agriculture.

However, in Egypt, the majority of the AHRs (or agricultural workforce), as a part of the rural population, is lagging behind the latest scientific developments related to technical agricultural innovations, and best-fit production and marketing practices that are essential to achieving Sustainable Agricultural and Rural Development (SARD).

The population of Egypt, in general, could be described as mainly rural and relatively young in age, since, according to CAPMAS (2022), the rural population represents around 57% of the total population, more than half (52.4%) of Egypt's population is under 25 years of age, and more than three-fourths (75.5%) are under 40 years of age. The old age category (60 years and above) represents a relatively small proportion of the total population (only 6.7%).

This young population, especially in rural communities, which is assumed to be an active and highly energetic category of the population, could be considered a positive input for an efficient and highly productive agricultural workforce, or AHRs. The effectiveness of the AHRs in promoting and sustaining the agricultural sector depends mainly on their current professional knowledge and skills, in addition to their ability to cope with the latest advancements in agricultural production and marketing technology and innovations.

Small land holdings are prevailing in Egypt's agriculture. As reviewed by Chaaban et al. (2018) the agriculture sector in Egypt is still characterized by small-scale farms that usually rely on traditional methods. Farmers also do not own physical capital, leading most farmers to rent tractors and pumps, while other farmers may resort to more traditional methods of relying on labor-intensive activities to compensate for the work of heavy machinery, which can be both time- and money-consuming.

These problems represent serious challenges that could be a cause, and probably an effect, of the prevailing poverty and illiteracy among Egyptians, especially in rural areas. As mentioned in the Human Development Report (2021), in 2018, around one third (32.5%) of Egypt's population is below the national poverty line, and illiteracy rates are prevailing among females (30.8%) and males (21.2%). Consequently, the relatively high poverty and illiteracy rates among farmers, especially small landholders, are seriously challenging their easy access to essential production and marketing assets and inputs, including sources of the latest scientific developments and innovations.

Additionally, this situation is articulated by other severe challenges facing Egyptian agriculture, among which the most important examples include:

1. The overpopulation challenge, since Egypt is the third most populated country in Africa (after Nigeria and Ethiopia). According to CAPMAS (Dec. 2022), the total population exceeded 107 million by the end of 2022 (December 21st), and expected to reach 121 million in 2030, and around 160 million in 2050. This challenge aggravates the current and potential critical situation of food security in the country due to the very limited available natural resources (especially cultivable lands and irrigation water), resulting in ever-increasing demands for basic food items.
2. The climate change challenge, since as a result of climate change's negative effects on agricultural production and the quality of available natural resources, as reported by the World Bank Group (2021, p. 13), Egypt's agricultural sector is particularly vulnerable to climate change, due to its dependence on the Nile River as its primary water source, its large traditional agricultural base, as well as the intensifying development and erosion along coastal areas. The country's water scarcity, dependence on the Nile River, and high temperatures make agriculture productivity increasingly vulnerable to climate variability and future projected climate change trends.
3. Moreover, hotter temperatures and reduced rainfall are projected to reduce agricultural productivity (including livestock rearing and fishing activities) by 15%–20% by 2050 (Kwasi et al., 2022, p. 31).
4. Emerging inconvenient threatening events, such as the COVID-19 pandemic and the Russian-Ukrainian conflict that add more pressures to the current and future situation of national food security and are particularly imperative since Egypt is one of the highest importers of important food stuffs (especially wheat and cooking oils) from these two countries.
5. The relatively low levels of personal qualifications of considerably high proportions of farmers and producers, especially small land holders, at the local rural community level (e.g., relatively high illiteracy rates or low levels of education, lack of technical knowledge, scientific experiences, and training opportunities, in addition to low access to key resources and assets (such as land, capital, information, agricultural credit, and production inputs).

Consequently, and considering all these serious challenges, achieving the SDGs 2030, especially the first two goals related to eradicating hunger and ending multi-dimensional poverty, necessitates the development and well-functioning of a high-quality, well-trained, and highly motivated AHRs or work force that can effectively perform their assigned tasks at all levels. This work force covers a wide variety of AHR categories that include:

A) Farmers and agricultural producers, living and working in local rural communities, especially small landholders (up to 1 hectare), who represent the critical mass (around 75% of the total Egyptian agricultural producers), As reported by FAO (2016, p. 3), small holdings characterize Egyptian agriculture, with about 50 percent of holdings having an area less than 0.42 ha (1 feddan).
B) Services' providers, working at different administrative levels to secure the timely provision of the services (especially production inputs), that are needed and demanded by farmers and agricultural producers.
C) Personnel from the Ministries of Agriculture and Land Reclamation (MALR) and Water Resources and Irrigation (MWRI) who work in the agricultural and irrigation water sectors. This category includes different administrative levels of agricultural and irrigation technicians and researchers, in addition to grass-root agricultural and irrigation extension workers and SARD specialists and agents, who facilitate the effective performance of all the management functions (based on the "POSDCORB" model) covering all agricultural and irrigation tasks and activities at all administrative levels. As reviewed by Chalekian (2013, pp. 9–15), these functions could be defined as follows:
    - Planning: "working out in broad outline the things that need to be done and the methods for doing them to accomplish the purpose set for the enterprise,"
    - Organizing: "the establishment of the formal structure of authority through which work subdivisions are arranged, defined, and coordinated for the defined objective."
    - Staffing: "the whole personnel function of bringing in and training the staff and maintaining favorable conditions of work."
    - Directing: "the continuous task of making decisions, embodying them in specific and general orders and instructions, and serving as the leader of the enterprise."
    - Coordinating: "the all-important duty of relating the various parts of the work."
    - Reporting: "keeping those to whom the executive is responsible informed as to what is going on, which thus includes himself and his subordinates, informed through records, research, and inspection."
    - Budgeting: "All that goes with budgeting in the form of fiscal planning, accounting, and control."

Therefore, CACs are highly and urgently needed in all the agricultural educational institutions of Egypt (agricultural high schools, technical high institutes, and faculties). CACs serve as a multi-functional platform for facilitating the planning, implementation, and evaluation of highly effective life-long learning opportunities and experiences through highly varied programs and activities provided for all relevant actors, partners, and stakeholders within the AHRs.

As reported by Swanson et al. (2007), establishing career resource centers on each university campus represents an institutional mechanism needed to assist students in securing private sector jobs. An important example of applying this mechanism is the project that was implemented in Egypt (during the period from 2003 to 2007) by the Midwest Universities Consortium for International Activities, Inc. (MUCIA), in collaboration with six U.S. Land Grant Universities. This project covered five Faculties of Agriculture (FOAs) at five universities in Upper Egypt, including Assiut, Cairo, Fayoum, Minia, and South Valley Universities, and sought to strengthen specific academic departments within each FOA (agricultural economics, livestock production, and horticulture), plus the Faculty of Veterinary Medicine at Assiut University (p. 333). Lessons learned from this project need to be reviewed.

OBJECTIVES AND FUNCTIONS OF THE CAC

It is widely accepted and well-known that the university, including all of its faculties, has three important functions: a) the academic function, which focuses on preparing students to be highly qualified as professionals (technicians and researchers) in various specializations; b) the research function, which involves conducting various types of scientific research (basic, applied, and adaptive) for resolving field problems and facing climate challenges; and c) the outreach extension education function to serve and develop the surrounding communities. In this function, the highly sophisticated but profitable and applicable technologies are treated, simplified, and made available in an easy, understandable, and applicable form to all the different categories of end users and beneficiaries.

Based on this view, the main objective of the CAC is to assume the third function or responsibility of the faculty concerning environmental development and serving the surrounding communities, especially rural communities. This responsibility could be effectively assumed through systematic and integrated lifelong learning programs and activities to build and develop personal and professional capacities. This, in turn, could result in better current and future careers for all individuals, groups, and organizations involved in the different areas and activities of the agriculture profession. This objective could be achieved through organizing and conducting regular, systematic, and integrated training and life-long learning opportunities, programs, projects, and activities for all categories of AHRs. These AHRs include faculty staff members, students, graduates, and administrative personnel in addition to the agricultural labor force in all areas of science-based, smart, and environment friendly agriculture.

The Specific Objectives and Functions of the CAC

1. Up-scaling and improving the personal capacities of all individuals involved in the AHRs, especially the capacities that are necessary to secure their effective and positive contribution to the achievement of the SDGs by 2030. Achieving this objective could effectively contribute to improving food security through increasing agricultural productivity, farm income, or net return, and, in turn, up-scaling the quality of rural life. All the faculty's out-reach extension educational activities and programs are targeting different categories of female and male farmers and rural household heads.

This objective could be achieved through developing programs, activities, and efforts for updating, upgrading, and improving the individual behaviors of all the AHRs involved in different areas related directly or indirectly to smart agricultural production and marketing. This behavior includes the main components of human behavior: knowledge, attitudes, aspirations, practices, and skills (referred to as KAAPS). These major components of individual human behavior (the KAAPS) could be viewed and thought of as similar to an iceberg (it could be called a human iceberg). In the human iceberg, both P and S are like the parts of the iceberg that are observable, since they are above the water level, while the K, A, and A are the parts of the iceberg that are unobservable and, in turn, need accurate assessment or measurement since they are hidden under water. The most important reflections, or implications, of this way of thinking concerning the human iceberg, include:

- The P and S could be easily observed and/or measured by outsiders. They tend to be highly predictable when compared to the prevailing local cultures or ways of life; however, in some cases, they can be choking, questionable, and appear strange when compared to the normal and prevailing ones.
- In order to understand and/or interpret the P or S of an individual, the hidden parts of the iceberg (K, A, and A) need to be assessed, measured, and understood.
- Changing or improving an individual's P and/or S requires an accurate assessment of the K, A, and A gaps, in addition to working on de-freezing the current conditions and then moving to better, more up-to-date, and more desirable conditions, and finally stabilizing, maintaining, continuing, and

sustaining the new ones to avoid discontinuance and/or returning to the old ones.

- Updating the individual's K, A, and A need high involvement and participation in the available self-capacity building or career development programs, efforts, and activities.

2. Encouraging individuals in the AHRs to cope with the ever-advancing agricultural knowledge, innovations, and technologies through the acquisition of the knowledge and skills essential for facing, positively handling, and adapting to current and potential climate change. Regular and ongoing campaigns and activities of awareness-raising related to personal self-development for enhancing current and future career development are required to achieve this goal.
3. Empowering all individuals in the AHRs to acquire the skills of preparing and implementing personal capacity-building action plans for effective self-career development. This objective could be achieved through encouraging, motivating, and providing technical assistance to all individuals in the AHRs to deeply think, decide, and apply the following suggested paradigm for developing their careers, through a self-capacity building approach:
    a. Conducting a simple self-SWOT analysis (analysis of personal Strengths, Weaknesses, Opportunities, and Threats);
    b. Diagnosing and prioritizing personal KAAPS gaps or problems,
    c. Preparing future career development self-action plan,
    d. Implementing the plan, and,
    e. Objective monitoring and evaluation of all the previous steps for reconsideration and ever-improving personal current and future capacities.
4. Achieving a more participatory, pluralistic extension educational approach for providing AHRs, particularly small farmers and producers, with high-quality extension and advisory services. Applying a Participatory Extension and Advisory Services (PEAS) in Egypt could significantly contribute to improving food security. A study conducted in 5 Governorates in Egypt by a multi-disciplinary research team (El-Shafie et al, 2022) confirmed the possibility of implementing a PEAS approach that involves all service providers in integrated efforts to provide Small Producers (SPs) with need-oriented advisory services. This study concluded that implementing this approach assisted SPs in proper and timely application of the best-fit practices for promoting their production, with positive impacts including SPs' strong willingness to be active partners through high participation in the implemented activities. Lessons learned from this

study include: a) the importance of bringing different services providers together, in collaborative work. b) encouraging each partner to effectively contribute to help SPs to understand and properly and timely apply the best-fit production and marketing practices, c) securing reciprocal communication and constructive dialogue among all partners.

This objective could be achieved by assisting the different actors in the AHRs in establishing and activating public-private partnerships (PPPs). These PPPs should involve representatives from all relevant SARD actors and stakeholders to provide the AHR (as individuals and/or groups) with timely and sufficient resources needed for facilitating training and capacity-building activities for self-career development. These PPPs, as well, need to be highly inclusive to attract the agri-business community to fulfil their social responsibilities through different types of assistance and support for conducting a wide variety of personal and career development programs, projects, and activities that cover different categories of AHRs.

5. Empowering all relevant SARD actors and stakeholders (on an individual or group basis) through capacity-building programs and activities in the areas closely related to personal growth and career development Examples of these areas include:

   - Effective communication skills
   - Leadership skills
   - Understanding group dynamics for better and more efficient group work
   - Problem diagnosis and prioritization
   - Planning, implementing, monitoring, and evaluating self-development activities and personal capacity-building efforts
   - Building, activating, and sustaining efficient teamwork
   - Understanding and applying different diagnostic analysis tools (e.g., SWOT analysis, stakeholder analysis, objective tree, or problem tree)
   - Designing and applying socio-metric tests for assessing social relationships among groups' and communities' members to understand and reform group structures
   - Critical and creative thinking for innovation
   - Assessment of individual KAAPS gaps or needs
   - Social learning as an approach to lifelong learning
   - Using the log frame approach in planning, implementing, monitoring, and evaluating SARD programs
   - Effective time management
   - Effective management of small income-generating projects

For effective management of the Faculty of Agriculture's CAC, the Egyptian educational institutions have their own laws, rules, and regulations for forming their Boards of Directors (BoDs). However, in addition to the Faculty Dean and Vice-Dean for Community Services and Environment, who serve as the BoD's top leaders, an acting director and assistant director are required to carry out detailed specific tasks and activities. Members of the Board of Directors must represent departments within the faculty that are closely related to the CAC objectives and activities, such as Human Agricultural Sciences (Agric. Economics, Agric. Extension, and Rural Development), various Technical Agriculture Sciences, and representatives from the Agri-Business Community and Civil Society Organizations who are interested in career development.

Based on the concept or view that "form follows function," which was articulated at the end of the 19th century by the American architect Louis Sullivan in 1896 (Environment and Planning, 2022), it is worth noting that there are some important units that are needed, within the CAC organizational structure, to assume the responsibility of performing its basic functions to secure the effective achievement of its objectives. As shown in Figure 1, the following two units are examples of the recommended units within the suggested organizational structure of the CAC.

A Sustainable Agricultural and Rural Development Information System (SARDIS) unit to provide farmers and producers, in addition to agricultural policymakers and decision-makers, with reliable information they need to make wise production and marketing decisions. Currently, information systems and technologies are imperative components of successful and competitive businesses. Information technologies consist of Internet-based information systems is playing a vital and expanding role in enhancing firm's economic growth. The experience of organizations' managers needs to be provided with the necessary information to reduce risks and make the most appropriate decisions, Chuma (2020).

Figure 1. Career Advancement Center

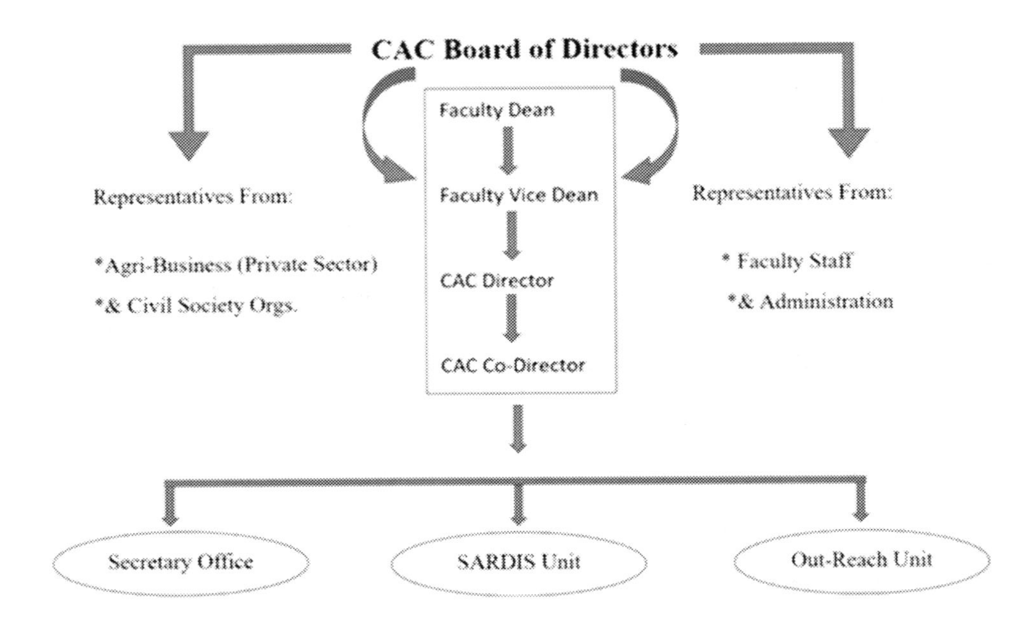

The importance of this unit is based on the fact that reliable, timely and easily usable information could be of high value to all categories of AHRs as essential input in making wise decisions related to agricultural production and marketing and other rural life activities. It is logical to recognize the fact that inserting unreliable, distorted, overestimated or underestimated data or information in any decision making or planning process will be eventually resulted in failure. Therefore, the well-functioning of this unit protects individuals, groups and organizations, within the AHRs, from falling in the "Garbage in Garbage Out, known as GIGO trap or syndrome".

The main functions of this unit will focus on:

- Identifying the types of data and information that are relevant to the most important agricultural production and marketing problems and issues.
- Collecting accurate, reliable, current, and sufficient data and information about past and current and expected SARD conditions
- Reviewing, verifying, processing, and analyzing the collected data in order to clearly present and display the results and conclusions

- Writing short and simple reports about the past and current situations to assist planning teams in planning effective programs
- Making written reports available to other SARD actors and stakeholder groups at various administrative levels;
  -Monitoring and evaluation of the whole process to secure the reliability and validity of the data and information upon which wise decisions could be made

Outreach, or EASs, unit, for expanding the advantages of applying the scientific research results and new agricultural technologies and innovations generated inside the faculty by making them accessible, understandable, and applicable to all interested categories of producers in the surrounding rural communities. This unit could effectively provide needed and desired EASs to farmers and producers in the surrounding rural communities.

Within this unit, a hotline telephone service needs to be developed for responding to farmers' and producers' requests or calls looking for how to handle or resolve any specific production or marketing problem. In addition to that, several mobile phone applications could be developed to establish groups of farmers who share the same interests, providing them with timely information, advice, and recommended practices to adapt to any sudden climate change. Best-fit production and marketing practices could also be delivered to the members of these groups.

It is worth noting, as reported by Morse (2020), that the USAID, in coordination with the Ministry of Higher Education and Scientific Research and the private sector, is establishing 21 sustainable career development centers at 13 public universities across Egypt. These centers will serve more than 70 percent of all public university students in Egypt, including specialized support for students with disabilities, by providing them with career mentoring, employability skills training, English language training, private sector networking, internships, and, ultimately, jobs.

The Career Development Center (CDC) of Cairo University's Faculty of Agriculture is one example of an active center that plans and implements training programs and efforts to improve its students' future careers by upscaling their personal skills and preparing them to face the challenges of current and potential employment opportunities. The faculty graduates, in addition to different professional individuals and groups within all categories of the AHRs, are also covered by highly specialized training programs, workshops, and discussion meetings. During these activities, all the knowledge and experiences related to the latest scientific and technological developments are delivered, shared, and exchanged through interactive interpersonal communication. Training areas cover a wide spectrum of topics and subjects, including, for example, food safety, the application of nanotechnology in agriculture, self-development for a changing labor market, and building effective teamwork.

In a win-win scenario, agri-business, or agricultural private sector companies, provide financial support for sending their young and middle-experienced employees to attend and learn.

CONCLUSION

- The agriculture sector is increasingly becoming very important for Egypt's economy. This is based on the critical challenges facing the country, including scarce natural resources; food insecurity; very fast growth and overpopulation; the negative impacts of climate change; and the COVID-19 pandemic, in addition to the Russian-Ukrainian war.
- For a more effective contribution to achieving the SDGs 2030 in Egypt, all categories of AHRs need professional capacity building and empowerment through quality training, adult education, and lifelong learning programs.
- Agricultural Educational Institutions (AEIs) are highly involved and responsible for this process of professional capacity building and empowerment of the AHRs through the active functioning of their three basic functions, namely the academic, research, and extension education or outreach functions.
- Establishing a CAC in each Egyptian AEI is necessary and urgent for activating and maintaining its extension education, development outreach function, and serving the surrounding communities, particularly rural communities. This function focuses on assuming the responsibility for achieving the objectives related to the three dimensions of sustainable development (the economic, the environmental, and the socio-cultural dimensions).

Recommendations

For achieving the SDGs 2030 and Egypt's Vision 2030, through developing more efficient AHRs and based on the previously presented ideas, the following recommendations could be suggested.

- Each and every AEI, especially in rural areas (large villages, districts, and regions), including faculties, higher education institutions, and high technical schools, needs to have a CAC or CDC. Establishing these centers is a participatory responsibility in which the government, civil society, and agribusiness or private sector projects and companies must be actively involved in establishing, funding, and sustaining the active functioning of these centers in order to assist rural people in positively dealing with the current and potential SARD challenges.

- The established and currently working career development and advancement centers need active support, especially financial and technical support, from all sustainable development actors and stakeholders to maintain their active functioning. They need regular and systematic meetings that could be conducted on a monthly basis at the location of the center.
- The Governor of each governorate, especially rural ones, and all his administrative personnel need to be highly involved in establishing and ensuring the effective functioning of the CACs or CDCs.

REFERENCES

Aquastat, F. (2016). Country Profile-Egypt. Rome, Italy: Food and Agriculture Organization of the United Nations (FAO).

Batty, M. (2022). The conundrum of ‘form follows function.’ *Environment and Planning B: Urban Analytics and City Science, 49*(7), 1815-1819. DOI: doi:10.1177/23998083221120313

Central Agency for Public Mobilization and Statistics (CAPMAS). (2022). *Egypt in figures*. Retrieved from https://www.capmas.gov.eg/Pages/Publications.aspx?page_id=51 04&YearID=23602

Chaaban, J., Chalak, A., Ismail, T., & Khedr, S. (2018). *Agriculture, Water and Rural Development in Egypt: A Bottom-Up Approach in Evaluating European Trade and Assistance Policies*. Academic Press.

Chalekian, P. (2013, December). POSDCORB: Core patterns of administration. In *Proceedings of the 20th Conference on Pattern Languages of Programs. The Hillside Group* (*Vol. 17*). Academic Press.

Chuma, L. L. (2020). The Role of Information Systems in Business Firms Competitiveness: Integrated Review Paper from Business Perspective. *International Research Journal of Nature Science and Technology, 2*(4). www.scienceresearchjournals.org

Demir, N. K. (2020). The Need of Adult Education and Training Administration in Lifelong Learning. *Mediterranean Journal of Social & Behavioral Research*, *4*(3), 41–45. doi:10.30935/mjosbr/9600

Egypt's Human Development Report. (2021). *Development, a right for all: Egypt's pathways and prospects.* Retrieved from https://www.undp.org/egypt/publications/egypt-human-development-report-2021

El-Shafie, E. M., Azam, A., & Ibrahim, R. H. (2022). *Participatory Extension and Advisory Services (PEAS) as an extension approach to achieve food security in Egypt: Lessons learned from 5 Governorates.* Retrieved from https://esciencepress,net/Journals/IJAE

FAO. (2012). *Greening the economy with agriculture.* FAO.

Kwasi, S., Jakkie C., & Yeboua, K. (2022). *Race to Sustainability? Egypt's Challenges and Opportunities to 2050.* Academic Press.

Ministry of Higher Education and Scientific Research. (2019). National Strategy for Science, *Technology, and Innovation.* Arab Republic of Egypt. Retrieved from http://mohesr.gov.eg/en-us/Documents/sr_strategy.pdf

Morse, L. (2017-2020). *UCCDs develop students' and graduates' job-seeking skills to strengthen lifelong career prospects.* USAID. Retrieved from https://2017-2020.usaid.gov/egypt/higher-education/university-centers-career-development

Oxford Business Group. (2022). *How Egypt is boosting food production and exports.* Retrieved from https://oxfordbusinessgroup.com/overview/seeds-growth-crop-exports-rise-research-and-training-programmes-seek-support-next-generation-farmers

Sadiku, M. N., Ashaolu, T. J., & Musa, S. M. (2020). Emerging technologies in agriculture. *International Journal of Scientific Advances*, *1*(1), 31–34. doi:10.51542/ijscia.v1i1.6

Shubenkova, E. V., Badmaeva, S. V., Gagiev, N. N., & Pirozhenko, E. (2017). Adult education and lifelong learning as the basis of social and employment path of the modern man. *Espacios*, *38*(25), 25–30.

Swanson, B. E., Barrick, R. K., & Samy, M. M. (2007, May). Transforming higher agricultural education in Egypt: Strategy, approach and results. In *Proceedings of the 23th annual conference of Association for International Agricultural Extension Education (AIAEE), Polson* (pp. 20-24). Academic Press.

World Bank. (2022). *Egypt - Country Climate and Development Report*. World Bank Group. Retrieved from https://documents.worldbank.org/curated/en/099510011012235419/P17729200725ff0170ba05031a8d4ac26d7

Chapter 9
Advancing Sustainable Development Through Teacher Professional Development

Melanie J. Miller Foster
https://orcid.org/0000-0002-0270-2431
The Pennsylvania State University, USA

Daniel D. Foster
https://orcid.org/0000-0002-2175-0623
The Pennsylvania State University, USA

ABSTRACT

Teachers are at the heart of education systems because they directly interact with and influence learners in communities around the world. Teachers should be supported through dynamic professional development to engage them as critical partners in sustainable development. Professional development should exhibit the research-based characteristics of a strong content focus, active learning, collaboration, and sufficient duration. Connecting to teacher professional identities and supporting the development of teacher efficacy can help inspire teachers to become change agents in their communities. Communities of practice are one example of professional development practice that bring these elements together. Teacher professional development is an important component of the transformation, but teacher professional development needs to be supported by the educational system to fully realize the potential for impact.

DOI: 10.4018/978-1-6684-4050-6.ch009

INTRODUCTION

Teachers are essential actors in preparing the next generation to address the world's most pressing issues and facilitating the transition to a more sustainable society. Teachers are at the heart of education systems as they directly interact with learners and community members on a daily basis. They are familiar with community challenges as well as community assets. Teachers can work with learners to shape the communities that learners want to live in, and influence mindsets toward a more sustainable future. Sustainable Development has been most commonly defined as "*development that meets the needs of the present without compromising the ability of future generations to meet their own needs*" (World Commission on Environment and Development, 1987, p. 41) What is needed is to harness the strategic positionality of teachers as community change makers and to invest in their professional development to advance progress on sustainable development issues.

To date, Education for Sustainable Development (ESD) has been seen as the way that educational systems must realign themselves to work toward sustainable development outcomes. The term Education for Sustainable Development emerged at the 21st Agenda of the UN in Rio de Janeiro in 1992 and has evolved to focus on transforming the educational systems worldwide to foster knowledge, skills and attitudes that will lead to a sustainable society (Leicht et al., 2018). The concept of sustainability has evolved to encompass not just environmental movements, but to recognizes the interconnectedness of the economy and society within the biosphere (United Nations, 2015).

For many teachers, ESD is an opportunity for professional growth because they must reorientate their teaching practice along educational principles such as holism *(that environmental, social, and economic systems should be viewed as interconnected taking into consideration time and space as well),* pluralism *(that instructional approaches should be open encouraging critical thinking and dialogue helping students form their own opinion)* and an action orientation (*that students should with the help of the teacher set realistic goals and measurable outcomes to influence change*) (Sinakou et al., 2019). Teachers also must attend to the emotional aspect of learning and help learners to develop a sense of possibility for the future and inspire a sense of wonder (Grund & Brock, 2020). To be active and effective partners in the complexity of ESD, teachers need holistic and interdisciplinary professional development (UNESCO, 2020).

Reorienting education toward sustainability can begin at the level of the individual classroom (UNESCO, 2020). Teacher professional development is one of the methods that can be utilized to transform teaching and advance sustainability mindsets by impacting what occurs in the individual classrooms. Teacher professional development encompasses "all types of learning undertaken by teachers beyond the point of their initial training" (Craft, 2000, p. 9). Teachers must be empowered with the necessary pedagogies, tools, partnerships, and resources to transform their teaching to align with sustainable development efforts.

This chapter will address Education for Sustainable Development in the Middle East and North Africa (MENA) region, the importance of teacher professional development, the characteristics of effective teacher professional development and examples of effective teacher professional development.

EDUCATION FOR SUSTAINABLE DEVELOPMENT IN THE MENA REGION

Education for Sustainable Development (ESD) has its roots in environmental education efforts, which took hold in the 1970s in the MENA region (Saab et al., 2019). As the understanding of sustainability grew to include social and economic pillars, so too did the participation of countries in the MENA region in the global ESD movement. Multiple countries in the region were involved with the Decade for Education for Sustainable Development spanning from 2005 to 2014 including: Algeria, Egypt, Jordan, Lebanon, Libya, Morocco, Oman, Palestine, Tunisia, and United Arab Emirates.

Considering the example where ESD has been a policy priority of one country in the region, Oman. Oman was one of the first to develop a ministry of environment and then coordinated an effort to integrate ESD themes into the national curricula in the subjects of life skills, science, mathematics, and Arabic (Mula & Tilbury, 2011). The concept of ESD was embedded into teacher education in the College of Education in Sultan Qaboos University, the only public university in the country (Ambusaidi & Al-Rabaani, 2009). Prospective teachers were found to have a high level of awareness or positive attitudes towards sustainable development concepts and were prepared to deliver these concepts to their future students (Ambusaidi & Al Washahi, 2016).

A second example, Qatar, has worked to shift its economy from natural resource base to a knowledge base (Richer, 2014). Qatar has expanded sustainable development research capacity in higher education programs (Saab et al., 2019). The most recent curricular reform, the Qatar National Curriculum Framework, went into place in 2016. The framework is modeled on the issues by the UN Sustainable Development Goals and focuses on social, economic, and environmental challenges related to the water, energy food nexus (Zguir et al., 2021). The Qatar National Curriculum Framework also aligns with the ESD movement by promoting active student engagement in these issues (2021). However, there is no known provision for teacher preparation programs or teacher professional development programs to address these changes.

Zguir et al. (2022) found that there is work to be done on translating the vision of the national curricular policy into the curricula, textbooks, and lesson plans. Time constraints and system inflexibility appear to be additional barriers to the holistic integration of ESD into teaching practice.

The case of ESD in Egypt has taken a very different path. Faced with a variety of economic, social, and environmental pressures, there was a need to incorporate ideas of sustainability into the educational system, but the examination-focused educational system would need a complete transformation to support ESD (Sewilam, 2015). In this case, the European Commission funded a project to promote Education for Sustainable Development in Egypt by partnering European and Egyptian higher educational institutions, schools, non-governmental organizations, and ministries. The project was thought to be successful because educators were ready and willing to implement new pedagogies and content into their classes, but in the absence of structural educational reform, they were still limited by large class sizes, few resources, and little support. Egypt developed a new Sustainable Development Strategy called Egypt Vision 2030, which called on the educational system to transform to become student-centered and linked to societal needs (Mostafa, 2021). The new educational strategy, "Education 2.0", began in the 2018 academic year.

Other countries in the region, such as Morocco (Babou et al., 2020), have implemented educational reforms consistent with ESD, while others have evidence of ESD integration only at the level of higher education such as Bahrain (Buong, 2020), Jordan (Abu-Hola & Tareef, 2009), Lebanon (Hammoud & Tarabay, 2019), and Libya (Mahmoud & SP, 2022).

The roadmap that the United Nations Educational, Scientific, and Cultural Organization (UNESCO) (2020) developed for ESD suggests that policy support for ESD is the first step in scaling up ESD in education systems, which then allows space for trained and empowered educators to facilitate learning focused on sustainable development. With so many countries in the region and around the world without ESD-aligned policies, the transformation of educational systems is by default left largely in the domain of individual educational institutions and teachers. The next section focuses on the importance of empowering educators to transform the educational system through high quality and impactful professional development.

IMPORTANCE OF TEACHER PROFESSIONAL DEVELOPMENT IN MENA

The importance of teacher professional development in the Middle East and North Africa (MENA) region cannot be understated. According to the World Bank (2022), poor learning outcomes contribute to human capital deficiencies in the region. Poverty is rising, and food insecurity issues continue to grow in the face of conflict and food price inflation. Schools in MENA were closed for an average of 170 days during the covid-19 pandemic, and on average, learning losses totaled to one learning-adjusted year of school. The World Bank's three-pronged approach involves focusing on human capital as a driver of a greener, more inclusive, and more resilient growth in MENA. Investing in the professional development of teachers can contribute to the goals of building human capital through increasing the quality of teaching and learning.

Teacher professional development is an opportunity to "create relevant, sustained change and renewal" (Posti-Ahokas et al., 2022, p 312), and in many cases, educational reform is synonymous with teacher professional development (Sykes, 1996).

This investment of time and resources into educator professional development is important because if properly designed and managed, the professional development can improve the quality of teaching and learning, and impact student learning outcomes (Yoon, Duncan, Lee, Scarloss, & Shapiro, 2008; Gess-Newsome et al., 2017; Fischer et al., 2018). In the model developed by Yoon et al. (2008), professional development enhances teacher's knowledge, skills, and dispositions, improves classroom teaching performance, and positively affects student achievement. Additional work has highlighted the non-linear aspect of this process and underscores how evidence of success may take long periods of time before becoming detectable (Guskey & Yoon, 2009).

While the literature relating specific professional development methods to student learning outcomes is relatively sparse, professional learning communities (Vescio et al., 2008) and teacher-centered collaborative learning Akiba & Liang (2016) have been found to be positively correlated with student learning outcomes. These methods allow for teachers to try new practices and reflect with their colleagues, which are important elements in changing teacher practice (Webster-Wright, 2009; Saric & Steh, 2017).

Professional development is thought to be a vital aspect to achieving Education for Sustainable Development because of the complexity of the transformation required. An interdisciplinary approach is required to tackle the complex issues of sustainable development (Annan-Diab & Molinari, 2017), and student-centered pedagogies are at the heart of Education for Sustainable Development. In essence, teachers need to move from being conveyors of knowledge to facilitators of learning working in concert with colleagues across disciplinary areas. Teachers need to be supported in this endeavor through high quality professional development if they are going to be able to implement new teaching practices and philosophies.

Professional Identity

Appealing to teacher professional identities is an important outcome of teacher professional development, and a first step toward transforming education toward sustainability. Erickson (1979) described identity as "being at one with oneself as one grows and develops: it also means, at the same time, a sense of affinity with a community's sense of being at one with its future as well as its history or mythology" (pp. 27–28).

In many contexts, ESD has relied on individual teachers (Bryan & Bracken, 2011). These champions of sustainable development are teachers who have a global orientation, practice a strong sense of empathy, and believe that teachers can make a difference in the world (Gerry & Quirke-Bolt, 2019; Agirreazkuenaga, 2019). Complementary personal characteristics such as passion, self-efficacy and persistence are also associated with ESD teachers (Timm & Barth, 2020).

Beijaard et.al (2004) describes teacher professional identity as including professional knowledge, attitudes, beliefs, norms, values, emotions, and agency. Teacher professional identity is a dynamic process of reflection and re-interpretation of experiences (Kerby, 1991), corresponding to the idea that teacher professional identity development is an ongoing process (Conway, 2001). Indicators of teacher professional identity are correlated with teacher satisfaction, collegial relationships, and self-efficacy (Canrinus et al., 2012).

When teachers feel professionally marginalized by definition of their role and status, when their contributions as teachers to sustainable development are overlooked, this feeling can have a negative impact on their sense of professional identity and contributions to ESD (Qi et al., 2021). Timm and Barth (2020) identified two categories of teachers as ESD change agents: those who function as change agents by interacting with students toward sustainable development goals, and those that work to incite institutional change to value the teaching of ESD within the system.

Professional development can provide opportunities for teachers to build professional identities through peer-to-peer learning (Mbowane et al., 2017), and provide a space for teachers to reflect on their professional role individually and collectively (Posti-Ahokas, et al., 2022). To transform society toward sustainability, professional development can catalyze teachers to self-identify and connect their professional identity as a teacher to serving as agents of change in their communities and regions.

While professional identity of what the job of a teacher is can be critical, professional development can also assist in increasing the teachers' belief that they can do the job at high level. Taking into consideration teacher self-efficacy is another vital element to constructing effective teacher professional development.

Teacher Self-Efficacy

Education for Sustainable Development (ESD) presents new challenges to teachers as they play key roles in reorienting education towards the empowerment of students to shape a sustainable future. Professional development can enhance teacher self-efficacy, which is a trait that is described by Bandura (1997) as the judgement of one's ability to plan, implement, and ultimately perform a task. Improving teacher self-efficacy can contribute to ESD goals because teachers need to feel confident and prepared to implement instruction on sustainable development concepts. How do we grow self-efficacity in educators? Bandura (1997) reported four main sources of efficacy:

1. **Master Experience** – the educator experiences success firsthand
2. **Vicarious Experience** – the educator sees success being modeled
3. **Social Persuasion** – the educator hears positive affirmation or encouraging dialogue from trusted sources
4. **Physiological Feedback** – the educator if confident and not anxious, thus feels excited about the task at hand.

Self-efficacy can profoundly impact everything from psychological states to behavior to motivation. Stated differently, a teacher's self-efficacy will determine the goals they pursue, how they accomplish those goals and how they reflect upon their own performance. People with low self-efficacy generally avoid challenging tasks, and the call for effectively engaging in ESD is certainly a challenging task demanding high self-efficacy. Research (Hajloo, 2014) suggest that self-efficacy predicts self-esteem. Teachers with low self-esteem typically are not retained in the profession (Kelchtermans, 2017).

To effectively engage with ESD, teachers need to have content knowledge in a variety of subjects because they must have the ability to identify linkages between disciplines (Timm & Barth, 2020). When teachers are more familiar with sustainable development content, they have a higher self-efficacy to focus on solutions (Akça, 2019). Self-efficacy in ESD is correlated with implementation of ESD practices, which is why it is important to have professional development opportunities that have active learning components (Boeve-de Pauw et al., 2022).

Illustration: Teachers are important agents of change in addressing food insecurity

Professional development can be an opportunity for professional renewal providing an avenue to build teacher identities and teacher self-efficacy. The World Food Prize Foundation Global Guides Program is a professional development program curated by the Global Teach Ag Network that empowers educators from any discipline to connect their learners to issues of food security in their local communities.

A clear value woven into the fabric of the program is that teachers are professionals that have valuable skills, knowledge, and experiences, and that they are uniquely positioned to identify and address issues in their local communities. This value is not only authentically expressed by the facilitators but is also intentional included in the design of the 9-month professional development program delivered with online and face-to-face components.

Professional development is not always inspiring and empowering. The World Food Prize Foundation Global Guides facilitators do not speak to the value of teachers, they demonstrate this by designing program activities that illuminate the knowledge, skills, and experiences that each cohort of teachers brings into the program. In this way, teachers learn to see the value not only in themselves, but also in the other members of the cohort.

As one Global Guide expressed, "Truly, the inspiration is something that I did not expect to find, at least not to the degree in which I was inspired."

A belief stressed during the World Food Prize Foundation Global Guides program is that sustainable development challenges will not be solved by one discipline alone. Teachers can be inspired to model interdisciplinary thinking when encouraged to analyze how their discipline relates to a specific global challenge such as food security. One Global Guide expressed, "My eyes were totally opened to the fact that it will take all of our voices from the different disciplines to tackle the problem."

World Food Prize Foundation Global Guides participants have expressed that they better understand how teachers can be involved in creating sustainable change in their communities and how this contributes to solving global issues. They engage deeply in the subject matter of food security and are tasked with teasing out how their discipline relates to aspects of this multi-faceted, complex issue. With an interdisciplinary cohort, it is a process of discovery, not simply sharing a prescribed connection. One Global Guide noted, "I feel energized and empowered to share the information with my students and colleagues. In the wake of climate change and food security issues presently, there is also a sense of urgency to get the message to my students."

Many World Food Prize Foundation Global Guides initially feel a sense of "imposter syndrome" or a feeling of internalized fear of having inadequate skills and talents and being exposed as fraud when interacting with researchers, development professionals, politicians and other individuals working in sustainable development. By giving teachers a seat at the table and the ability to engage deeply and meaningfully with individual working in these areas, teachers can begin to see their role in addressing sustainable development issues such as food insecurity. Knowing that they have valued skills, knowledge and experiences leads to a sense of efficacy and confidence. As one Global Guide stated, "I feel more confident in myself as an educator in general. I feel better equipped with resources to discuss global food security but overall, the confidence gained was for teaching overall."

When teachers develop a strong sense of identity fueled by a well-developed sense of self-efficacy and feel empowered to make change, they can transform into teacher leaders. Many <name redacted> share program messages with teacher colleagues in their schools and professional organizations. One Global Guide noted that, "I have already volunteered to do a school-wide training on global competence for my colleagues."

The World Food Prize Foundation Global Guides professional development program is an opportunity for professional renewal as teachers are called upon to remember why they were called to the profession and understand the important role that they play in communities around the world. As one Global Guide noted, "I was counting the days to retirement next year. But after participating in this program, I think I'll continue teaching so I can implement what I've learned about food security."

World Food Prize Foundation Global Guides Program is a research-based high quality professional development program that facilitates a process of discovery on how educators can play an important role in sustainable development and works to inspire educators to meaningfully engage their learners in issues of food security in their local communities. World Food Prize Foundation Global Guides have expressed that have a better understanding of their role in sustainable development, that they are inspired to engage their students in food security, and that they experience a sense of excitement and renewal in their profession.

CHARACTERISTICS OF EFFECTIVE TEACHER PROFESSIONAL DEVELOPMENT

There is an array of teacher professional development engagement strategies that have been implemented over time, across contexts, and across geographies. Previous technical-rational-top-down strategies have focused on the deficits of educators and provided "one-shot" professional development sessions to fill in gaps (Caena, 2011). Practitioner theory is the theoretical underpinning of such top-down approaches, and implies that as new policy initiatives are implemented, teachers need to be "up-skilled" to rise to the challenge (Svendsen, 2020). Governments and universities are positioned as the experts, and professional development is something that happens to teachers, not something in which they have ownership.

Teacher professional development has since shifted to more cultural-individual interactive approaches, driven by adult learning, and situated cognition theories (Caena, 2011; Creemers, et al., 2013). This more progressive model considers teachers as professionals, and rests on the notion that teachers are active partners in their own professional development. The result is that change is more likely to be innovative and enduring (Svendsen, 2020).

Dynamic Teacher Professional Development

Creemers et al. (2013) propose a model of dynamic teacher professional development that can be utilized by policy makers and practitioners to positively impact student learning outcomes and transform educational systems. The eight components of a dynamic teacher professional development program are presented in Table 1.

Table 1. Components of dynamic teacher professional development (Creemers, et al, 2013)

Component	Definition
Evidence-based	*teacher professional development should be focused on developing skills associated with positive learning outcomes.*
Comprehensive	*teacher professional development should focus on groups of related skills rather than isolating specific skills.*
Individualized	*teachers will have different areas of improvement and needs for professional development.*
Actively involved	*teachers should know how professional development outcomes will impact student learning.*
Integrated	*teacher professional development should not be proscriptive nor leave development entirely up to individual teachers. teachers should be offered the opportunity to identify improvement needs and create an action plan to address the identified needs.*
Supported	*teachers should be supported through data collection to evaluate progress.*
Formative evaluation	*teachers critically reflect on progress and improve on their action plan.*
Summative evaluation	*measure the impact on teaching and learning.*

The dynamic model combines elements from different teacher professional development paradigms to capitalize on strengths and minimize weaknesses, with a particular focus on improving pedagogical content knowledge (Creemers et al., 2013). The dynamic model holds promise for transforming education to meet sustainable development challenges, because a holistic change encompassing learning content, pedagogy and the learning environment will be required (UNESCO, 2020).

Additional Characteristics of Effective Teacher Professional Development

Teacher professional development should have a strong content focus. To effectively engage in ESD, teachers need to have content knowledge in a variety of subjects because they must have the ability to identify linkages between disciplines and engage in cross-curricular connections (Timm & Barth, 2020). For example, Boeve-de Pauw et al. (2022) found teachers in their study were more comfortable with the environmental aspects of sustainable development, but were challenged when including economic, political and international elements.

Teacher professional development should be practice-based. Professional development programs should include opportunities to implement and reflect on new knowledge and skills (Boeve-de Pauw et al., 2022; Sims & Fletcher-Wood, 2019, Popova et al. 2022). A longitudinal study by Boeve-de Pauw et al. (2022) found that in a three-year professional development program, self-efficacy in ESD was boosted early in the program but fell back to initial levels as teachers struggled to implement new practices.

Teacher professional development should include collaborative elements. Collaboration can occur on several scales including in partners, in small groups, in whole-school settings, in professional organizations and beyond. Vygotsky (1978) discusses the power in social learning in that collaboration supports the development of collective knowledge that extends beyond individual experiences. According to Garmston (2007) collaborative cultures allow space for the exchange of ideas and opportunity for feedback relevant to practice. It can also increase understanding, commitment, and follow-through. A collaborative culture can also assist individuals in developing measurable goals. Sprott (2019) found that collaborative professional development can help to develop enduring professional relationships.

Teacher professional development must be of sufficient duration to better support teachers through the transformation process (Sims & Fletcher-Wood, 2019). One shot professional learning, no matter how dynamic, is not sufficient to support meaningful learning and transformation of teaching practice (Darling-Hammond, 2010).

Illustration: The World Food Prize Foundation Global Guides Program

The World Food Prize Foundation Global Guides Program is a teacher professional development experience designed using the dynamic model with a focus on research-based best practices for teacher professional development that was developed in 2018. The program utilizes a cohort model, has a duration of nine months, and is implemented in a blended format with digital and face-to-face interactions. Educator participants are from different countries, teach at all levels of education, and teach in a wide variety of subject areas. The common thread is that all participants have a desire to integrate food security topics into their teaching practice and connect their students to this real-world issue.

Evidenced-based: The World Food Prize Foundation Global Guides Program is based on teacher-centered collaborative learning, which has been shown to positively impact student learning outcomes. Using a cohort-based model, the knowledge, skills, and experiences of the participating teachers are highlighted and celebrated. Teachers are encouraged to share and collaborate with each other throughout the program, including workshops and the action project.

Comprehensive: The content areas of focus are food security and how it is embedded into the interconnected Sustainable Development Goals. Teachers are encouraged to deeply engage with the content area, identify linkages to their disciplinary area, and develop strategies for engaging learners. Project-based learning and storytelling are two approaches that are shared as complimentary skill sets that can supplement new knowledge of food security and sustainable development.

Individualized: Each participating educator develops their own action plan for engaging learners in the topic of food security. Individualized plans allow for teachers to take into account the learning context and the specific needs and interests of their learners. Additionally, an individualized plan allows them to work within the limitations imposed by time, resources, and other factors, while also leveraging community assets.

Actively involved: The facilitators of the World Food Prize Foundation Global Guides Program model active learning strategies that teacher participants can utilize with their students. Strategies are debriefed and underlying theories shared in order that teachers understand the science behind why the strategies are effective at improving student outcomes.

Integrated: An explicit value of the World Food Prize Foundation Global Guides Program is that the teacher participants are active partners in working to shape the future of teaching and learning of food security, and food insecurity solutions.

While guidance is provided on identifying high quality data sources, accessing food security experts, and sharing research-based best practices in teaching and learning, no "right answers" are presented. Teacher participants are explicitly invited as co-creators in teaching transformations required to improve society.

Supported: Teachers are supported throughout the entire program by a team of facilitators. Additionally, they are connected with a variety of experts in a range of areas that they can tap into for specialized assistance. Finally, they are invited into a community of practice consisting of educators from around the world who have an interest in teaching food security. A main value of the World Food Prize Foundation Global Guides Program is to provide spaces for educators to share ideas, best practices and ask for assistance, while also linking to high quality relevant information and resources.

Formative evaluation: During the stages when teachers are developing and executing their action plan, participants are provided repeated opportunity for iterative feedback process from peers, external stakeholders, and facilitators, who have open office hours where teachers can share progress, ask questions, and get feedback.

Summative evaluation: The final stage of the World Food Prize Foundation Global Guides program is the development and public sharing of a Reusable Learning Artifact (RLA). The RLA is based on their action project and created in such a manner that other teachers can utilize the ideas and information in their own context. A Celebration of Excellence is the culminating event to an authentic audience of peer educators with an instilled sense of professional responsibility to contribute to the collective knowledge and capacity of the group.

The program evaluation of the World Food Prize Foundation Global Guides Program by participating teachers is overwhelmingly positive. Many noted that the immersion was transformative and that they gained new knowledge and skills related to food security education. Participants have indicated that they are confident in their ability to grow their students' global competence and engage them in topics related to food security. They indicated that they have begun integrating the Sustainable Development Goals into their classroom directly and in leading school-wide efforts to address global competence. As one participant noted, "The Foundation <name redacted> Immersion has been the most meaningful professional development I have attended as an educator."

EXAMPLES OF EFFECTIVE TEACHER PROFESSIONAL DEVELOPMENT

While the Sustainable Development Goals provide a guideline for how global issues are complex and interconnected, sustainable development implementation strategies are currently evolving. Sustainable development is "holistic and transformational", and so too must sustainable development education (UNESCO, 2020). UNESCO (2020) stresses that sustainability education requires that we "rethink what, where and how we learn", and in this sense, teachers are challenged to develop students' critical thinking, collaboration, and problem-solving skills. There are no established answers, but rather students need to be empowered to envision, take responsibility for, and take action towards a sustainable future. Teacher professional development for sustainability education must address these complex needs. This section highlights promising teacher professional development strategies focused on a specific way teachers collaborate: communities of practice.

Teacher Communities of Practice

Communities of Practice (CoP) are defined by Wenger et al. (2002) as "groups of people who share a concern, a set of problems, or a passion about a topic, and who deepen their knowledge and expertise in this area by interacting on an ongoing basis" (p 4). Communities of Practice can provide a positive learning environment where teachers can reflect, share, provide and receive peer feedback, and develop new teaching methods with colleagues (Kelchtermans, 2006; Vangrieken, et al., 2015).

While CoPs are relatively common in western countries, Gitsaki et al. (2012) trialed and studied a CoP in United Arab Emirates. Teachers trialed new practices introduced via the CoP and self-reported a positive impact on their teaching practice. A study conducted of three CoPs in MENA by Johnson & Khalidi (2005) revealed that two of three CoPs studied were successful in meeting goals within two years. The study found that CoPs were most likely to be successful if there was a felt need of the members, if there is funded and dedicated leadership, and when they established linkages to relevant organizations.

In the seminal text on CoPs, Wenger et al. (2002) described seven principles for cultivating CoPs that are presented in Table 2. The seven principles from Wenger et al. (2002) for cultivating healthy Communities of Practice center around the notion of nurturing community "aliveness" or vitality, which the authors describe as the interactions that take place in a community that engender excitement and demonstrate value to members. While aliveness cannot be contrived, it can be designed for by utilizing the seven principles for cultivating CoPs.

Table 2. Seven principles for cultivating communities of Practice (Wenger, et al., 2002).

Principle	Definition
Design for evolution	*design elements should facilitate the evolution of the community over time*
Open a dialogue for inside and outside perspectives	*community design should bring information from outside the community into the dialogue about what the community could achieve*
Invite different levels of participation	*members should be able to participate at a level that feels comfortable to them*
Develop both public and private community spaces	*communities should center on the relationships between members and facilitate different spaces in which relationships can develop and thrive*
Focus on value	*members need to find value in a community because participation is usually voluntary*
Combine familiarity and excitement	*communities should cultivate stability, but also a sense of adventure and excitement*
Create a rhythm for the community	*the frequency of events and interactions should reflect the norms of the community and the needs of the members*

Illustration: The Global Teach Ag Network Global Learning in Agriculture Community

The Global Teach Ag Network comprising over thirty academic, non-governmental organization and private entities, collaborate to empower educators to connect their learners to global issues in agriculture and the environment. One way this goal is achieved is through high quality professional development for educators through the Global Learning in Agriculture Community. The evolution of this Community of Practice demonstrates Wenger et al. (2002) seven design principles.

Design for evolution: *The Global Teach Ag Network based at The Pennsylvania State University grew out of a movement to globalize agricultural curriculum and promote multidisciplinary global food security education. In 2014, a workshop on global learning in agriculture was held at The Pennsylvania State University as a culminating grant activity for a United States Department of Education Fulbright-Hays project. Since that initial face-to-face event, the Global Learning Conference become an annual event that is held online so that a greater number of educators can participate from across the U.S. and around the world. Thousands of educators from around the world have joined the conversation through the conference, listserv, social media, and face-to-face professional development events. The Global Learning in Agriculture Conference then transitioned into the Global Learning in Agriculture Community to better reflect the goals of a year-round community of practice to support educators in their professional development journey.*

Open a dialogue for inside and outside perspectives: *The <Network> comprises academic, private, public, non-governmental, and industry partner organizations. The purpose of these partnerships is to bring relevant information to educators for purposes of engaging their learners in current and authentic global issues. Additionally, partners can engage educators as key stakeholders in their work by developing connections and identifying educator needs.*

Invite different levels of participation: *Some members are extremely active and take on leadership roles, while others are more occasional contributors. All are valued members of the community. Specific examples included the opportunities for educators to be program manager of complimentary affinity groups and programming efforts in the community while other educators serve as volunteer committee members or participants.*

Develop both public and private community spaces: *Within the digital community of practice there are spaces that are open to all members of the community, while some working groups maintain a closed affinity space in which they can conduct their business only among a select group of members.*

Focus on value: *Participation in the Global Learning in Agriculture Community is completely voluntary, meaning that educators access the community because they find it valuable in some way. The real value of participation in the community may not be apparent immediately, as it takes time for teachers to reflect and integrate new knowledge and ideas into their teaching.*

Combine familiarity and excitement: *The Global Learning in Agriculture Community provides spaces for educators to bring their ideas to other group members. One member brought the idea of writing poetry about food security topics to process the emotions that arise when discussing food insecurity, and to encourage creativity around the issue. For some educators in the community this was an idea that they had never considered before, and it gave members an opportunity to practice open-mindedness.*

Create a rhythm for the community: *The rhythm for the Global Learning in Agriculture Community followed the evolution from a conference to a community. The conference had previously taken place in February, so the community has come to expect a catalyzing event at that time of year. Other "seasons" in the community include an annual book study and a poetry contest. Members are otherwise active throughout the year with activities such as addressing questions and sharing resources in community forums.*

The Global Learning in Agriculture Community has provided space for hundreds of hours of professional development events and thousands of points of contact between members. For some members, this is the only access they have to teacher-oriented professional development while for others, this is just one opportunity of many. The Global Learning in Agriculture Community has organically grown and developed over the years with resources dedicated by organizational partners of the Global Teach Ag Network.

SUMMARY, CONCLUSIONS, AND RECOMMENDATIONS

Teachers are at the heart of education systems because they directly interact with and influence learners in communities around the world. Teachers can be engaged as partners in sustainable development if they are valued as professionals due to their specialized knowledge, skills, and experiences.

Teachers should be engaged in dynamic professional development that exhibit the research-based characteristics of a strong content focus, active learning, collective participation, coherence, and sufficient duration. Communities of Practice are one example of professional development practice that bring these elements together.

Recommendations for the MENA region include following the roadmap for ESD developed by UNESCO (2020) which lays out a plan for advancing policy, transforming learning environments, building capacities of educators, empowering, and mobilizing youth, and accelerating local level action. Teacher professional development is an important component of the transformation but needs to be supported by the entirety of the educational system for impact to be fully realized.

Teacher professional development should be high quality and research based. This chapter emphasized a dynamic professional development model in which teachers are partners in their own learning and can support the holistic change needed in teaching practice to support ESD. Specific characteristics of professional development supported by the literature include a strong content focus, active learning, collective participation, coherence, and a sufficient duration. Appealing to teacher professional identities and supporting the development of teacher efficacy can help to inspire teachers to become change agents in their communities.

Communities of Practice are one research-based example of a professional development practice that encourages collaboration among teacher colleagues. Communities of Practice can provide a positive learning environment where teachers can reflect, share, provide and receive peer feedback, and develop new teaching methods with colleagues. As demonstrated with the example of the Global Learning in Agriculture Community curated by the Global Teach Ag Network, CoPs can lend themselves well to the transformation of teaching and learning needed to achieve ESD.

REFERENCES

Abu-Hola, I. R., & Tareef, A. B. (2009). Teaching for sustainable development in higher education institutions: University of Jordan as a case study. *College Student Journal*, *43*(4), 1287–1306.

Agirreazkuenaga, L. (2019). Embedding sustainable development goals in education. Teachers' perspective about education for sustainability in the Basque Autonomous Community. *Sustainability*, *11*(5), 1496. doi:10.3390u11051496

Akça, F. (2019). Sustainable development in teacher education in terms of being solution oriented and self-efficacy. *Sustainability*, *11*(23), 6878. doi:10.3390u11236878

Akiba, M., & Guodong, L. (2016). Effects of teacher professional learning activities on student achievement growth. *The Journal of Educational Research*, *109*(1), 99–110. doi:10.1080/00220671.2014.924470

Ambusaidi, A., & Al-Rabaani, A. (2009). Environmental education in the Sultanate of Oman: Taking sustainable development into account. In *Environmental education in context* (pp. 37–50). Brill.

Ambusaidi, A., & Al Washahi, M. (2016). Prospective teachers' perceptions about the concept of sustainable development and related issues in Oman. *Journal of Education for Sustainable Development*, *10*(1), 3–19. doi:10.1177/0973408215625528

Annan-Diab, F., & Molinari, C. (2017). Interdisciplinarity: Practical approach to advancing education for sustainability and for the Sustainable Development Goals. *International Journal of Management Education*, *15*(2), 73–83. doi:10.1016/j.ijme.2017.03.006

Babou, A. I., Jiyed, O., El Batri, B., Maskour, L., El Mostapha Aouine, A. A., & Zaki, M. (2020). Education for sustainable development and teaching biodiversity in the classroom of the sciences of the Moroccan school system: A case study based on the ministry's grades and school curricula from primary to secondary school and qualifying. *Brock Journal of Education*, *8*(2), 13–21. doi:10.37745/bje/vol8.no2.pp13-21.2020

Bandura, A. (1997). *Self-efficacy: The exercise of control*. Freeman.

Beijaard, D., Meijer, P. C., & Verloop, N. (2004). Reconsidering research on teachers' professional identity. *Teaching and Teacher Education*, *20*(2), 107–128. doi:10.1016/j.tate.2003.07.001

Boeve-de Pauw, J., Olsson, D., Berglund, T., & Gericke, N. (2022). Teachers' ESD self-efficacy and practices: A longitudinal study on the impact of teacher professional development. *Environmental Education Research*, *28*(6), 867–885. doi:10.1080/13504622.2022.2042206

Bryan, A., & Bracken, M. (2011). *Learning to read the world? Teaching and Learning about Global Citizenship and International Development in Post-primary Schools*. Irish Aid.

Buong, A. (2020). Achieving SDGs through higher educational institutions: A case study of the University of Bahrain. In *Sustainable Development and Social Responsibility—Volume 2* (pp. 19–23). Springer. doi:10.1007/978-3-030-32902-0_3

Caena, F. (2011). Literature review Quality in Teachers' continuing professional development. *European Commission*, *2*, 20.

Canrinus, E. T., Helms-Lorenz, M., Beijaard, D., Buitink, J., & Hofman, A. (2012). Self-efficacy, Job Satisfaction, Motivation and Commitment: Exploring the Relationships between Indicators of Teachers' Professional Identity. *European Journal of Psychology of Education*, *27*(1), 115–132. doi:10.100710212-011-0069-2

Conway, P. (2001). Anticipatory reflection while learning to teach: From a temporally truncated to a temporally distributed model of reflection in teacher education. *Teaching and Teacher Education*, *17*(1), 89–106. doi:10.1016/S0742-051X(00)00040-8

Craft, A. (2000). *Continuing professional development: A practical guide for teachers and schools*. London: Routledge/Falmer.

Creemers, B., Kyriakides, L., & Antoniou, P. (2013). *Teacher Professional Development for Improving Quality of Teaching*. Springer. doi:10.1007/978-94-007-5207-8

Darling-Hammond, L. (2010). Teacher education and the American future. *Journal of Teacher Education*, *61*(1/2), 35–47. doi:10.1177/0022487109348024

Erikson, E. H. (1979). *Dimensions of a new identity*. WW Norton & Company.

Garmston, R. J. (2007). Results-oriented agendas transform meetings into valuable collaborative events. *The Learning Professional*, *28*(2), 55.

Gerry, J., & Quirke-Bolt, N. (2019). Teachers' professional identities and development education, *Policy & Practice: A Development. Education Review*, *29*, 110–120.

Gitsaki, C., Donaghue, H., & Wang, P. (2012). *Using a community of practice for teacher professional development in the United Arab Emirates. In The international handbook of cultures of professional development for teachers: Comparative international issues in collaboration, reflection, management and policy*. Analytrics.

Grund, J., & Brock, A. (2020). Education for sustainable development in Germany: Not just desired but also effective for transformative action. *Sustainability, 12*(7), 2838. doi:10.3390u12072838

Guskey, T. R., & Yoon, K. S. (2009). What works in professional development? *Phi Delta Kappan, 90*(7), 495–500. doi:10.1177/003172170909000709

Hajloo, N. (2014). Relationships between self-efficacy, self-esteem and procrastination in undergraduate psychology students. *Iranian Journal of Psychiatry and Behavioral Sciences, 8*(3), 42–49. PMID:25780374

Hammoud, J., & Tarabay, M. (2019). Higher Education for Sustainability in the Developing World: A Case Study of Rafik Hariri University1 in Lebanon. *European Journal of Sustainable Development, 8*(2), 379–379. doi:10.14207/ejsd.2019.v8n2p379

Johnson, E. C., & Khalidi, R. (2005). Communities of practice for development in the Middle East and North. *Africa KM4D Journal, 1*(1), 96–110.

Kelchtermans, G. (2006). Teacher collaboration and collegiality as workplace conditions. A review. *Zeitschrift fur Padagogik, 52*(2), 220–237.

Kelchtermans, G. (2017). 'Should I stay or should I go?': Unpacking teacher attrition/retention as an educational issue. *Teachers and Teaching, 23*(8), 961–977. doi:10.1080/13540602.2017.1379793

Kerby, A. (1991). *Narrative and the self.* Indiana University Press.

Leicht, A., Heiss, J., & Byun, W. J. (2018). *Issues and trends in education for sustainable development* (Vol. 5). UNESCO Publishing.

Mahmoud M, D., & SP, B. (2022). Science and Quality Education for Sustainability Development in Libya. *INTI Journal, 2022*(18).

Mbowane, C. K., De Villiers, J. J., & Braun, M. W. (2017). Teacher participation in science fairs as professional development in South Africa. *South African Journal of Science, 113*(7-8), 1–7. doi:10.17159ajs.2017/20160364

Mostafa, Y. (2021). Educational Reform Movement in Egypt towards 2030 Vision: Learning from History to Incorporate New Education. *Journal of School Improvement and Leadership*, *3*, 115–125.

Mulà, I., & Tilbury, D. (2011). National journeys towards education for sustainable development: reviewing national experiences from Chile, Indonesia, Kenya, the Netherlands, Oman. In United Nations Decade of Education for Sustainable Development 2005-2014. UNESCO.

Popova, A., Evans, D. K., Breeding, M. E., & Arancibia, V. (2022). Teacher professional development around the world: The gap between evidence and practice. *The World Bank Research Observer*, *37*(1), 107–136. doi:10.1093/wbro/lkab006

Posti-Ahokas, H., Idriss, K., Hassan, M., & Isotalo, S. (2022). Collaborative professional practice for strengthening teacher educator identities in Eritrea. *Journal of Education for Teaching*, *48*(3), 300–315. doi:10.1080/02607476.2021.1994838

Qi, W., Sorokina, N., & Liu, Y. (2021). The Construction of Teacher Identity in Education for Sustainable Development: The Case of Chinese ESP Teachers. *International Journal of Higher Education*, *10*(2), 284–298. doi:10.5430/ijhe.v10n2p284

Richer, R. A. (2014). Sustainable development in Qatar: Challenges and opportunities. *QScience Connect*, *2014*(1), 22. doi:10.5339/connect.2014.22

Saab, N., Badran, A., & Sadik, A. K. (2019). *Environmental Education for Sustainable Development in Arab Countries*. Arab Forum for Environment and Development.

Saric, M., & Steh, B. (2017). Critical reflection in the professional development of teachers: Challenges and possibilities. *CEPS Journal, 7*(3), 67-85.

Sewilam, H., McCormack, O., Mader, M., & Abdel Raouf, M. (2015). Introducing education for sustainable development into Egyptian schools. *Environment, Development and Sustainability*, *17*(2), 221–238. doi:10.100710668-014-9597-7

Sims, S., & Fletcher-Wood, H. (2021). Identifying the characteristics of effective teacher professional development: A critical review. *School Effectiveness and School Improvement*, *32*(1), 47–63. doi:10.1080/09243453.2020.1772841

Sinakou, E., Boeve-de Pauw, J., & Van Petegem, P. (2019). Exploring the concept of sustainable development within education for sustainable development: Implications for ESD research and practice. *Environment, Development and Sustainability*, *21*(1), 1–10. doi:10.100710668-017-0032-8

Sprott, R. A. (2019). Factors that foster and deter advanced teachers' professional development. *Teaching and Teacher Education*, *77*, 321–331. doi:10.1016/j.tate.2018.11.001

Svendsen, B. (2020). Inquiries into teacher professional development – What matters? *Education*, *140*(3), 111–131.

Sykes, G. (1996). Reform of and as professional development. *Phi Delta Kappan*, *77*(7), 464.

Timm, J. M., & Barth, M. (2021). Making education for sustainable development happen in elementary schools: The role of teachers. *Environmental Education Research*, *27*(1), 50–66. doi:10.1080/13504622.2020.1813256

United Nations. (2015). *Sustainable Development Goals*. Retrieved from: https://sdgs.un.org/goals

United Nations Educational, Scientific and Cultural Organization. (2020). *Education for sustainable development: A roadmap*. Retrieved from: https://unesdoc.unesco.org/ark:/48223/pf0000374802.locale=en

Vangrieken, K., Dochy, F., Raes, E., & Kyndt, E. (2015). Teacher collaboration: A systematic review. *Educational Research Review*, *15*, 17–40. doi:10.1016/j.edurev.2015.04.002

Vescio, V., Ross, D., & Adams, A. (2008). A review of research on the impact of professional learning communities on teaching practice and student learning. *Teaching and Teacher Education*, *24*(1), 80–91. doi:10.1016/j.tate.2007.01.004

Vygotsky, L. S. (1978). *Mind in society: The development of higher psychological processes* (M. Cole, V. John-Steiner, S. Scribner, & E. Souberman, Eds. & Trans.). Harvard University Press.

Webster-Wright, A. (2009). Reframing professional development through understanding authentic professional learning. *Review of Educational Research*, *79*(2), 702–739. doi:10.3102/0034654308330970

Wenger, E., McDermott, R. A., & Snyder, W. (2002). *Cultivating communities of practice: A guide to managing knowledge*. Harvard Business Press.

World Bank. (2022). *Human capital plan for the Middle East and North Africa*. Retrieved from: https://www.worldbank.org/en/region/mena/publication/2022-human-capital-plan-for-the-middle-east-and-north-africa

World Commission on Environment and Development. (1987). *Our Common Future: A Report from the United Nations World Commission on Environment and Development*. WCED.

Yoon, K. S., Duncan, T., Lee, S. W. Y., Scarloss, B., & Shapley, K. L. (2007). Reviewing the evidence on how teacher professional development affects student achievement. *Regional Educational Laboratory Southwest (NJ1)*.

Zguir, M. F., Dubis, S., & Koç, M. (2021). Embedding Education for Sustainable Development (ESD) and SDGs values in curriculum: A comparative review on Qatar, Singapore and New Zealand. *Journal of Cleaner Production*, *319*, 128534. doi:10.1016/j.jclepro.2021.128534

Zguir, M. F., Dubis, S., & Koç, M. (2022). Integrating sustainability into curricula: Teachers' perceptions, preparation and practice in Qatar. *Journal of Cleaner Production*, *371*, 133167. doi:10.1016/j.jclepro.2022.133167

KEY TERMS AND DEFINITIONS

Community of Practice: For this chapter, a community of educators is a group of professionals who come together to share their experiences, knowledge, and practices related to a specific topical area, in this case agricultural education.

Education for Sustainable Development: Education for sustainable development is a term originating from the United Nations in 1992 to describe an approach to learning that aims to equip individuals with the knowledge and skills to create a more sustainable future for all.

Global Teach Ag Network: The Global Teach Ag Network is a collaboration coordinated by Penn State University in the US focused on providing educators professional development opportunities and access to communities of practice to advance global learning in food, fiber, and natural resources.

Professional Development: Professional development refers to activities or programs that help educators improve their skills, knowledge, and performance.

Teacher Association: A teacher association is a professional organization often with a membership structure, that supports and advocates for teachers, providing them with opportunities for networking, learning, and collaboration.

Teacher Professional Identity: Teacher professional identity refers to a teacher's sense of self and purpose as a professional educator, encompassing their beliefs, values, and attitudes about teaching and learning.

Teacher Self-Efficacy: Teacher self-efficacy refers to a teacher's belief in their ability to successfully carry out specific teaching tasks or goals, even in the face of challenges or obstacles.

REVIEW QUESTIONS

1. What are two examples shared in the chapter of Education for Sustainable Development from the MENA region?
2. According to the chapter, how could investment in teacher professional development yield positive outcomes for the MENA region?
3. How would you describe the professional identify of a teacher from your past educational activities? What do you wish one of your past educators knew about how you view them?
4. What are the potential risks of teachers feeling marginalized in our society?
5. Why would teacher self-efficacy be critical to success in efforts of education for sustainable development?
6. What would be a specific example of an activity from each of the four sources of efficacy that could positively impact the self-efficacy of a teacher towards education for sustainable development?
7. What is imposter syndrome?
8. What is an example of an effective professional development intervention you have participated in? Please describe specifically how this program reflects the characteristics of effective teacher professional development presented in the chapter.
9. What is an active "Community of Practice" you are currently a member of? Describe three specific elements that make it affective according to Wenger, et al. in the chapter.
10. If resources were not a concern, what would be an example of an effective teacher professional development program to advance education for sustainable development you would bring to your community?

Chapter 10
Improving Teaching and Learning of Ornamental and Medicinal Plants in the Arab Countries

Safia Hamdy El-Hanafy
Cairo University, Egypt

ABSTRACT

This chapter sheds light on the main features of education in the universities of Arab countries. Limitations and obstacles were highlighted, with presentation of non-controversial keys for their resolution. The language used in education is largely discussed to reveal the great significance of using the Arabic language in the Arabic world. The discussion reveals clearly how this issue can guarantee easy and perfect education, keeping in mind that this is the situation in all advanced countries who are sticking to their national languages in education. The chapter demonstrates the nature of ornamental and medicinal plants as an application for the intended aim, together with highlighting their significance for human being. Expansion of essential knowledge in the field of ornamental and medicinal plants was also demonstrated. The economic significance of these plants was also mentioned.

INTRODUCTION

Education is fundamental to development and growth. The human mind makes possible all development achievement from health advances and agricultural innovations to efficient public administration and private sector growth (King, 2011). Agricultural

DOI: 10.4018/978-1-6684-4050-6.ch010

education takes place in all the colleges of agriculture all over the Arabic world. The Faculty of Agriculture (FOA) in Cairo University is the oldest higher agricultural education institution in the Arab world. It was established in1889. Today the FOA presents a big number of programs such as Animal Production, Plant Production, Commercial and Social Agricultural Sciences, Soils and Waters, Food Science, Plant Protection, Agricultural Engineering, and Biotechnology. The Department of Floriculture is an important component of the program of Plant Production at the FOA, Cairo University. This, together with the Department of Vegetables, Department of Pomology and Department of Crops represent the four departments forming the Program of Plant Production.

Undergraduates are granted teaching in all the four departments of the program of Plant Production to get their Bachelor degree in science (BS), without being specialized in any. A graduate has the knowledge and skills of horticulture crop production, management and marketing. Their career prospects range from production of field crops for small-scale farmers and herb crops for export, management of landscape business for recreation to marketing of fruits and vegetables. A postgraduate student can, however, get the master degree (M.Sc.), then the Philosophy Doctor degree (Ph.D.) of either of the four above mentioned departments. The Cairo University Department of Floriculture offers postgraduate degrees that prepare for a myriad of diverse professional opportunities.

The Department of Floriculture is concerned with the study of the subjects of ornamental plants, medicinal plants, flower designs, flower shows, aromatic plants, forests and trees, landscape, interior scape, tissue culture, plant reproduction, plant production, post-harvest handling of ornamental, medicinal and aromatic plants and the subject of breeding.

The mission of the department of Floriculture is to:

- Prepare students for professions with a broad base of ornamental and medicinal plant knowledge
- Develop research-based knowledge for efficient and profitable production of ornamental and medicinal plants
- Improve the competitiveness of the ornamental and medicinal plant industry in Egypt
- Increase the quality, variety, and availability of ornamental and medicinal products
- Deliver research-based knowledge about the ways ornamental and medicinal plants improve our environment and serve as a source of personal enjoyment.

Expansion of essential knowledge in the field of ornamental and medicinal plants is of paramount importance for development of education in this field. Continuous updating of the educational content is essential in teaching. Fertilization practices in cut-rose production have for example evolved significantly in the last century, going from a mostly organic and rudimentary methodology based on mineral soils to highly technical, computerized nutrient delivery systems based on soilless media or hydroponics (Cabrera and Solis-Perez, 2017). Another example is education of technological methods of pathogen detection which improve the ability for managing diseases of glasshouse ornamental crops (Dinesen and van Zaayen, 1996).

Ornamental plants are primarily decorative and, as such, are purchased and appreciated according to their visual characteristics. Beyond this, the other senses may be also concerned. This multi-sensorial repercussion justifies the use of sensory methodologies to study their characteristics (Symoneaux, Segond, and Maignant, 2022). Given the immense range of varieties of different colors and sizes of flowers, breeders do their best to improve the existing assortment by producing a large diversity of several thousands of seedlings and by trying to find the few elite plants which are worth being tested to create one new variety (Chaanin, 2003). The industry of drug production involves a large number of non-stopping research activities on the way of further active ingredient extraction from medicinal plants.

PROSPECTS OF EDUCATION ENHANCEMENT

Teaching and learning strategies and methods in the field of ornamental and medicinal plants.

Teaching of ornamental and medicinal plants should be put in suitable theoretical and practical frameworks that guarantee achievement of the aspired targets of good understanding of the concerned data, making proper connection between them, having clear imagination of their existence and application, and having enhanced mental ability for their adequate storage and recall on a practical basis.

Vast fields, farms and greenhouses are available for experimental and productive scales of activities, together with whatever labs required. Teaching classrooms, conference halls and libraries of different specialties are also available. There are about thirty faculty members in the Department.

The objectives of a course including indoor plant siting, flower arrangements and ornamental plant exhibitions should provide an understanding of the scientific principles and associated costs involved in choosing plants and flowers.

The teaching and learning methods should include lectures, class meetings (where students may be asked to write papers on relevant subjects), seminars, workshops and field trips.

Large groups of students are supposed to attend theoretical discussions in the form of lectures on a regular basis, whereas small groups have external visits to farms, export companies or exhibitions, where some lessons should be discussed while actual procedures and arrangements are being done.

Students in small groups are also distributed at laboratories, greenhouses and farms for the purpose of demonstration and application of practical steps of plant reproduction, cutting and preservation of flowers, collection and preservation of seeds, landscaping, collecting medicinal and aromatic plants which are dried and preserved, oil extraction from parts of such plants.

The topics involved in the subject of decoration using ornamental plants may include identification of indoor plants used in interior decoration, discussion of the suitable environmental conditions for indoor plants, basis of interior decoration of lobbies, other applications of interior decoration and studying decoration maps.

The subject of flower arrangements should highlight different types of flowers and cut foliage used in arrangement, containers in use and available holders and equipment and different designs of flower arrangement (eastern, European, American, Japanese ...etc.).

The assessment of student learning is a dynamic process that includes practical exams to evaluate the professional and practical skills gained by students, written exams to assess the students' knowledge and understanding of information provided to them in the lectures, viva exams to assess the students' intellectual skills and creativity, quizzes on readings and lecture material being frequent and unannounced, discussions to assess the students' comprehension of their reading assignments and evaluation of papers prepared by students on relevant subjects to weigh their ability to gather information and solve problems.

The Facilities Required for Teaching and Learning

Those may include visual and audio teaching aids (like slide sets, video tapes, data show, projectors and whiteboard), plant material (cut flowers, pot plants, medicinal and aromatic plants ... etc.), course notes and handouts, books, atlas of flowers, web sites, appropriate Laboratory with chemicals and equipment necessary for practical classes, library with access to sources of up-to- date information in the field of interior decoration, computers with access to the internet and visits to farms and flower shows.

Auditing

A process of self- assessment from the side of the mentor at the end of each teaching task can be concluded from the answers of the following self-questioning:

- What did I want to accomplish today?

- What did I want students to learn?
- What went well in what I did?
- What did students seem to learn?
- What material did students seem not to understand?
- What could have been done differently?

At the side of the students by the end of the semester, they are expected to give a feedback including their comments on:

- Course management.
- Mentor characteristics and style.
- Student outcomes.
- Instructional environment.
- Student preferences for instruction/learning style
- Instructional settings.

A peer review is essential for any course including:

- Peer observation of teaching.
- Peer review of course material.

Fruitful Teaching Techniques

Whatever and wherever education is concerned, the use of a foreign language in either teaching or training decreases the outcome of the process of education. That is simply because a foreign language needs extra time and effort, is less clear, less amenable for storage in memory or recall from it, and is less flexible for mental association, integration and elaboration that arrives at recent conclusions. A foreign language learner will never manage it for the rest of his life like an original speaker of that language. As a tool for knowledge acquirement, an original speaker will get optimum benefit from using his mother tongue in education, whereas a student who learned it, will never get the same level of benefit (Mahmoud, 1999). For teaching Floriculture, this by no means is different.

Some techniques of questioning may enhance student understanding and improve their capability of active sharing in the process of education and learning. Examples of such techniques of active learning are:

1. Brainstorming exercises, like asking the students about the mistakes in the process of flower arrangement which lead to early welting.

2. Case studies that suggest the presence of some problem during a practical situation and asking the student to give a solution for the suggested obstacle. A case study may for example put forward the possibility that a student is working at one department of a flower exhibition and is suffering the problem of receiving a limited number of visitors. The student is then asked to define the possible etiological factors responsible for the problem, and accordingly propose a solution.
3. Concept mapping is another modality of enhancing active learning. The students at the end of a course are encouraged to go into the depth of the different thoughts discussed throughout the course and try to suggest concepts related to the thoughts.
4. Jigsaw procedure is also a method of active learning, where different items of a course content are assigned to a number of students each of whom will search, read and prepare his specified bit of material, and also review that of others.

Examples of reading assignments given to individuals within a group are:

- Types of flowers which can be dried
- Methods of desiccation of flowers

Designs of Flower Arrangement

5-A learning cycle is a set of steps concerned with active learning. It starts with the phase of exploration, where the students are asked to explore some sort of a phenomenon, like for example something that may be noticed when cut flowers are held in a container under unfavorable conditions. The second phase of the learning cycle is to encourage the students to present their observations of abnormalities in the form of concepts, e.g., vase solution turbidity, flower desiccation, cut flower neck bending, failure of bud opening of the cut flowers or loss of cut flower petals. The third and last phase of the cycle is the stage of concept application, where the candidate is encouraged to fashion an experiment that tests the possible etiology underling an abnormality. The experiment at the end may or may not be able to explain the etiology of the concerned phenomenon.

Significance and Examples of Ornamental Plants

Ornamental plants were known to be a pleasure maker since the forest man had first noticed their presence until now. They are either used indoor for houses, hotels, hospitals, malls or offices, or otherwise are used outdoor for house gardens, public gardens, squares or streets. Cut flowers are a mobile form of pleasure offering plants, being nationally and internationally marketed.

The courses of ornamental plants should include the study of trees and palms, shrubs, perennial and annual plants, ornamental bulbs, edging plants, turf grasses, hedges and vines, aquatic and semiaquatic plants, succulent plants and cactus, and pot plants. Ornamental trees may give shade, wind buffering or flowering, which all help in designing gardens and ornamenting roads. Ficus, cupressus and poinciana are well known examples of ornamental trees. Ornamental palms offer beauty, fruits or by-products. Examples are pritchardia, cocos and phoenex. Shrubs are used for designing hedges, ornamenting small gardens and flowering. Hibiscus, tecoma and plumeria are examples of ornamental shrubs. Perennial plants give long standing ornamentation. They may be in the form of shrubs like rose, herbs like Carnation or bulbs like Canna. Annual herbs offer beauty on a short-term basis, like Calendula and Helianthus. Ornamental bulbs give the most beautiful flowers known in the market like Gladiolus, Dahlia and Tulipa. Edging plants define the outlines of lawns. Artemisia, alernanthera and verbena are some of their examples. Turf grasses make a nice green land cover, like aptenia and wedelia. Hedging plants may be in the form of shrubs like duranta or vines like bougainvillea and arum. Aquatic plants make like green cabs on the water surface (like nymphaea), while semiaquatic plants jacket the stream banks like bambusoideae. Ornamental pot plants may be essentially having foliage like epipremnum, syngonium and dracaena or may have both foliage and flowering like anthurium, gardenia and spathiphyllum. Ornamental plants may stay at their cultivation territories, or else are harvested for national or international marketing. Specific procedures may then be required for their preservation and shipping.

Cut flowers are characterized by their keeping viable for a relatively long period after being cut, which enables their being used in preparation of bouquets and vases. Cut flowers are used either nationally, or they are exported. Special criteria for cut flower quality control should be present, including full flower maturity before being cut, flower intactness and freedom from scratches, mechanical harming or diseases; with guaranteed tolerance for manipulation and transportation that enables their arrival to the using hands in perfect quality.

Flower cutting should take place at an adequate stage of their opening, during the recommended day hours, and via the recommended mode of cutting, so as to guarantee their good quality and post-cutting longevity.

Manipulation of cut flowers includes a number of steps starting with harvesting which is followed by immersing in preservative solution, cooling, assortment and grading, encasement and storage, or otherwise transportation and marketing. Great care should be taken during each of these stages to protect cut flowers from deterioration or derangement.

Cool chains, or what is called "Door to Door Cooling" is a golden rule in dealing with cut flowers which should be always preserved in cool chambers in between different manipulation procedures, until they reach their target place of using. This is extremely essential as it minimizes flower respiration, decreases ethylene gas tissue production, lowers the rate of water loss via transpiration, increases flower resistance to mechanical harm and generally keeps biological tissue activity of the flower at its lowest level. This policy of cold preservation and transportation guarantees adequate prolongation of the flower's way towards senescence.

The production of ornamental plants for use in homes, gardens and outdoor amenity areas is a commercial discipline that can be further divided into specialist sectors: nursery stock, house plants and bedding plants (Cameron and Emmett, 2003).

Significance and Examples of Medicinal Plants

This is a critical topic, unlike any other topics of agricultural science, as it is concerned with the study of plants used for maintaining health and/or treating specific ailments. Such plants are used in a plethora of ways in both allopathic and traditional systems of medicine in countries across the world.

Even people using only allopathic medicine throughout their lives are likely to be somewhat medicinal plant reliant as 20-25% of drugs prescribed are plant derived (Rates, 2001).

The World Health Organization has estimated that 80% of the world's population relies solely or largely on traditional remedies for health care (Bannerman, 1982) and there is speculation that more than two billion people may be heavily reliant on medicinal plants (Lambert, Srivastava, and Vietmayer, 1997). Although considerable uncertainty surrounds these often-cited figures, there is no doubt that medicinal plants play an important role in the livelihoods and welfare of a vast number of people in both developed and developing countries. The importance of medicinal plants in health care is increasingly recognized in the health sector as exemplified by discussions of the role of traditional medicine in contributing to achieving the Millennium Development Goals (MDG), three of which are directly health related (UN, 2002), and by work towards European harmonized criteria for the assessment of herbal medicinal products (Steinhoff, 2005; *Action plan for herbal medicines 2010–2011,* 2010).

The subject of medicinal plants should discuss specific plants containing actively acting substances which may have an impact on human health. The medically active ingredients in a medicinal plant may be better concentrated at the roots as in case of glycyrrhiza, at the rhizomes like zingiber, at the bark like in case of cinnamomum and salix, at the leaves as seen in mint, thymus and ocimum, at the flowers as in matricaria, at the fruits of coriandrum or at the seeds like the case of nigella. Such plants may be devoted for this single purpose, or else are required to satisfy other human needs.

The target medicinal plant parts may be dried to give a chance for better concentration of the component precious substance or else to enable their proper manipulation throughout marketing. Otherwise, fresh parts are directly involved in procedures of active ingredient extraction. Each medicinal plant has its specific recommended time for raw material collection, its preferred method of drying (natural or artificial techniques) and its reference procedure of extraction of the component active ingredients. Packing is one of the most important manipulations for medicinal plants, whether they are in the form of dry parts, derivatives or extracted oils.

The industry of active ingredient extraction from medicinal plants has a wide range of different procedures, either in field or after harvesting. This can also take place regionally, nationally or internationally.

A suitable packing method is essential for each product to guarantee easy transportation, storage and marketing without any change in the physical or chemical characteristics of its ingredient components.

The storage duration for medicinal plant derivatives differs with different ingredients at different physical circumstances, being clearly defined in the pharmacopoeia. storage in general should protect against invasion by rodents, insects or microorganisms. It should also guarantee against enzymatic reactivation that can spoil the component ingredients.

Medicinal plants have been known since the very ancient history for their great role in treatment of human and animal illness, and are still the main raw material for the international pharmaceutical industry. Various interesting possible applications of essential oils replace synthetic drugs to circumvent the increasing resistance of some pathogens (Gudrun and Gerhard, 2011).

Teaching medicinal plants, should therefore throw the light on examples of drugs derived from plants, like Acetyldigoxin (carditonic, derived from the plant Digitalis lanata), Adenoside (cardiotonic, derived from Adonis vernalis), Aescin (antiinflamatory, derived from Aesculus hippocastanum), Aesculetin (antidysentery, derived from Frazinus rhychophylla) and Agrimophol (anthelmintic, derived from Agrimonia supatoria) (Anne, 2020).

Environmental and Commercial Aspects for Teaching Ornamental and Medicinal Plants

Specific features characterize ornamental plants, making them different from other members of the plant kingdom. They constitute natural living beauty that enriches the quality of human life and represent an important economic grouping within the ornamental industry (Bautista-Baños, Romanazzi, & Jiménez-Aparicio, 2016).

Not only this, but also that the late climate changes which are becoming increasingly prominent on our planet made it necessary to give the greatest care for extending the green blanket of the earth, a matter that if made by ornamental plants, will get the double target of amendment climate aberration and augmentation of the human spiritual satisfaction.

Ornamental plants are largely not harvested, being non-consumption plants. This makes them ideal as a persistent green cover. Also, their being widely cultivated in urban sectors of the world makes them the suitable alternative green cover in such areas, whereas rural sectors of our planet are almost always covered by consumption crops.

Cover plants offer a natural and inexpensive solution for climate derangement through their ability to capture atmospheric carbon dioxide (CO2) into soils. Cover plants not only remove CO2 from the atmosphere but also offer the soil better health and make the plants more resilient to a changing climate. Healthy soil has better water infiltration and water holding capacity and is less susceptible to erosion from wind and water. Cover crops also trap excess nitrogen – keeping it from leaching into groundwater or running off into surface water – releasing it later to feed growing plants. This saves water and fertilizer and makes plants more able to survive in harsh conditions (Greene, 2022).

Cover plants absorb carbon dioxide through photosynthesis and store the carbon in the soil, helping to mitigate climate change. It is estimated that 20 million acres of cover plants can sequester over 66 million tons of carbon dioxide equivalent per year, equal to the emissions of about 13 million vehicles (Bertrand, Roberts, and Walker, 2022).

Ornamental plants contribute immensely to the quality of life. Wood and wood fibers are important to human survival and comfort. The production of ornamental and conifer trees is limited in unique ways by competition from both native and introduced wild vegetation (LeBaron, McFarland, Burnside, 2008)

Enhancing awareness of the commercial significance of ornamental and medicinal plant production is significant, where it can be done by inviting policy makers, business and community heads for workshops. Establishment of projects for production of such plants can represent a cooperation between the teaching authority and the investors, so that the scientific authority has the role of supervision and student training in the cultivated fields. This saves some of the costs of educational training and some of the costs of workforce. The expansion of business activities in the aspect of ornamental and medicinal plant production allows for the availability of adequate jobs for the agricultural graduates, and in general for a promising career for such specialized trainees.

An inter-relation exists between ornamental and medicinal plants, where we sometimes get one plant that can be described as an ornamental, as well as a medicinal plant, examples are nerium and ocimum. Medicinal plant products like essential oils (e.g., mint oil and thyme oil) are also famous for their antimicrobial activity which can be beneficial in improving the characteristics of vase solution of cut flowers (El-Hanafy, 2007)

The Arabic world has a vast farm of ornamental and medicinal plants, being produced in open field or at green houses. Different processes are applied for their reproduction including vegetative or seed propagation.

Obstacles and Solutions for Education in the Arabic Area

Agricultural education in the Arabic region, like other types of education suffers a myriad of defects, not unlike other modalities education having a scientific nature. Since successful education always leads to performance of distinct research activities. The lack of original research is a prominent witness of an almost friable process of education that is in bad need of editing and enhancement.

Research Activities

True research almost always starts with a scientific questioning that has no answer. The researcher then suggests an answer based on scientific facts and goes through specific procedures which at the end prove or disprove his suggestion. The researcher should support the idea representing the pivot of his research by presenting a review of publications at or near the same field of his research with discussing the similarities and differences between his research and their research, and should highlight clearly the area of innovation in the concept presented or the results arrived at.

Another modality of research activity is an extensive collective statistical effort by going through a big amount of previously recorded data to arrive at a new conclusion. True research is therefore always representing a necessity, as it adds something new to the knowledge concerned with the scientific fields in vogue, and their ecological repercussions.

What actually happens in the Arabic countries, and is wrongly called research, is in most of the situation just a repetition of some experimental activity done before, whether internationally or even nationally. This odd behavior going on in the Arabic area, is nothing but a waste of time, effort and money.

The reason behind this problem of repetition is the presence of laws in the universities which hold up promotion of any member of the teaching board until he is able to present a certain number of research works. There is also a requirement in the form of a research thesis for a postgraduate student to fulfil his master or philosophy university degree. With education in the Arabic countries being through a foreign language, there is significant limitation of the scientific capability of the members of the university teaching panel, as well as of their postgraduate students (Mahmoud, 2022).

The Language used in the Process of Education

Throughout the last 50 years, teaching in the Arabic world has largely turned from being through the national into being through foreign languages in many university specialties, claiming that this change is a sort of development for education. This, of course, is not development as extra effort and more time is needed to study, fogging of data is an association, together with less understanding, compromised memory, and marked limitation of the capability of mental association, integration and elaboration of thoughts. This is then a grave civilizational fall that underlies the scarcity of new concepts or innovated ideas. The lack of true reaction and emotional integration of the Arabic researcher with his scientific specialty, is a natural result of using a foreign language in education, with marked limitation of his knowledge depth and capacity both quantitatively and qualitatively. Teaching and studying through the Arabic language in the Arabic countries is therefore the golden key to guarantee a thorough vast and deep understanding of every bit of science within an optimally short time, with maximized ability to keep information in mind, and to have adequate mental processing of such information. This improves ability to recall information, to notice links between different data, to analyze, criticize and discuss every detail at a very high level of accuracy and clearance. This spontaneously enables creation of true scientific sensation and reaction with knowledge, and establishment of a scientific base in the society of schools and universities, as well as in the whole society. Then, and only then, Arabs can produce true research and enjoy its benefits. The wisdom that should never be forgotten here is the "the national language is the most perfect tool to manipulate education". Application of this wisdom was the golden key for building old civilizations of the ancestors of humans, with translation being the normal channel of transmission of information from generation to generation. The same wisdom is still being applied by all advanced nations of today, being spontaneously applied as a biological necessity without any hesitation. Inter-translation between the distinct nations of today is also still the logic pathway for mutual exchange of everything new. Replacement of languages is therefore the wrong choice, as it represents suicide of civilization, with its associated prominent scientific and cultural failure witnessed by any wise observer. Imitation of successful nations means to do what they do, which means to study through national language like what they do, not to study through another nation's language. To understand this clearly, imagine the possibility that one of the today's advanced nations changed their language of teaching from the national to a foreign one, and subsequently imagine how poor they will be in knowledge. This is actually the shortest way for failure (Mahmoud, 1998 and 1999).

A Student's Self-confidence

This is a big issue in the field of enhancement of education. A student gets self-trust if he feels at any stage of his education, and since his very early childhood, that he understands and feels the meaning of material discussed in no time, keeps it in mind properly and remembers it clearly. Language is defined as the tool of understanding and the vessel of culture. To guarantee self-confidence, there is, therefore, no alternative to the use of mother tongue in education. On the other hand, loss or weakness of self-trust throughout the process of education weakens the sentimental relation between a student and knowledge. Consequently, education becomes a burden, with tendency for escaping confrontation. Once self-esteem deteriorates, staggering will be the rule, and hatred will be the hidden feeling towards education. Nobody likes to find himself less intelligent than he should be, whereas being indulged in learning through a foreign language creates this disagreeable feeling deeply within a student. Those who are keen to fulfil human rights should consider it an exclusive priority for every human on our planet to get his education through his own language. A student with deranged self-confidence will end with a graduate suffering the same defect in the field of knowledge throughout the rest of his life. The loss is really scaring, being a sure way of loss of capability of any country to develop its human resources (Mahmoud, 2016).

The Amount of Data Included in a Course

Minimization of the volume of information required to get a university degree is a prominent mistake that was practiced by the educational bodies which used a foreign language in place of Arabic. No comparison can actually be held between the rich information present in the Arabic curriculum applied in a similar college, and that very concise content studied in a foreign language. The reason for this, is the inability of the Arabic students to study a big amount of details through such a foreign language. On the contrary, an Arabic content is very clear and easy to understand within a short time, and keeps stable in mind. This pitfall is then a matter of deformation of education, not a sort of its enhancement, and adds much to the educational problems in the Arabic area.

Having limited capability in foreign languages in comparison to the national language, is actually not a problem in the Arabic student nor in his mentor, but it is simply a fact in all humans who have different languages, as well as different colors. Anyone, therefore, raised on his native language should have the privilege of being educated through it, which is true in all countries exporting recent civilization, and was also true throughout the whole human history. This grave mistake of changing the language of education from Arabic to a foreign language in the Arabic area with its gloomy upheavals, should attract our attention to understand the fact that "change does not always mean development, but may sometimes mean derangement". A sort of defective handling of the problem of incapability in foreign language, is to feel ashamed, and consequently to try defending its use in education, just for avoidance of being stigmatized with inefficiency, which is unacceptable by a mentor (Mahmoud, 2022).

Should a Course Content be Fully Written?

It is wise to have much interest in fully writing the whole amount of knowledge in each subject, devote enough time to discuss each bit of such knowledge face to face with undergraduate or postgraduate students, and make sure that every bit of information is simply clear and well understood. This means abandoning the method of just telling candidates the names of topics and let them search for the content and read it.

The principle of offering the student a definite source including all the required details of a subject, should be largely considered. Since knowledge is unlimited, the opposite party imagines that we should not hinder the student's capability by defining the amount of information we want him to know. This poor alternative of giving the student just titles and letting him collect the information, has the drawbacks of losing the advantage of preparing every detail of the subject by the highly experienced members of the teaching panel, having a lazy student who may not be keen to collect a good amount of information and may not collect whatever he does not understand, losing the chance of having detailed class discussions for every information of a unified source and losing the benefit of having timely assessment for knowledge gained in the absence of a definite single source. In addition, the absence of a definite source gives the sensation (at the side of the student, as well as at the side of his mentor) of lack of clear idea of what is truly required to know, gives an impression of missing the road map of education, increases the level of indifference for both the student and the mentor, and finally ends with a poor level of knowledge (with of course, little exceptional students who happen to be knowledge lovers). Another claim of the other party who prefer leaving a student without detailed written knowledge, is the rapid rate of addition of recent knowledge. This is simply argued by confirming the significance of detailed studying of well-settled basic facts, regularly updating the content of a course, and leaving a limited space for student activities in gathering recent data, which should be discussed with them and submitted for assessment (Mahmoud, 2023).

A Paper Book or an Electronic One?

The true companion of a student throughout his endless learning journey is a true paper book. An actual relation of love and respect develops between the student and his true book. The features of the front paper and every corner of every paper of a student's book have their mental pictures that make him feel scientifically and culturally strong with his paper sources. The electronic pages should be auxiliary, but not the main source of the student (Mahmoud, 2023).

Direct Education or On-Line Communication?

In the process of education, as well as on training, sight, ear and heart are the truly effective tools for transmission of clear knowledge. A mentor should mix with his students, attract the sight of one student at a time, and gather the ears of all students all through. This, together with the internal feelings of a mentor, confirm to every student that the mentor is very keen to make every student understand every detail he personally understands, and enables a mentor to simplify any complicated bit of information. Continuous questioning of this and that student at every point being discussed raises the level of arousal for all attendees and confirms that everything is well-understood. Brain storming by suggesting difficult cases and getting quick variable answers enhance the feeling of responsibility and improves self-esteem of the students. This direct and very private relationship of love and respect between a mentor and his students is essential "to make today's students tomorrow's scientists". Personal questioning from the side of students to the mentor, as well as among themselves by the end of this direct cession, adds much to the aspired clarity and well-understanding of the required knowledge and techniques. The outcome of telecommunication in this regard, is very disappointing and inefficient as it simply lacks attendance in person, with loss of all functions mentioned above. An on-line student, is at best, just a watcher of the process of education, not a part of its team (Mahmoud, 2023).

One Subject or Multiple Subjects at a Given Time of the Academic Year?

An idea that may improve the outcome of any subject being taught, is to allow the student to study only one subject till its end, and then move to another subject until he finishes the number of subjects required for this year. This gives the school student (excluding young children of the primary school) or university student an ideal chance to go into the depth of the subject, and makes it easier for him to feel that he is getting good experience in the subject. The last month of the academic year should however be devoted for reviewing all the subjects of the year (but again subject by subject, each being within a few days) with an exam at the end of the last day of every subject reviewing.

The application of the idea of one subject at a time (not sharing a number of subjects at the same time of the academic year) enhances the desire of the student to widen his knowledge beyond the defined syllabus. This results from his gained firm knowledge in the subject he is devoted to. Settings should be specified to discuss whatever students collect from electronic or paper sources beyond the curriculum, and a specified assessment should be held at the end of each setting (Mahmoud, 2022).

Distribution of Assessment

To complete the picture of a sound process of education, assessment can be held at the last hour of each academic day, whether at schools or universities. This should be completed by weekly and monthly settings of assessment, each being held at the last hour of a regular academic day, and is crowned by a final assessment at the last hour of last day of a course. Distribution of the assessment marks among such frequent settings nullifies the significance of what is called an exam day, and maximizes the significance of understanding and keeping in mind every bit of information. Holding the settings of assessment always at the end of a regular teaching day (whether they are daily, weekly or monthly assessment, or even assessment at the end of the course) confirms the concept of learning, and only learning first, not just getting a certificate (Mahmoud, 2022).

Certificates of Education, Are They Significant?

Another problem affecting the process of teaching and assessment in Arabic countries is the present status of giving much care for getting some certificate of education, without giving the same level of care for the process of getting knowledge. Teaching and learning are consequently negatively affected at the expense of caring much for the exams, their timing, their models of questions, the criteria for success in them......etc. A great difference actually exists between a student who succeeded in exams, and that who has got a good amount of cumulative knowledge throughout his academic journey.

What is originally called education is to give much interest for achievement of a proper scientific and cultural level of knowledge, both quantitatively and qualitatively, that really suits the targeted rank. Exams and consequently certificate offering should therefore be a secondary less significant target within the insight of the teaching or learning individuals (Mahmoud, 2022).

The Types of Questions Which Appear in an Exam, and the Fad of Question Banks

Omission of any rules controlling the types and forms of questions at any exam at any level confirms more and more that the issue is to learn, not to get a certificate. An undergraduate or a postgraduate student should therefore never be given or told about how the questions will appear at any exam. There is continuously ongoing controversy as to the preference of long essay questions, short questions, filling in the spaces, enumeration, requesting comments, multiple choice questions....etc. That is simply because any of these modalities has its advantages and disadvantages in the process of evaluation. A wrong perception is considering some modality as the more recent or the better, and a wrong policy is to put regulations for the types of questions to be used. A correct behavior is then to give the mentor full freedom to choose whatever modality of questions at any given procedure of assessment, and to encourage the students to know how to manage all possible modalities. The spread of the phenomenon of question banks led to deviation of interest of both students and teaching staff from being directed to proper learning and teaching, to being directed towards preparation for the exams. The better policy that improves learning is to encourage students to make questions on every bit of information they study, with the aim of intensifying their mental arousal and deepening their understanding of the given data. A deep stab at the heart of education is to forget that everybody should learn to change from an ignorant to a cultured person enjoying the knowledge he knows and the cleverness he got from training, even in the absence of a certificate indicating his knowledge or experience (Mahmoud, 2023).

The Value of Devoting the Last Year Before Graduation to Reviewing all the Subjects Studied Since the First Day of College

The last year of the university can be devoted for reviewing all the subjects studied throughout all the years spent in the university. This can be done by dividing the last year into short time sectors, one being specified for each subject, with an exam at the last hour of the last day of each subject. This idea guarantees a state of continuous learning at the side of the student who will be graduated at an optimum level of knowledge and experience in his field (Mahmoud, 2022).

Optimization of the Level of Knowledge of a Mentor

To enhance the level of knowledge of the members of the university's teaching panel without obliging them to present repeated useless mock research, a new system should be applied for their evaluation and subsequent promotion. This system should intend at enhancing continuous fruitful reading, being witnessed by presenting a sort of knowledge creation or translation at a specified amount and on defined regularity. The staff members should give much care for the subject of research methodology including teaching and training through research models. There should be also continuous encouragement of the staff members to do research only when they have innovated ideas that can change or add facts. Better rewarding should always await whoever presents an original research work, whereas useless works should be rejected (Mahmoud, 2022).

Scientific Journals

Each Arabic country has its own scientific journals where postgraduates publish their research works in different fields of knowledge. The filling of such journals with imitated, feeble and useless publications made bad reputation for such journals. This gradually led to the craze of recommending publishing in internationally known journals. Those well-known foreign journals have strict criteria for acceptance of publications. This story at the end, leads to progressive deterioration of the Arabic journals and progressive improvement of the well-known foreign journals. This odd situation means that the Arabic regulations for evaluation of publications, which give a higher rank for a paper published in foreign journals, are directly sharing in stoppage of development of the Arabic scientific journals, and are consequently putting more obstacles in the way of enhancement of Arabic education, including agricultural education.

The proper solution of this problem is to omit regulations which encourage publication in foreign journals, and alternatively establish a high committee that supervises all scientific and cultural journals and puts strict regulations for accepting only original invaluable papers for publishing. This, on one hand will make research papers at the aspired optimum level, and on the other hand will improve the reputation of Arabic scientific journals. It is important to point here that these Arabic journals should publish in Arabic, but they should have on the internet an Arabic site as well as a foreign language site, which enables publishing every paper in both languages. With improvement of the credibility of these journals they will be internationally well-known, and may accept papers for publication from non-Arabic countries (Mahmoud, 2022).

The Issue of the Significance of an ISO Certificate

The international organization who develops and publishes international standards is called ISO. ISO does not perform certification, but only develops international standards such as ISO 9001. The process of accreditation and certificate offering is actually performed by external certification bodies. However, ISO's committee on conformity assessment (CASCO) has produced a number of standards related to certification process, which are used by certification bodies. Once an educational establishment gets certified, the impression will be that this school or institute has perfect quality management standards, or that it is in conformity to management system standards (iso.org, 2015).

Certification can be a useful tool to add credibility. However, the standards used for assessment of institutes do not measure the scientific or cultural knowledge of the graduates, whether quantitatively or qualitatively. It subsequently gives a false impression that this school or that institute is capable of having highly efficient graduates at their specified fields of knowledge. Such believes are sometimes nothing but mirage, and therefore should not be given more than their true size, if the Arabic world is working seriously for enhancement of education (Mahmoud, 2023).

Twinning of Universities

An issue related to credibility of teaching establishments, is their being eager to make some sort of relation with an internationally well-known educational body, to make use of its name in convincing candidates to join them, in spite of lacking similarity at most. Other Arab areas reached in this aspect up to building complete educational bodies, carrying the same name of internationally well-known bodies, and even brought foreigners to teach and to do research activities. Their candidates however, are Arabic students whose mother tongue is the Arabic language, and whose performance can be optimum only through their mother language. The research activities performed in such hired institutes are claimed to be Arabic research activities, while they have nothing to do with Arabs, as they are actually suggested and carried out mainly by the hired guests. Much care is also given in such situations to offering the students certificates carrying the name of the internationally known educational body, a situation that means retreating back to the square of giving much care to getting a certificate, and less care to getting truly useful education.

The sound consideration, or logic concept in this issue, is to hold a twinning agreement between an Arabic and a foreign university on the basis of exchanging knowledge and training programs, and transmission of innovated policies and successful systems of management and teaching. However everything should be explored with a critical eye, as not everything doing well there, will do well here. Unifying the language of education in both universities is not, and cannot be a part of the twinning agreement at any scale (Mahmoud, 2022).

The Inter-Reaction Between an Institute and Its Incubating Environment

Environment and knowledge finders are companions since the early start of creation. In this concern, environment inspires a human endless observations, facts and findings, and directs his attention to difficulties and obstacles. This makes a human always look at his environment, contemplate, understand and extrapolate. Environment is therefore the human's attendant mentor, and is the uterus incubating him, being the immortal source of his mental nutrition. Those concerned with knowledge and education, lose their compass for development once they ignore this biological relation. The other face of this relation, is the human touch, innovating and producing recent equipment, giving the environment a new reality that inspires subsequent generations. This is the scene of how human knowledge and sciences are being developed. Academies were subsequently established, keeping in pace with environment, and always having mutual enrichment. No success is therefore expected from a system of education that disregards local environment, or else, regional, international or cosmic environment. This is true, whether being knowledge-wise, literary-wise and sentimental-wise (Mahmoud, 2022).

Agricultural education is logically intimately related to environment. Education and research should always reveal this fact, and standards of evaluation should always consider it. The use of national language in the whole of this scenario cannot be overemphasized.

The Triangle of Education

All advanced nations, without exception, use their national languages in teaching their students, at all levels and within all grades. This is always their way in introducing scientists to the international society, although the population in some of these advanced countries is little. This actually induces life in science by using a national language and induces life in a national language by studying science through it. The sound vessel for a human's culture is really his mother tongue, not any other alternative. Those who try to learn through a foreign language are just like those trying to hold water within an outstretched hand. This simply explains why every advanced nation defends the use of its national language in all aspects of life, particularly in the process of education. This does not mean, by any way, neglecting teaching a foreign language to the grown-up students, as it enables communication with others, translation from and to other languages, as well as attending international conferences. Yes, advanced nations teach their students foreign languages, but they never teach them through these foreign languages.

The basis for a successful process of education, can therefore be imagined in the form of a triangle, with its three angles. The first angle is to perform the whole process of teaching and learning, at all levels, through the national language. This enables getting the greatest personal cumulative amount of knowledge at its highest quality, which means that a future scientist or expert is being introduced to the national and international society. The second angle is augmentation of the national language of the students at all levels, as it is their perfect tool in the struggle for earning knowledge. The third angle is to teach the student a foreign language, including its general as well as its professional aspects. That professional side of a foreign language should be specific for the student's field of knowledge (Mahmoud, 1999 and 2005).

Enhancement of education and training in the Arabic world requires confirmation of the use of Arabic language at every bit of the process. This gives the best level of accurate understanding at the shortest time, allows for maximum capacity of storage of data and maximum ability for their recall. It also augments the highest level of mental processing of the studied data to conclude new facts in the concerned field. Having a look on the whole world, all countries sharing effectively in innovation of international knowledge are using their own languages in all aspects of education without any exclusion. The exchange of recent knowledge among them is always going on through the logic channel of translation. For the Arabic world, education will give its adequate outcome if Arabs apply the triangle of education. The foreign language chosen should be only one, not more, so as to save the time and effort of students who should be devoted to the scientific knowledge they are supposed to know. The logic foreign language for the eastern part of the Arabic world is English, whereas it is French for Tunisia, Algeria, Morocco and Mauritania.

REFERENCES

Action plan for herbal medicines 2010–2011. (2010). European Medicines Agency. Retrieved from https://www.ema.europa.eu/en/documents/other/action-plan-herbal-medicines-2010-2011_en.pdf

Anne, M.H. (2020). List of medicines made from plants. *Science, Tech, Math.*

Bannerman, R. H. (1982). Traditional medicine in modern health care. *World Health Forum, 3*, 8–13.

Bautista-Baños, S., Romanazzi, G., & Jiménez-Aparicio, A. (Eds.). (2016). *Chitosan in the preservation of agricultural commodities*. Academic Press.

Bertrand, S., Roberts, A.S., & Walker, E. (2022). *Cover Crops for Climate Change Adaptation and Mitigation.* Agriculture and Climate Series, Environmental and Energy Study Institute (EESI).

Cabrera, R. I., & Solis-Perez, A. R. (2017). *Mineral nutrition and fertilization management*. Academic Press.

Cameron, R. W. F., & Emmett, M. R. (2003). Production Systems and Agronomyl Nursery Stock and Houseplant Production. Encyclopedia of Applied Plant Sciences, 949-956.

Chaanin, A. (2003). Breedingl Selection Strategies for Cut Roses. Encyclopedia of Rose Science, 33-41.

Dinesen, I. G., & van Zaayen, A. (1996). Potential of Pathogen Detection Technology for Management of Diseases in glasshouse Ornamental Crops. *Advances in Botanical Research*, *23*, 137–170. doi:10.1016/S0065-2296(08)60105-6

El-Hanafy, H. S. (2007). Alternative additives to vase solution that can prolong vase life of carnation (Dianthus caryophyllus) flowers. *J. Product Dev*, *12*(1), 263–276. doi:10.21608/jpd.2007.44957

Greene, M. G. (2022). *Cover Crops Play a Starring role in Climate Change mitigation*. U.S. Department of Agriculture. Retrieved from https://www.farmers.gov/blog/cover-crops-play-starring-role-in-climate-change-mitigation#:~:text=Cover%20crops%20offer%20agricultural%20producers,resilient%20to%20a%20changing%20climate

Gudrun, L., & Gerhard, B. (2011). A Review on Recent Research Results (2008-2010) on Essential Oils as antimicrobials and Antifungals. A review. *Flavour and Fragrance Journal*, *27*, 13–39.

Iiso.org. (2015). The International Organization for Standardization.

King, E. (2011, Jan. 28). Education is fundamental to Development and Growth. *Education for Global Development*.

Lambert, J., Srivastava, J., & Vietmayer, N. (1997). *Technical Paper no. 355. Medicinal plants – rescuing a global heritage.* Washington, DC: World Bank.

LeBaron, H., McFarland, J., & Burnside, O. (2008). The Triazine Herbicides. 50 years Revolutionizing Agriculture. Academic Press.

Mahmoud, M. E. K. K. (1998). *Arabization and Westernization between what is Common and what is Required (translated from Arabic)*. Published in the Policy Newspaper.

Mahmoud, M. E. K. K. (1999). *The Linguistic Concept for keeping up with civilization*. Conference of Cairo University, Egypt, for Development of University Education. Vision to the Future University. Republished in the site: www.alhiwartoday.net

Mahmoud, M. E. K. K. (2005). The Language of a Nation is Their Tool for science. Conference of the Society of the Tongue of Arabs, held in The League of Arabic States, Cairo, Egypt.

Mahmoud, M. E.K.K. (2016). *The Way to Get out of The Crisis of Deterioration of Education in Egypt, with a Guarantee of The Occurrence of a Distinct Scientific Renaissance within a Few Years*. A letter for state stakeholders in Egypt.

Mahmoud, M. E. K. K. (2022). *A Call towards Amendment of Scientific Research and Education in the Modern Republic*. A Letter for state stakeholders in Egypt.

Rates, S. M. K. (2001). Plants as source of drugs. *Toxicon*, *39*(5), 603–613. doi:10.1016/S0041-0101(00)00154-9 PMID:11072038

Steinhoff, B. (2005). Laws and regulation on medicinal and aromatic plants in Europe. *Acta Horticulturae*, (678), 13–22. doi:10.17660/ActaHortic.2005.678.1

Symoneaux, R., Segond, N., & Maignant, A. (2022). Sensory and consumer sciences applicated on ornamental plants. In *Nonfood Sensory Practices* (pp. 291–311). Woodhead Publishing. doi:10.1016/B978-0-12-821939-3.00007-5

UN. (2002). *The millennium development goals and the United Nations role*. Fact Sheet. United Nations Department of Public Information. https://www.un.org/millenniumgoals/MDGs-FACTSHEET1.pdf

Chapter 11
Food Security and the Future Role of Agricultural Extension in Building Agricultural Human Capital in the Arab Countries

Mohamed Samy
https://orcid.org/0000-0003-3522-0989
Independent Researcher, USA

ABSTRACT

Most Arab countries have been struggling to deal with the global food crisis and the shortage of food commodities. In spite of their efforts, they face a critical food insecurity crisis of increasing the production of food commodities less rapidly than their consumption. This chapter presents efforts of three Arab countries—Egypt, Saudi Arabia, and Morocco—in tackling the food insecurity crisis and the role of public and private agricultural extension in building agricultural human capital and in turn increasing agricultural productivity. These countries varied on their reliance on public agricultural extension to educate and train farmers, especially small-scale farmers, who are the majority of food crop producers. This chapter identifies challenges facing public agricultural extension and provides guidelines to reform and improve extension services to be able to assist in building agricultural human capital, increase productivity, and achieve acceptable levels of food security.

DOI: 10.4018/978-1-6684-4050-6.ch011

INTRODUCTION

From the beginning of the twenty-first century, much of the developing world, especially the Arab countries have been struggling to deal with the global food crisis and the shortage of major food commodities. In spite of their efforts to increase agricultural productivity and production of major food commodities during the past two decades, the 22 Arab countries, without exception, have been facing a critical food insecurity crisis of increasing the production of major food commodities less rapidly than their consumption. Several factors have contributed to this food insecurity crisis including low production and stagnate productivity of major food commodities, and the realities of rapidly increasing population, food consumption and food imports, and changing global markets. In addition, the impact of international political conflicts, the COVID-19 pandemic, supply chain disruption, inflation, and natural calamities caused by climate change have all aggravated the Arab food insecurity crisis. Moreover, most Arab governments are very concerned about the effect of rising global food prices on their economy (The Economist, 2022).

Achieving food security is an essential goal of the United Nations (UN) 2030 Agenda for the Sustainable Development. The UN Sustainable Development Goals call on achieving food security, ending hunger, improving nutrition, and promoting sustainable agriculture. Food security exists "when all people, at all times, have physical, social and economic access to sufficient, safe and nutritious food to meet their dietary needs and food preferences for an active and healthy life" (UN, 2022). Food security has four dimensions: availability, access, utilization and stability; it can be evaluated at individual, household, national, regional or global levels (FAO, 2021a). The availability of food examines the supply side of food security as it is mainly concerned with issues related to food production, food trade, food distribution, in addition to the availability of resources such as land and water for agriculture.

The Arab governments in general and the ministries of agriculture in particular are responsible for making food available; and their efforts to achieve the UN 2030 Sustainable Development Goals reflect their commitment to tackle the food insecurity crisis. The 2021 Regional Overview of Food Security and Nutrition Report provided an update on the progress made in the Arab countries towards food security and nutrition, set by the World Health Assembly (WHA). Findings presented in this report showed that the Arab countries face serious challenges in ensuring regular access to sufficient, safe and nutritious food for all Arab people (Reliefweb, 2023). The deteriorating food security situation in the Arab countries indicates that a greater number of people face difficulties in accessing a healthy diet, (Reliefweb, 2023). An estimated 32.3 percent of the 527 million Arab population did not have access to adequate food in 2020, with 10 million more people reporting food insecurity than the previous year (TRT, 2022). The IMF Managing Director, Kristalina Georgieva, said in October 2022 that 141 million people across the Arab region are exposed to food insecurity (ASHARQ AL-AWSAT, 2022).

The prospective of reaching the UN goal of achieving acceptable level of food security in the Arab countries greatly depends on the ability of their governments to build agriculture human capital to be able to increase agricultural productivity and sustainably manage natural resources. Building human capital focuses on educating and training farmers, especially small-scale farmers who constitute the majority of food crop producers in the Arab countries, and their families. Adopting the goal of building and strengthening agricultural human capital should be a top priority of the Arab countries in the next decades. To achieve this goal, the Arab governments should utilize all forms of public and private, formal and non-formal agricultural education to educate and train Arab farmers and their families.

The primary public institutional actors in the Arab countries designated to educate and train farmers and their families are the Ministries of Agriculture by means of non-formal education in the forms of public agricultural extension and advisory services (AEASs) and Ministries of Education by means of formal agricultural education. Engaging non-governmental organizations (NGOs) and the private sector in providing education and training to farmers is essential to complement government efforts in building agricultural human capital. Private sector providers of AEASs include farm input supply firms, private consultants and advisory services, supply chain and food processing factory advisors, and producer and user associations.

OBJECTIVES

This chapter presents examples of the efforts of the Arab countries in tackling the food insecurity crisis and the role of non-formal agricultural education in the forms of AEASs in improving agricultural productivity in three Arab countries; Egypt, Saudi Arabia and Morocco. These three countries represent different approaches of investing and using public AEASs in the Arab region to increase production of major food commodities. These examples include activities and programs, challenges, and degree of effectiveness and success of these extension organizations in educating and training farmers to increase agricultural productivity. In addition, these three examples show the role and scope of activities of private firms and NGOs in providing AEASs. Other chapters in this book discuss in details how the Arab countries can improve their formal agricultural education systems.

The objectives of this chapter are:

- To present efforts of Arab countries to increase agricultural productivity and to tackle the food insecurity crisis using AEASs during the past two decades

- To identify challenges facing public AEASs in the Arab countries in providing effective educational and training services to farmers
- To provide agricultural leaders, policy makers, and planners of rural development in the Arab countries with guidelines and recommendations to reform and improve public AEASs to be able to assist in building agricultural human capital and achieve acceptable levels of food security.

Greater emphasis should be given to small-scale farmers, who constitute the majority of major food crop producers, to enable them improve their productivity and production, and to sustainably manage agricultural resources, especially water. This would require innovative approaches to reform and improve the public AEASs to be able to provide more effective programs that meet the current and future educational and training needs and interests of Arab farmers and their families in the next decades. The Arab countries should have the vision to invest, reform and revitalize public AEASs to assist in building agriculture human capital. Public AEASs can play very important role in educating and training small-scale farmers and their families and enable them improve their knowledge, skills, and capabilities, increase their productivity, sustainably manage natural resources, and, in turn, achieve acceptable levels of food security.

AGRICULTURAL HUMAN CAPITAL (AHC) AND THE AVAILABILITY OF FOOD IN THE ARAB COUNTRIES

The Arab region includes 22 countries and territories stretching from Iraq to Morocco and from Syria to Somalia (see Map 1). The agriculture sectors in the Arab countries supported a population of 453 million in 2022, with an annual population growth rate of 1.9 percent, among the highest in the world (Statista.com, 2022), over 40 percent of them live in rural areas (The World Bank. Data, 2022). Of these, about 84 million are dependent on agriculture, including fishing and livestock for their livelihoods (The World Bank. Data, 2022). The Arab region covers an area of 13.1 million square kilometers and includes a diversity of farming systems. However, arid and semiarid areas with low and variable rainfall predominate. The main rainfed crops are wheat, barley, legumes, olives, grapes, fruits and vegetables; they are grown during the winter period. In addition, the large-scale irrigated areas, representing only 2 percent of total land area, produce grains, fruits and vegetables year-round.

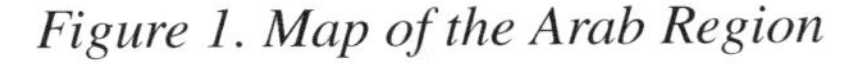

Figure 1. Map of the Arab Region

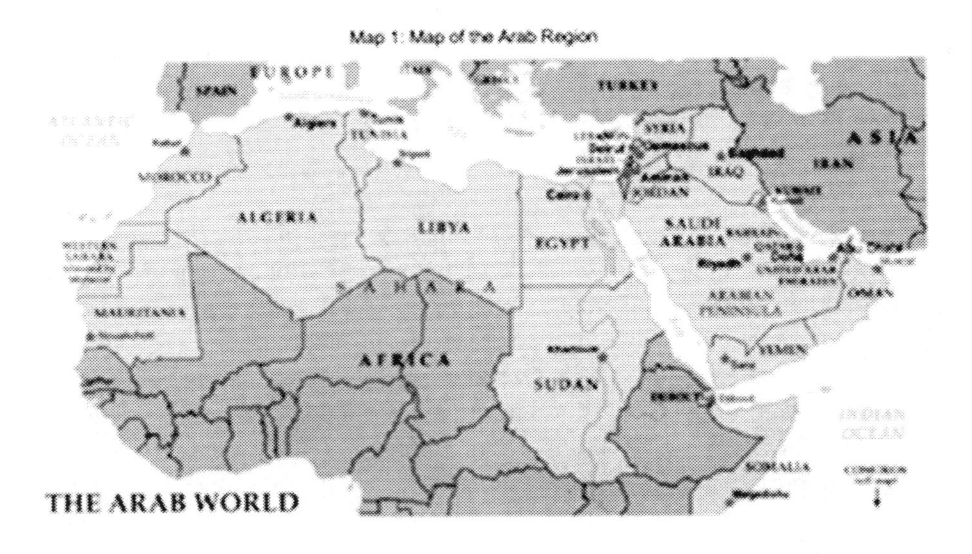

Most of the Arab countries have invested in agricultural development projects during the last two decades to increase agricultural productivity and production. For example, Egypt implemented several agricultural projects to increase the crop production areas in Toshka near Aswan in Upper Egypt, and in Northern Sinai. These projects added thousands of hectares to its wheat and other food crop production areas. Saudi Arabia has continued to make intensive efforts to develop the agricultural sector as part of its "Vision 2030 Program" (Arab News, 2022). This program was initiated in 2016 to foster sustainable agriculture practices and better management of natural resources, especially water (Kingdom of Saudi Arabia, 2022). Morocco also implemented the Green Morocco Plan (GMP) in 2008, to improve the performance of the agriculture sector, that included investments on construction of dams, expansion of irrigation and the conversion to crops better suited to the climate. In 2015, Morocco also launched the National Development Strategy for Rural and Mountainous Areas 2020-2030 to promote sustainable development and build human capital by investing in projects that support rural development efforts and food security (Amanah et al. 2021).

Despite all these agricultural development efforts and investments over the last twenty years, the agriculture sectors in the Arab countries have grown slowly and production of major food commodities, such as wheat, maize and rice has increased annually at no more of 3 percent during the last 10 years (USDA, 2022). In addition, there are great difficulties in expanding the cultivated areas or in increasing the availability of water for agriculture, which make the food insecurity crisis in the

Arab countries exceptionally severe. Furthermore, the rapid population growth, combined with increased economic growth, in some Arab countries has made the food insecurity problem more complicated because the demand for food is rising at an annual rate of approximately 5 percent in recent years (FAO, 2021-b). As a result, the discrepancy between the consumption and production of major food commodities has widen significantly since 2000 in most Arab countries, and these Arab countries continue to import increasing quantities of their consumed food commodities every year. For example, Egypt is the largest importer of wheat in the world at about 11.5 million tons in 2022, representing around 60 percent of its wheat consumption, Egypt also imported 10.2 million tons of maize in 2020-21. For the same year, Morocco imported 7.5 million tons of wheat, becoming the 11th largest importer of Wheat in the world, and imported 2.5 million tons of maize; and Saudi Arabia imported 3 million tons of wheat, and 4 million tons of maize. This inability of the Arab countries to produce enough food, and the consequent growth of food imports, puts more pressure on the already unbalanced Arab economies (USDA, FAS, 2022).

With the deteriorating situation of the economy in most Arab countries, due in large part to the outflow of capital (foreign currency) to import food, agriculture has become a central component in their efforts to tackle the food insecurity crisis. Clearly, the success of these development efforts to increase agriculture productivity and production of major food commodities, and sustainably manage natural resources in the Arab countries will be determining factors in achieving acceptable levels of food security. Several crucial factors contribute to increasing agricultural productivity and production in the Arab countries including agricultural policy, availability of land and water, utilization of appropriate and improved genetic, chemical, mechanical and management technology, and access to credit and markets; but the contribution of agricultural human capital and skilled workforce are unquestionably of great importance. Investing in farmers; what is known as 'agriculture human capital' (AHC) is essential to address the challenges facing the Arab food production systems, and to sustainably feed the growing population with food that is safe and healthy (FAO, 2021a). Investing in farmers is just as important as investing in infrastructure and other physical capital (FAO, 2022).

Agricultural human capital is the set of qualifications and capabilities that an individual possesses and can apply to successfully perform a farm task or manage an agribusiness enterprise. Human capital in agriculture also refers to both individual and group organization skills and capabilities and include education, skills, expertise, talents and other relevant knowledge, as well as self-esteem, empowerment, creativity, increased awareness and mindsets (Davis et al., 2021). "Investing in AHC is crucial to addressing challenges in agri-food systems" (Davis et al., 2021). Czyżewski et al., (2021) indicate that education and training are the main components of building

AHC; and they also add that human capital explains about a half of the economic performance of farms, while the other half relies on efficiency of farm assets.

Furthermore, Huffman & Orazem (2007) state that economic growth cannot occur in developing countries until farmer productivity is improved through education, training and utilization of improved technology. They add that building AHC is as important as developing and transferring improved agricultural technology. In addition, Djomo & Sikod (2012) show that the results obtained from their research in Cameroon indicate that an additional year of education or work experience increases agricultural productivity. They add that governments should invest more in AHC to provide solution to food insecurity problems. The case study of "Rural Empowerment and Agricultural Development Scaling-up Initiative in Indonesia" provides perspectives on how investment in AHC increases agricultural productivity. This case study shows that training and coaching using lead farmers, crop doctors and farmer field schools increased knowledge and improved practical skills in farming, including technical skills, and soft skills and empowered farmers, leading to better productivity and crop quality (Davis et al., 2021).

Building and strengthening AHC is a key to increase agricultural productivity as well as sustainably manage natural resources. Most Arab countries have a large share of their labor force in agriculture, the majority of the labor force is small-scale farmers, and economic growth cannot occur until these countries strengthen their AHC and increase the productivity of small-scale farmers. By building and strengthening AHC; the Arab countries will be better equipped to increase agricultural productivity and production, reduce food imports, sustainably manage agricultural resources, and achieve acceptable levels of food security. In other words, the future of food security in the Arab countries lies in the skills, capacities and talents of small-scale farmers, who constitute the majority of the Arab farmers.

EFFORTS OF ARAB COUNTRIES TO BUILD AGRICULTURAL HUMAN CAPITAL THROUGH AGRICULTURAL EXTENSION AND ADVISORY SERVICES

Recent developments in the Arab agricultural policies to tackle the food insecurity crisis have re-emphasized the importance of reaching the UN 2030 Sustainable Development Goals of achieving food security, as well as sustainably manage agricultural natural resources ((Ministry of Agriculture and Land Reclamation MALR (2009), The Ministry of Environment, Water and Agriculture, Saudi Arabia, (2012); and The Ministry of Agriculture, Rural Development, and Maritime Fisheries, Morocco, (2022)). Furthermore, these polices have underlined the vital role of education and training in achieving these goals, and the importance of the public-private partnership in allowing private firms and NGOs to effectively participate in building AHC by educating and training famers and their families.

The agricultural extension organizations (AEOs) in the Arab countries are the primary institution actors designated to provide education and training services to farmers and their families. These public AEOs were introduced to the Arab countries and developed based on modified versions of those in the U.S. and European countries. Furthermore, these AEOs have been supported by international donor agencies that helped finance and shape their goals and activities during the past few decades. The goals of these AEOs in the Arab countries have been usually associated with the government development goals of achieving national food security and improving rural income; and in some cases, promoting sustainable agricultural practices. Resources and activities of the AEOs have been focused on disseminating knowledge and transferring agricultural technology and production practices that would increase the productivity of major crops and livestock production.

Since the beginning of this century, the support of Arab governments for both public and private AEOs have varied considerably from country to country. On one hand, the government support for public AEOs has been greatly influenced by their past performance, as well as the impact of their activities on increasing agricultural productivity and improve quality of life in rural areas. This government support was expressed by the amount of financial resources, number of extension agents and types and scope of extension activities allowed for these public AEOs. However, in most Arab countries, these public AEOs have been unable to cope with rapidly rising demand for agricultural knowledge and technology (El-Shafie, 2009; Al-Zahrani, et al. 2017; and Al Balghiti & Mouaaid, 2010), and also influenced by the government bureaucratic culture towards small-scale farmers and rural families. On the other hand, the scope of work and size of business allowed for the private firms and NGOs have also varied among Arab countries depending on the recognized level of participation of the private sector in agricultural development, and the tolerated level of financial and political benefits.

The following are three examples of the roles and activities of the public and private AEOs in assisting governments tackle the food insecurity crisis in three Arab countries: Egypt, Saudi Arabia and Morocco during the past two decades. These three examples describe the scope of activities and extension programs, degree of success and effectiveness of public AEOs in building AHC and in increasing agricultural productivity through AEASs to farmers. In addition, these three examples present the role and scope of activities of private firms and NGOs in providing AEASs in each of these three countries. These examples illustrate that these three Arab countries have shared the same challenges of low or stagnate agricultural productivity and scarcity of water and lands resources; however, they have varied on their responses of using public and private AEOs to build AHC and tackle the food insecurity crisis. The volume of wheat production and rates of wheat yield growth/productivity are used as indicators for the performance and success of the agricultural sectors in general, and AEOs in particular in increasing agricultural production and productivity in these three Arab countries.

Example 1: Agricultural Extension and Advisory Services in Egypt

Agriculture is a key sector in the Egyptian economy, providing livelihood for approximately 55 percent of the 104 million Egyptian people in 2022, and directly employing around 25 percent of the labor force in 2021 (The World Bank. Data. 2022). Although its share of gross domestic product (GDP) has fallen from 30 percent in 1981 to about 11 percent in 2021; farming is still a vital source of exports and foreign exchange, accounting for 20 percent of export revenue (IFAD, Egypt. 2022). The area of agricultural land in Egypt is confined to the Nile Valley and Delta, with a few oases and some arable land in Sinai. The total cultivated area is 9.5 million feddans (1 feddan = 0.42 ha), representing only 4 percent of the total land area; however, the total crop area is about 11.5 million feddans (The World Bank Group, 2022). The entire crop area is irrigated, except for some rainfed areas on the Mediterranean coast. The Nile River is the main source of surface water in Egypt, and contributes 77% of agriculture's annual water supply. The efficiency of Nile irrigation system is low due to the use of flood surface irrigation. (MALR, 2009). Scarcity of water is a key constraint to agricultural development and food security in Egypt.

Since its establishment in 1953, the public AEO has stated its goal of carrying research results, in the form of technical recommendations, to farmers and helping them increase their agricultural productivity and production. Extension programs have focused on two main activities: 1) various forms of field demonstrations, including farmer field schools to disseminate knowledge of new plant varieties and management practices; and 2) distribution of extension publications. Small-scale farmers control about 80-85 percent of the total cultivated area, with the average farm size of 2.5 feddans. The majority of the small-scale farmers have very limited access to education and they use traditional farming practices that do not meet global standards (USAID/Egypt, 2022). These small-scale farmers still use flood surface irrigation techniques, resulting in high losses of very limited water resources. Major crops grown in Egypt are wheat, rice, sugarcane, beet, cotton, fodders, clover, vegetables, and fruits such as citrus and grapes (Ministry of Agriculture and Land Reclamation, Egypt. 2021).

From early 2000s, Egypt has invested in several agricultural development projects to increase the crop production areas in Toshka near Aswan, and in Northern Sinai;

but these projects were implemented without real involvement of the public AEO in educating or training farmers to improve their production, irrigation and management practices. As a result, these projects succeed in adding thousands of hectares to the wheat and other food crop production areas; but the yield growth/productivity has remained stagnate during these years. To illustrate, the wheat production has increased from 8.1 million tons in 2015-2016 to 9.8 million tons in 2022-2023 (chart 1). While the wheat yield growth/productivity remained stagnate during the same period at 6.4 tons per hectare (Chart 2) (USDA, FAS. 2022).

Figure 2. Wheat Production in Egypt from 2012-2013 to 2022-2023

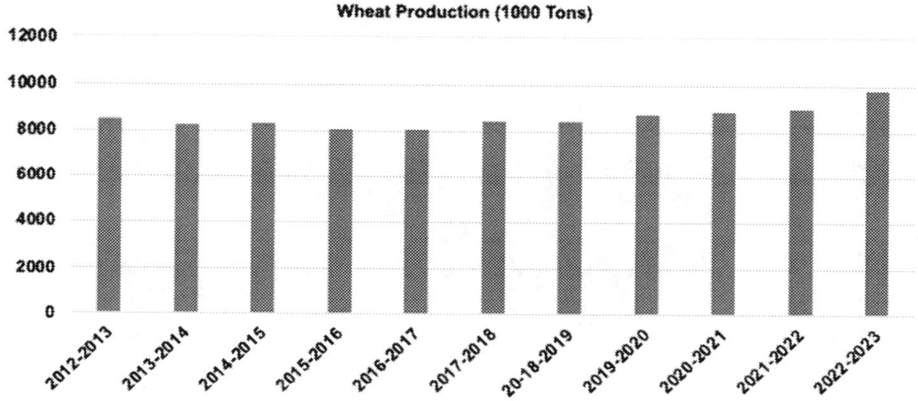

Wheat Production (1000 Tons)

Source of Data: FAS, USA, 2022

Figure 3. Wheat Yield Growth in Egypt from 2012-2013 to 2022-2023

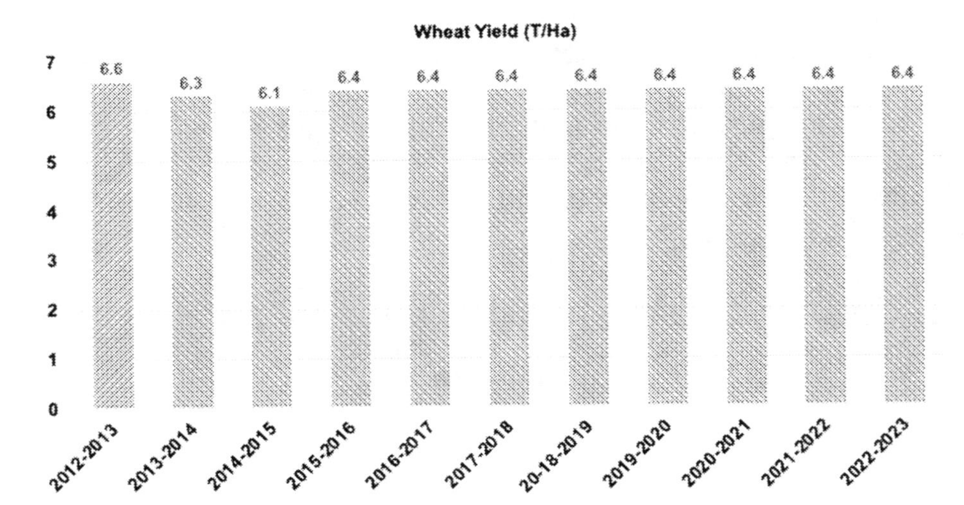

Source of Data: FAS, USA, 2022

Meantime, the impact of the government no-hiring policy and the decreasing of funds allocated to the public AEO in Egypt have resulted in severe cutback of extension activities and a dramatic reduction of the number of extension agents from 9,658 agents in 2007 to 2,503 agents in 2018. The majority of these agents have a secondary school diploma, and most of them near retirement age (Elmenofi et al., 2015). These extension agents are responsible for providing education and training services to more than 20 million farmers in 4,200 villages (CAAEE, 2018). Swanson & Davis (2014) describe the public extension in Egypt as it is in a dire need for reorganization, and farmers rated it as the least useful of all information providers. They added that most of the extension agents are poorly trained and if well-trained at some point in their careers, they are now sadly out of date and nearing retirement (Swanson & Davis, 2014). In addition, El-Shafie, (2009), Abdel-Ghany and Diab (2013) and Elmenofi et al. (2015) indicate that public AEO in Egypt suffers from several problems, including lack of resources, lack of institutional organization and coordination, ineffective and irrelevant extension activities and programs, and extension agents lack proper qualifications. It is clear that the role of public AEO in Egypt has been significantly diminished and is for the most part ineffectual (GFRAS, 2013). Small-scale farmers, who represent the majority of the producers of major food crops including, wheat, maize and rice have been have been left without any viable public extension services to help them improve their productivity or utilize new irrigation techniques.

The private AEASs in Egypt have filled the gap created by the diminishing role of public extension services since 1990s, and started to provide their services to various groups of farmers in Egypt. Today, three different types of private firms and NGOs provide their services to farmers and producers: 1) farm input supply firms, 2) private consultants and advisory services, and 3) producer and user associations.

Farm input supply firms sell fertilizers and nutrients, seeds and seedlings, agro-chemicals, feeding stuff, machines and other farm supplies. These input suppliers have expanded their networks of dealers and representatives to cover the main agricultural regions in Egypt, along the Nile valley and the Delta. They provide various groups of farmers with information and advice on how to successfully use their products. They usually recover all or part of the cost of their education and advisory services from the sales of their products. In addition, they also adapt to small-scale farmers' cash flow and allow for credit repayment at harvest. However, in some cases, they have provided small-scale farmers with lower quality inputs, or incorrect information; they also operate under minimal government supervision.

Private consultants and advisory service agencies have provided technical assistance and technology transfer services to large-scale commercial horticultural farms in the Delta and then to other regions in Egypt. These advisory agencies provide farm inputs, farm management practices, and improved technology, including new fruit and vegetable varieties. In some cases, these agencies have provided technical advisors to supervise the production and harvesting of export crops during the production season. The services of these advisory agencies are provided to commercial farms that export their horticultural products to international markets, especially European markets. Some of these advisory services are branches of international and European agencies, and multinational agricultural companies.

Producer and farmer associations and water user associations have provided technical support, market information, financial assistance and advice to their members. These associations include The Horticulture Export Improvement Association (HEIA), The Potatoes Producer Association, Water Users' Associations, and CARE-supported small-scale horticulture associations. With the exception of the Potatoes Producer Association, all these associations have been established with the support of international donors including European Union (EU), USAID/Egypt, German Development Agency (GIZ), UNDP, World Bank and International Fund for Agricultural Development (IFAD). During the past 20 years, these international development organizations have helped in establishing many small-scale farmer associations for horticulture production and water user associations, especially in Upper Egypt. In addition, the World Bank (WB) and International Fund for Agricultural Development (IFAD) have been funding producer associations to assist in improving farm infrastructure through loans and technical assistance.

These internationally supported associations generally work in horticulture, not major food crops, have access to some agricultural information, but they vary widely in the knowledge and skills required to deliver good extension work. Christiansen et al. (2011) adds that it is unclear how new information is adapted locally to various farming systems in Upper Egypt without knowledgeable public research and extension education. In addition, the serious challenge facing these internationally supported associations working in Egypt, and other Arab countries, is how to sustain their efforts once the projects are concluded and the funds are dried.

In summary, although the Egyptian government has invested in several development projects to increase the production of major food crops to be able to achieve food security, the yield growth/productivity of these major food crops have remained stagnate, and Egypt continues to import more food and feed commodities. Most of the investment in agriculture has been directed to infrastructure and other physical capital, not to build AHC. It is clear that the role of the public AEO in Egypt has been significantly diminished; and the private extension and advisory services have filled part of the gap by providing advice and training to farmers, who are able to pay for the services. However, millions of small-scale farmers, who are the major producers of food crops, have lift without any help to improve their farming and managerial skills.

Example 2: Agricultural Extension and Advisory Services in Saudi Arabia

The Kingdom of Saudi Arabia is one of the largest Arab countries, covers approximately an area of 2,15 million square kilometers (The World Bank, 2015). The climate is typically desert, with the exception of the southwest region, which features a semi-arid climate with nominal yearly rainfall. Only 1.5% of the land area is arable. Agriculture is an important sector of the Saudi economy and it provides employment to nearly 6.7 percent of the workforce in 2021 (The World Bank, 2022). Currently, agriculture contributes by 3.4% to the GDP. The agriculture sector supported population of over 36 million in 2022; with an annual population growth rate of 1.59 percent. The agriculture sector plans to provide between 19-30 percent of the total food available for consumption (The Ministry of Environment, Water and Agriculture (MEWA), 2020).

Pursuing the policy of increasing agricultural production and achieving food security in wheat during the 1990s came at a very hefty price for the Kingdom. The government realized that agriculture consumed more than 87 percent of the available fresh water, more than 50 percent came from non-renewable fresh aquifers (UNDP, 2010); as a result, the fresh water resources were depleting rapidly due to inefficient irrigation techniques and mismanagement of water (Strategic Media,

2009). In addition, the intensive use of imported chemical fertilizers and pesticides had led to adverse environmental effects, including soil erosion, and deterioration of soil fertility. Furthermore, most of large-scale agricultural projects depended on foreign workers. For all these reasons, the kingdom became more dependent on foreign inputs to achieve national food security.

In early 2000s the General Directorate of Agricultural Extension (GDAE), within the MEWA started to assist in implementing a new sustainable agricultural policy to increase the efficiency of the limited fresh water and land resources while increasing agricultural production (GDAE, 2022). This new policy resulted in a decline in volume of wheat production and other food commodities (SAMIRAD, 2005; Al-Shayaa et al., 2012; Baig & Straquadine, 2014). The new sustainable agricultural development program introduced major changes to farming practices, including better management of water and soil; and also required the adoption of sustainable agriculture technologies and management practices. In support for this policy, the GDAE implemented several programs to educate and train farmers about sustainable agricultural practices, and on how to sustainably manage the limited natural resources, including water by adopting water-saving irrigation technology. The role of the GDAE, with a total number of 250 extension agents in 2014, has become more challenging (Al-Zahrani et al., 2017). These extension agents carried out extension activities in all agricultural directorates of the Kingdom (Al-Zahrani et al.,2017), 35 percent of them held Bachelor degrees, most of them specialized in crop production and protection. Less than a fifth (19 percent) of the extension agents held higher degrees (Al-Zahrani et al., 2017).

The Saudi government has continued to make intensive efforts to develop the agricultural sector as part of its "Vision 2030 Program". This program was launched in 2016 to continue fostering sustainable agriculture practices, provide locally-produced safe food, and sustainably manage natural resources, especially water (MEWA, 2020). Today, the goal of the GDAE is to support the adoption of sustainable agricultural technology and practices and increase the efficiency of the use of the limited natural resources, while achieving acceptable level of food security. In 2018, MEWA has launched "the Sustainable Agricultural Rural Development Program 2018-2025" supporting the Kingdom's food security strategy (MEWA, 2020). The program is designed to help small-scale farmers increase their productivity while adopting sustainable agricultural practices (The Poultry Site, 2019). The extension agents have encouraged farmers to participate in alternative sustainable agricultural activities, such as greenhouse farming, implementing advanced drip irrigation practices, and producing fruits and vegetables (Modern Intelligence, 2022).

These agricultural development projects have promoted sustainable agriculture practices and provided locally-produced fruits and vegetables; but the production and yield growth/productivity for major food crops have remained stagnate during these years. For example, Chart 3 shows that wheat production has increased from 0.8 million tons in 2012-2013 to 1.2 million tons in 2022-2023; while Saudi Arabia's imports of wheat reached 3 million tons in 2021-2022. In addition, Chart 4 shows that wheat yield growth/productivity decreased from 6.5 tons per hectare in 2012-2013 to 6.1 tons per hectare in 2022-2023 (USDA, FAS. 2022).

Figure 4. Wheat Production in Saudi Arabia from 2012-2013 to 2022-2023

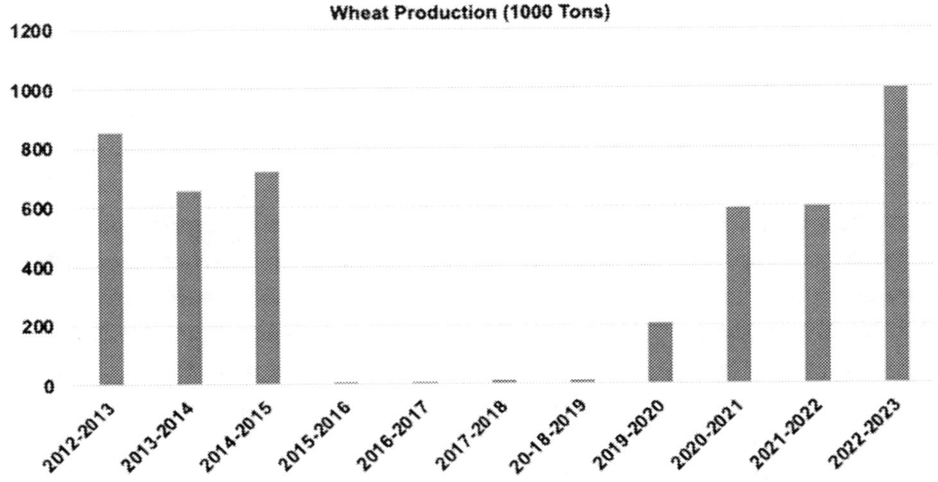

Wheat Production (1000 Tons)
1200
1000
800
600
400
200
0
2012-2013
2013-2014
2014-2015
2015-2016
2016-2017
2017-2018
20-18-2019
2019-2020
2020-2021
2021-2022
2022-2023

Source of Data: FAS, USA, 2022

Figure 5. Wheat Yield Growth in Saudi Arabia from 2012-2013 to 2022-2023

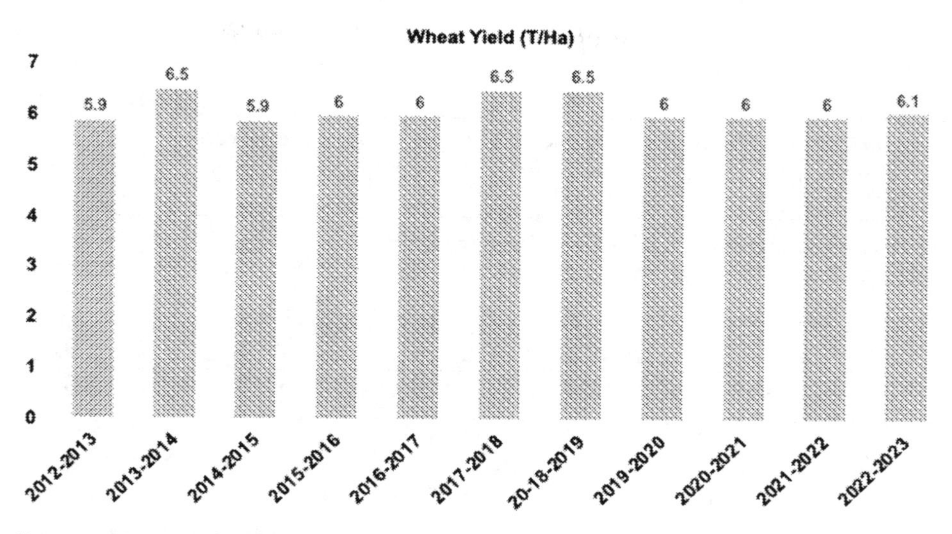

Source of Data: FAS, USA, 2022

With regard to the role of NGOs in advising and training farmers, there are very few farmer organizations that offer advisory services to its members, however, most of them are supported by the government. For example, Al-Butain Agricultural Cooperative Association in Al-Qassim region has over 200 members, it offers technical consulting services to its members, in addition to supply of inputs, maintenance of farm machinery, and marketing services (GFRAS, 2013). Moreover, there are hundreds of commercial farms and food processing facilities in Saudi Arabia; most of them employ foreign workers; managers of these farms and facilities obtain technical information and technology from national or international consultants, advisors and private firms (GFRAS, 2013).

In spite of these intensive development efforts, the impact of the Saudi public AEO in educating and training farmers and building AHC was not noticeable. There are several challenges that have affected its ability to achieve its goals of assisting farmers increase agricultural productivity and adopt sustainable agricultural practices. These challenges include inadequate number of qualified extension agents and subject matter specialists, ineffective extension contents, insufficient funds for activities, and poor linkages with agricultural research centers (Al-Zahrani et al., 2016). In addition, Muneer (2014) states that most activities and services provided to date palm producers in 7 agricultural regions by agriculture extension were "inadequate and provided low degree of benefits". Al-Zahrani et al., (2016) also point out that farmers from Al-Qassim and the Al-Kharj regions were not satisfied with the activities of the agricultural extension services regarding the information and technology they received to be able to make sound and timely decisions regarding using sustainable production practices in their farms. More than half (51.6%) of the farmers in these two regions reported that the quality of extension services they received was low. Al-Zahrani et al., conclude that these challenges would have negative impact on the quality and results of the extension activities in the Kingdom (Al-Zahrani et al., 2021).

Example 3: Agricultural Extension and Advisory Services in Morocco

Agriculture has always been of great importance to Morocco, as it accounts for 13% of GDP and 20 percent of the country's exports in 2022 (International Trade Administration, U.S., 2022). Population is expected to reach 38.2 million by 2023 (Statistics times, 2022). More than 40 percent of Moroccans live in rural areas and depend, directly or indirectly, on agriculture for their livelihood (IFAD, Morocco, 2021). The total arable land is 7.9 million hectares and the output of the agriculture sector is highly variable from year-to-year due to fluctuating annual rainfall. In addition, climate change and population growth are adding more pressure on water and land resources in Morocco (The World Bank, 2022).

The Division of Agricultural Extension and Advisory Services (DAEAS), within the Ministry of Agriculture, Rural Development, and Maritime Fisheries (MARDMF), is responsible for providing public AEASs to farmers in both rainfed small-scale farms, traditional sector and irrigated large-scale farms, modern sector. At the local level, AEASs are provided through Agricultural Extension Centers. In rainfed areas, there are 122 Extension Centers (CT) providing extension services to small-scale farmers, mainly on production and pest management practices. These small-scale farmers control 80 percent of the total agricultural land, and continue to use traditional production practices and have limited access to education and training, and limited access to farm inputs including fertilizers, pesticides, and mechanization. They produce cereal, legume, and livestock for local and domestic markets (International Trade Administration, U.S., 2022). In irrigated areas, there are 185 Development Centers (CMV) and Centers of Agricultural Development (CDA), that provide extension and advisory services to large-scale commercial and export farms which controls 16 percent of the agricultural land, and growing fruits and vegetables for export and domestic markets (GFRAS, 2013). Extension services include information on production, marketing, packaging, processing, and government incentive programs (Al Balghiti & Mouaaid, 2010; GFRAS, 2013).

The Green Morocco Plan (GMP), launched in 2008, was the Morocco's first strategic policy to improve the performance of the agricultural sector, including the agricultural extension service (Oxford Business Group, 2012). An essential element of the GMP was reforming the public AEO and connecting it with other agricultural institutions, including agricultural research centers, farmer organizations, and cooperatives. In addition, the GMP included investments on construction of dams, expansion of irrigation and the conversion to crops better suited to the climate. The GMP was designed to change Morocco's AEASs from focusing only on production and pest control practices, to a more market-oriented services that integrate all aspects of production, management and marketing, including information technology (IFPRI, 2012).

In 2010, the Education, Research and Development Directorate was established, within the MARDMF, that was an important step in developing more effective public AEASs. Today, it is called Education, Training and Research Directorate (DEFR) (Al Balghiti & Mouaaid, 2010). In 2013, the National Office for Agricultural Advisory Services (ONCA) was established (El Bilali et al., 2013). In 2015, Morocco launched the National Development Strategy for Rural and Mountainous Areas 2020-2030 to promote sustainable development and build AHC by investing in projects that support rural development efforts and food security (IFAD, Morocco. 2021).

The Government of Morocco launched its second strategic plan for agricultural development in February 2020 to deal with water shortages due to less rainfall and more extreme weather events, as well as population growth (The World Bank, 2018). The new plan, called "Generation Green," sets out an agricultural development strategy through 2030 to "strengthen sustainable agriculture by streamlining climate-smart practices" and adopting sustainable agricultural technologies; and to build AHC (The World Bank, 2018). The role of public agricultural extension is not clearly defined in implementing the activities of this strategy. In addition, starting 2022, the World Bank is funding a new irrigation project "Resilient and Sustainable Water in Agriculture". The project is design to enhance the governance of water and improve quality of Irrigation services; as well as increase access to advisory services and to modern on-farm irrigation technologies (The World Bank, 2022).

In spite of all the national and international efforts to support and reform public AEO in Morocco during the past two decades, the impact of extension activities on the production and productivity of major food crops as well as food security situation was not clear. For example, Chart 5 shows that wheat production was stagnate or decreased during the last 5 years; while Morocco imports of wheat increased to 7.5 million tons, to become the 11th largest importer of Wheat in the world in 2021-2022. Moreover, Chart 6 shows that wheat productivity/yield growth has decreased by 45 percent during the same period. The public AEO in Morocco has been facing several challenges affecting its ability to help farmers increase agricultural productivity during the past decade. These challenges include insufficient number of qualified agents, lack of funds for activities, inadequate staff training, and incentives, ineffective or non-operational linkages with research institutes and farmers organizations, and gender biased activities. (Al Balghiti and Mouaaid 2010).

Figure 6. Wheat Production in Morocco from 2012-2013 to 2022-2023

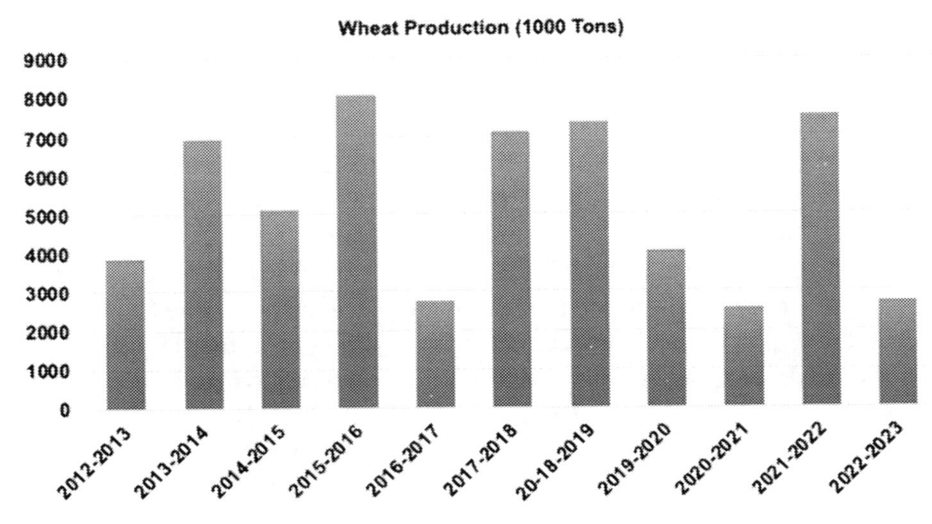

Figure 7. Wheat Yield Growth in Morocco from 2012-2013 to 2022-2023

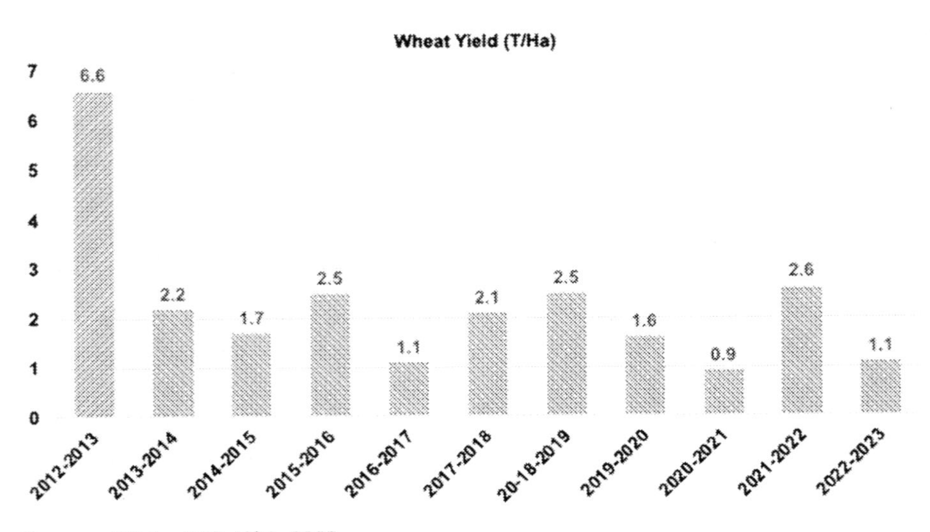

Furthermore, there has been a limited support for rural women, that was due to a gender imbalance among extension agents (Al Balghiti & Mouaaid, 2010; Rivera et al. 2005). Financial resources devoted to the public AEO remain inadequate, taking into consideration its mission and functions. Rivera et al. (2005) point out that the reform of the Moroccan agricultural extension system should have promoted demand-driven extension services, provided education and training for farmers, especially small-scale farmers; promoted gender equality activities; designed effective institutional cooperation; built more effective linkages with research institutes, and promoted the use of communication technologies (Qamar, 2005).

Besides the public AEASs, there are several agribusiness firms and producer associations in Morocco that provide information and technology to farmers. Agribusiness firms sell farm inputs and equipment and provide farmers with information and training on how to use them. These farm inputs and equipment include irrigation equipment, agricultural machinery, chemicals, fertilizers and biological products. Furthermore, there are several producer associations and cooperatives in Morocco that provide technical and marketing support to their members, especially for fruits and vegetables producers. An example of farmers' associations and agricultural cooperatives include "Fruits and Vegetables Producers and Exporters Association" (APEFEL) that provides information, training, and marketing services to its members (APEFEL, 2022).

CONCLUSION

To summarize the efforts of Egypt, Saudi Arabia and Morocco in reaching the UN 2030 Sustainable Development Goal of achieving food security during the past two decades, these three Arab countries have implemented several agricultural development projects and programs and invested in building agricultural infrastructure and physical capital, as well as in educating and training farmers to build AHC. However, they used different approaches to increase agricultural production and productivity; and they also varied on their reliance on public AEOs to educate and train farmers, especially small-scale farmers. Each one of these three countries represents different farming systems and could be used as an example for other Arab countries in their attempts to tackle the food insecurity crisis.

For example, Egypt invested in building agricultural infrastructure and physical capital, but diminished the role of public extension and allowed the private sector to provide information and advisory services to farmers who can afford to pay for these services. In addition, it allowed internationally-supported farmer associations and projects to help farmers getting organized, produce and export vegetables and fruits to international markets, but not producing wheat and other food crops for domestic markets. Meantime, Saudi Arabia invested in agricultural infrastructure and water saving and sustainable agricultural technology. Public extension programs encouraged farmers to produce more fruits and vegetables and less of other food crops. In addition, a sample of Saudi wheat producers considered extension services to be of low quality. While, Morocco, with the support of international donors, has invested in building irrigation infrastructure, and in adopting water saving irrigation technology, especially in irrigated areas. The public AEO in Morocco has also assisted farmers to produce horticultural crops for export markets; in contrast, small-scale farmers in rainfed areas, who mainly produce grains, wheat and barley, for domestic market were left with very little extension help to increase their production. The impact of the investment on AEASs in Saudi Arabia and Morocco has not been noticeable, especially on wheat production and productivity during the past decade.

The three public AEOs in Egypt, Saudi Arabia and Morocco are facing similar, but very serious challenges affecting their abilities to provide effective education and training programs to farmers and rural families. These challenges include lack of vision, insufficient funds for activities, rigid centralized extension structure, inadequate number of qualified extension agents and subject matter specialists, ineffective extension contents, inadequate staff training and incentives, ineffective or non-operational linkages with research institutes and farmers organizations.

Most of the small-scale farmers, who constitute the majority of the producers of major food crops including, wheat, maize and rice; indicated that their benefits from public extension services are minimal and extension activities are considered the least useful source of information. Private sector firms, including input suppliers, advisors and consultants, provide their services to farmers who can afford their products and services. Small-scale farmers in these three Arab countries, as in other Arab countries, are left without any viable public AEASs to help them improve their productivity of food crops or utilize new sustainable agricultural practices.

The Future Role of Agricultural Extension and Advisory Services in The Arab Countries and Food Security

As the World Economic Forum Executive in 2013 Chairman Klaus Schwab notes that "human capital is all the more valuable" in an era of global competition and multinational corporations and global supply chains; and as producers and farmers across the world are increasingly competing to increase productivity and production of food and fiber commodities (The World Economic Forum, 2013). In the Arab countries, the key for achieving acceptable levels of food security and sustainably mange agricultural resources lies in the skills, capacities and talents of the Arab farmers and producers (The World Economic Forum, 2013). Investing in building "agricultural human capital" – the potential of rural individuals – is going to be the most important long-term investment any Arab country can make for its food security and its rural people's future prosperity (The World Bank, 2018).

The Arab countries must invest in educating and training farmers and their families to develop their skills and capacities, and in turn increase their agricultural productivity. The majority of these farmers are small-scale farmers, who continue to use traditional production practices and have very limited access to education and training opportunities, limited access to farm inputs, and involving predominantly in grains and livestock production for domestic markets. The Arab governments must have the vision to act now, and reform and revitalize public AEOs to be able to build AHC to achieve acceptable levels of food security. The public AEOs are uniquely suited to be the primary institutional actors for building AHC and increasing agricultural productivity in the Arab countries. The future of Arab public AEOs crucially depends on how they are able to assist in building AHC, increasing agricultural productivity and in achieving food security goals.

The future and successful Arab agricultural extension organization will be able to:

- Develop clear well-defined vision and strategy for better governance, funding, and implementation of education and training services.
- Provide communication, educational, and training services to farmers and rural families to access information, knowledge and technology and to develop skills and capacities, and then, assist them in adopting and utilizing these knowledge, skills and technology to increase their productivity.
- Move beyond non-educational activities as well as centralized top-down transfer of knowledge and technology, towards more decentralized, demand-driven extension services
- Incorporate the present and future educational and training needs and interests of farmers and their families into extension activities and programs
- Identify skills gaps and advance agricultural workforce development by providing training on organizational, managerial and marketing skills; as neither agribusiness employers nor rural people themselves are able or willing to invest in skills development training (Hatch et al., 2018).
- Participate with the private sector in transferring improved and appropriate agricultural technology; as the private sector firms have the incentives to transfer genetic, chemical and mechanical technology, and usually recover all or part of the cost of their services from the sales of their products; while it is the responsibility of the public extension to transfer organizational, managerial, marketing and sustainable agricultural knowledge, practices, skills and technology, especially to small-scale farmers who need them the most.

- Strengthen operational linkages with national and international research centers
- Support farmer organization and producer associations to empower their members to participate in the formulation and implementation of agricultural policies and development projects.
- Advance the use of information and communication technology to improve extension program contents, programs delivery, and to reach greater number of rural people
- Ensure that there are sufficient resources, and sufficient numbers of qualified extension agents, subject matter specialists, and master trainers.
- Promote gender equality activities and programs
- Coordinate with NGOs, farmer organizations, private sector firms and international donors: funding, activities and programs to increase food production, especially grains, and to avoid duplications of efforts.
- Ensure that there is a sound monitoring and evaluation system in place able to provide feedback and assessment to agricultural leaders, policy makers, and planners of rural development.

REFERENCES

Abdel-Ghany, M., & Diab, M. (2013). Reforming agricultural extension in Egypt from the viewpoint of central level extension employees. *Arab Univ. J. Agric. Sci.*, *21*(2), 143–154. doi:10.21608/ajs.2013.14816

Al Balghiti, A., & Mouaaid, A. (2010). Country profile report on agricultural research, extension and information services: the case of Morocco. Sub-Regional Workshop on Knowledge Exchange Management System for Strengthening Rural Community Development, Cairo, Egypt.

Al-Shayaa, M., Baig, B., & Straquadine, G. (2012). Agricultural extension in the Kingdom of Saudi Arabia: Difficult present and demanding future. *Journal of Animal Plant Sci.*, *22*(1), 239–246.

Al-Zahrani, K., Aldosari, F., Baig, M., Shalaby, M., & Straquadine, G. (2016). Role of agricultural extension service in creating decision-making environment for the farmers to realize sustainable agriculture in Al-Qassim and Al-Kharj regions. *J. Anim. Plant Sci.*, *26*(4), 1063–1071.

Al-Zahrani, K., Aldosari, F., Baig, M., Shalaby, M., & Straquadine, G. (2017). Assessing the competencies and training needs of agricultural extension workers in Saudi Arabia. *Journal of Agricultural Science and Technology*, *19*, 33–46.

Al-Zahrani, K., Baig, M., Russell, M., Hasan, A., Abduaziz, H., Dabiah, M., & Al-Zahrani, K. A. (2021, May). Biological yields through agricultural extension activities and services: A case study from Al-Baha region – Kingdom of Saudi Arabia. *Saudi Journal of Biological Sciences*, *28*(5), 2789–2794. doi:10.1016/j.sjbs.2021.02.009 PMID:34012320

Amanah, S., Seprehatin, S., Iskandar, E., Eugenia, L., & Chaidirsyah. *(2021). Investing in farmers through public – private – producer partnerships – Rural Empowerment and Agricultural Development Scaling-up Initiative in Indonesia*. FAO Investment Center Country Highlights, No. 7. FAO and IFPRI.

APEFEL. (2022). *APEFEL: Moroccan Association of Producers and Exporters of Fruits and Vegetables*. https://www.agroligne.com/news-entreprises/14674-apefel-association-marocaine-des-producteurs-et-producteurs-exportateurs-des-fruits-et-legumes.html

Arab News. (2022). *Saudi Arabia takes important steps in securing farming sector*. https://www.arabnews.com/node/1597711

Asharq Al-Awsat. (2022). *IMF: 141 million people across Arab world are exposed*. https://english.aawsat.com/home/article/3910051/imf-141-million-people-across-arab-world-are-exposed-food-insecurity

Baig, M. B., & Straquadine, G. (2014). Sustainable agriculture and rural development in the Kingdom of Saudi Arabia: Implications for agricultural extension and education. In M. Behnassi (Eds.), *Vulnerability of agriculture, water and fisheries to climate change: Toward sustainable adaptation strategies*. Springer. DOI doi:10.1007/978-94-017-8962-2_7

CAAEE. (2018). *The Annual Report*. The Central Administration for Agricultural Extension and Environment. http://www.caaes.org/

Czyżewski, B., Sapa, A., & Kułyk, P. (2021). Human Capital and Eco-Contractual Governance in Small Farms in Poland: Simultaneous Confirmatory Factor Analysis with Ordinal Variables. *Agriculture, MDPI*, *11*(1), 1–16. doi:10.3390/agriculture11010046

Davis, K., Gammelgaard, J., Preissing, J., Gilbert, R., & Ngwenya, H. (2021). *Investing in farmers agriculture human capital investment strategies*. Food and Agriculture Organization of the United Nations. https://www.fao.org/family-farming/detail/en/c/ 1470656/

Djomo, N., & Sikod, F. (2012). The Effects of human capital on agricultural productivity and farmer's income in Cameroon. *Journal of International Business Research., 5*(4). Advance online publication. doi:10.5539/ibr.v5n4p149

El Bilali, H., Driouech, N., Berjan, S., Capone, R., Abouabdillah, L., Azim, K., & Najid, K. (2013). *Agricultural extension system in Morocco: problems, resources, governance and reform*. https://www.researchgate.net/publication/260890282

El-Shafie, E. (2009). *Improving agricultural extension in Egypt, a need for new institutional arrangements*. 19th European Seminar on Extension Education, Theory and practice of advisory work in a time of turbulences, Assisi, Italy.

Elmenofi, G., El Bilali, H., & Berjan, S. (2015). Contribution of extension and advisory services to agriculture development In Egypt. *Sixth International Scientific Agricultural Symposium Agrosym*. 10.7251/Agsy15051874e

Food and Agriculture Organization of the United Nations (FAO). (2021a). *Workforce development*. https://www.ruralhealthinfo.org/toolkits/sdoh/2/economic-stability/workforce-development

Food and Agriculture Organization of the United Nations (FAO). (2021b). *Regional Overview of Food Security and Nutrition; Statistics and Trends*. Food And Agriculture Organization of the United Nations.

Food and Agriculture Organization of the United Nations (FAO). (2022). *The State of Food Security and Nutrition in the World*. Food And Agriculture Organization of the United Nations.

Global Forum for Rural Advisory Services (GFRAS). (2013). *Morocco*. https://www.g-fras.org/en/world-wide-extension-study/africa/northern-africa/morocco.html#extension-providers

Hatch, C., Burkhart-Kriesel, C., & Sherin, K. (2018). Ramping up rural workforce development: An extension-centered model. *Journal of Extension, 56*(2).

Huffman, W. E., & Orazem, P. (2007). The role of agriculture and human capital in economic growth: Farmers, schooling, and health. Staff General Research Papers Archive 12003, Iowa State University, Department of Economics.

IFAD. (2021). *Country document*. https://www.ifad.org/en/web/ operations/w/country/morocco

IFPRI. (2012). *Agricultural extension and advisory services worldwide: Morocco*. International Food Policy Research Institute (IFPRI). http://www.worldwide-extension.org/africa/morocco

International Trade Administration. (2022). *Morocco-Country Commercial Guide*. https://www.trade.gov/country-commercial-guides/morocco-agricultural-sector#:~:text=Agriculture%20contributes%20almost%2013%25%20to,about%2031%25%20of%20Morocco's%20workforce

Kingdom of Saudi Arabia. (2022). *Vision 2030*. https://www.vision2030.gov.sa/v2030/ overview/

Mordor Intelligence. (2022). *Agriculture in Saudi Arabia - growth, trends, covid-19 impact, and forecasts (2023 – 2028)*. https://www.mordorintelligence.com/industry-reports/agriculture-in-the-kingdom-of-saudi-arabia-industry

Muneer, S. (2014). Agricultural extension and the continuous progressive farmers' bias and laggards blame: The Case of date palm producers in Saudi Arabia. *International Journal of Agricultural Extension*, *2*(3).

Oxford Business Group. (2012). *Economic update. Morocco: Boosting agriculture*. http://www.oxfordbusinessgroup.com/economic_updates/maroc-dynamiser-lagriculture#englishMorocco: Boosting agriculture

Qamar, K. (2005). Modernizing national agricultural extension systems: a practical guide for policy-makers of developing countries. Food and Agriculture Organization (FAO).

Reliefweb. (2023). *Regional Overview of Food Security and Nutrition; Statistics and trends*. Food And Agriculture Organization of the United Nations. https://reliefweb.int/report/algeria/near-east-and-north-africa-regional-overview-food-security-and-nutrition-2021

Rivera, W., Qamar, K., & Mwandemere, H. (2005). *Enhancing coordination among AKIS/RD actors: an analytical and comparative review of country studies on agricultural knowledge and information systems for rural development*. FAO.

Statista.com. (2022). *Population growth*. statista.com.

Statistics Times. (2022). *Demographics of Morocco*. https://statisticstimes.com/ demographics/ country /morocco-population.php

Strategic Media. (2009). *The process of Improving Farming in Saudi Arabia*. https://borgenproject.org/farming-in-saudi-arabia/

Swanson, B., & Davis, K. (2012). *Status of agricultural extension and rural advisory services worldwide: Summary Report*. https://www.g-fras.org/en/knowledge/gfras-publications.html?download=391:status-of-agricultural-extension-and-rural-advi sory-services-worldwide

The Economist. (2022). *Arab governments are worried about food security*. https://www.economist.com/middle-east-and-africa/2021/04/15/arab-governments-are-worried-about-food-security

The International Fund for Agricultural Development (IFAD). (2022). *Country documents*. https://www.ifad.org/en/web/operations/w/country/egypt

The Ministry of Agriculture, Fisheries, Rural Development, Water and Forests, Morocco. (2022). https://panorama.solutions/en/organisation/moroccan-ministry-agriculture-fisheries-rural-development-water-and-forests

The Ministry of Agriculture and Land Reclamation Egypt (MALR). (2009). A strategy for sustainable agricultural development up to 2030. Ministry of Agriculture and Land Reclamation, Agricultural Research and Development Council.

The Ministry of Agriculture and Land Reclamation Egypt (MALR). (2021). *Projects and Initiatives*. https://moa.gov.eg/en/projects-and-initiatives/

The Ministry of Environment, Water, and Agriculture (MEWA). (2012). *A Glance on Agricultural Development in the Kingdom of Saudi Arabia*. https://www.moa.gov.sa/files /Lm_eng.pdf

The Poultry Site. (2019). *King Salman unveils major Sustainable Agricultural Rural Development Program*. https://www.thepoultrysite.com/news/2019/01/king-salman-unve ils-major-sustainable-agricultural-rural-development-program me

The Saudi Arabian Market Information Resource (SAMIRAD). (2005). *Agricultural Developments in Saudi Arabia.* http:// www.saudinf.com/main/f41.htm.http: //www.saudinf.com/ main/f41.htm

The World Bank Data. (2015). https://data.worldbank.org/indicator/AG.SRF. TOTL. K2?locations=SA

The World Bank. (2018). *Growing Morocco's Agricultural Potential.* https://www.worldbank.org/en/news/feature/2016/02/18/growing -morocco-s-agricultural-potential1

The World Bank Data (2022). https://data.worldbank.org/indicator/SP. RUR.TOTL. ZS?locations=1A.

The World Bank Group. (2022). *Overview.* https://www.worldbank.org/en/country/ egypt

The World Economic Forum. (2013). *The Human Capital Report.* Prepared in collaboration with Mercer. Error! Hyperlink reference not valid.

TRT World. (2022). *Nearly one-third of the Arab world is experiencing food insecurity.* https://www.trtworld.com/magazine/nearly-one-third-of-the-ar ab-world-is-experiencing-food-insecurity-52711

United Nations. (2022), *Sustainable Development.* https://www.un.org/sustainable development/. https://www.un.org/sustainabledevelopment/development-agenda /

United Nations Development Program (UNDP). (2010). *Saudi Arabia progress towards environmental sustainability.* United Nations Development Program. http: //www.undp.org/energyandenvironment/sustainabledifference/PDFs/ ArabKingdomKingdoms/Sa udiArabia.pdf

USAID/Egypt. (2022). *Agriculture and Food Security.* https://www.usaid.gov/ egypt/ agriculture-and-food-security

USDA Foreign Agricultural Service (FAS). (2022). *Data and Analysis.* https:// www.fas.usda.gov/data

Chapter 12
The Perspectives of Higher Agricultural Education in the Gulf Cooperation Council Countries

Eihab Mohamed Fathelrahman
https://orcid.org/0000-0002-4818-7041
College of Agriculture and Veterinary Medicine, United Arab Emirates University, UAE

Paula E. Faulkner
North Carolina Agricultural and Technical State University, USA

Ghaleb A. Al Hadrami Al Breiki
United Arab Emirates University, UAE

ABSTRACT

This chapter offers a perspective on how agricultural education, research, and extension services address educational challenges and foster food security and sustainability in the Gulf Cooperation Council (GCC) countries. In the last two decades, significant progress has been made in Saudi Arabia, Oman, and the United Arab Emirates regarding agricultural education infrastructure and human capital development. However, agricultural education in the GCC countries face challenges, including stagnant enrollment, the need to offer more diverse programs, and enhanced education quality to satisfy the job market demand for qualified professionals. There is a need to formulate policies and initiatives to address the challenges facing agricultural education programs, research, and agricultural extension services. Graduates of agricultural higher education institutions must be trained and learn to be capable leaders to satisfy the needs for agricultural development, achieve food security goals and sustainability, and be industry entrepreneurs.

DOI: 10.4018/978-1-6684-4050-6.ch012

INTRODUCTION

The Gulf Cooperation Council (GCC) countries are capital-rich nations. However, the countries face challenges in combating the desertification of arable land and water scarcity. It is necessary to optimize agriculture production under such scarce resource constraints. The impacts of climate change in the GCC agriculture sector occur through the rise in temperature, the decline in rainfall, and an increase in evapotranspiration. Given these existing and predicted challenges, it is apparent that it would be challenging for the GCC countries to achieve food security locally unless there are considerable technological innovations in agriculture and food research (e.g., biotechnology and molecular biology) to boost domestic production. Without advances in agricultural technology, food imports will continue for many years ahead. Food import has various aspects of benefits and impacts on economies. Food imports allow trading with other countries and build up foreign direct agricultural investments. However, long-term food import concerns include significant financial obligations, capital flow from food-importing countries, which can affect national economies, lesser control on food quality production, and high risks of food insecurity during the war and when food import demand increases. Worldwide, almost all countries are importing food to various extents. However, the current food import by GCC countries showed an overall average range from 70 to 90% of the total food supply. Self-sufficiency has been reported to be in the range of 10 to 30% in dairy and poultry products in some of the GCC countries (Sahid &Ahmad, 2014).

This chapter provides the reader with a perspective on how agricultural education, research, and extension addresses agricultural and food security in the Gulf Cooperation Council (GCC) countries. The authors discuss the challenges of historical agricultural education development in the GCC countries. The chapter also provides an overview of the quality of agricultural higher education institutions and their various characteristics, such as degree programs offered, enrollment, policies, curricula, assessment processes, student opportunities with internships and professional training, and graduation statistics. The aim is to allow the reader to understand how higher education institutions focus on agriculture's contribution to human capital development. Human capital development in agriculture sciences is necessary so educators and students can address challenges of food security and sustainability in the GCC countries. The chapter presents current, and future growth workforce needs in the region; as such, the needs for various food and agriculture specializations are not fully known. The targeted audience of this chapter is higher education educators, policymakers, students, agricultural and food business employers and community leaders, and administrators at ministries of education. The targeted audience will also include universities, agricultural technical schools, deans of colleges of agriculture, ministries of agriculture environmental and sustainability agencies in the GCC countries, and faculty members of agricultural education programs in the Middle East and North Africa (MENA) countries.

BACKGROUND

The Gulf Cooperation Council (GCC) countries include the Kingdom of Saudi Arabia (KSA), United Arab Emirates (UAE), Sultanate of Oman, Kuwait, Qatar, and Bahrain. The GCC countries are capital-rich and have no foreign exchange limitation for food imports. The countries benefit from increased oil and gas prices to support economic development and stability. However, all GCC countries are net food importers, which adds more risk and uncertainty regarding food availability and price stability. The share of the agriculture sector to Gross Domestic Product (GDP) in the GCC countries in 2018 were 0.3, 0.5, 0.7, 2.2, 2.4, and 12.2% in Bahrain, Kuwait, UAE, Saudi Arabia, Qatar, and Oman, respectively -Table 1. In the GCC countries, the value of food imports increased from 8 billion U.S. Dollars in 2002 to 53 billion U.S. dollars in 2020. The GCC countries experience physical water scarcity, and high soil and land salinity levels, which necessitates efficient water resource management, as water extraction and use have exceeded sustainable limits. These facts indicate a need for agricultural education, research, and extension of institutions' services to address water scarcity, food security risk, and overall food and agricultural sustainability challenges in these countries (Faures et al., 2012).

Table 1. Demographic and macroeconomic characteristics of the GCC countries, agriculture sector contribution to the GDP, and food security status in 2018

Number	Country	Population in Millions	Gross Domestic Product in Billion International Dollars	Per Capita Gross Domestic Product in Thousands of International Dollars	The Agriculture Sector Contribution to the Gross Domestic Product (%)	Global Food Security Index
1	Saudi Arabia	33.7	1,604	47.6	2.2	70
2	United Arab Emirates	9.6	645	67.0	0.7	68
3	Oman	4.8	138	28.6	12.2	70
4	Kuwait	4.1	209	50.5	0.5	71
5	Qatar	2.8	253	91.0	2.4	70
6	Bahrain	1.6	73	46.2	0.3	65
	Total/ Average	56.6	2,922	55.2	3.1	69

Source: The Economist Group, 2021 and The World Bank Group, 2021.

Alston, Norton, and Pardey (1995) have indicated that growth in agricultural production is essential for improved welfare and economic development in developing and developed countries. When addressed, it helps keep food prices down, generates foreign exchange, and improves competitiveness in the world food market. Agricultural research, education, and extension are major sources of growth in production and income distribution among different income classes, geographic regions, and different types of producers and consumers. Most GCC countries experienced a growth recovery in 2018 as higher than anticipated oil prices provided additional fiscal budgets that helped rebuild the countries' account balances and foreign exchange reserves. The World Bank's Gulf Economic Monitor highlights that "GCC countries must pursue their national reform agendas with attention to institutional capacity building, extracting efficiencies in public spending on health and education, and fostering human capital formation" (Varma et al., 2019).

A study by Shahid and Ahmed (2014) discussed changes, climate change, sustainability challenges, and impacts on agriculture, food production, and consumption in the GCC countries. The authors used the Ecological Footprint (EF) approach to assess sustainability challenges. Ecological Footprint (EF) measures how much biologically productive land and water a population or activity requires to produce all the resources it consumes, measured in Global hectares (Gha). Biocapacity (BC) is the ecosystem's capacity to absorb the waste it generates. The authors found that per capita EFs in 2012 were 11.7, 9.7, 8.9, 5.7, and 4 Gha in Qatar, Kuwait, UAE, Oman, and Saudi Arabia, respectively, some of the highest worldwide. Meanwhile, per capita BCs were found to be 2.2, 2.1, 0.7, 0.6, and 0.4 Gha in Oman, Qatar, Saudi Arabia, UAE, and Kuwait, respectively, some of the lowest worldwide. Such high per capita Ecological Footprint (FT) and low Biocapacity (BC) indicate a higher role to be played by agricultural education and research in the GCC countries to address such environmental sustainability challenges.

Education for sustainable development (ESD) is a catalytic process for social change that seeks to promote the values, behaviors, and lifestyles needed to address challenges such as food security. ESD involves learning how to make decisions on long-term issues regarding the economy, the environment, and the well-being of all communities. Thus, agricultural education and training for college graduates, agribusiness professionals, and agricultural and food sector representatives are extremely important to support sustainable development (Medeiros, 2017).

CHALLENGES FACING AGRICULTURAL HIGHER EDUCATION IN THE GCC COUNTRIES

The following are the challenges agricultural education, research, and extensions services face in the GCC countries:

- Few strategies for developing education and research specify measurable targeted economic and societal agricultural development and growth goals.
- Limited means for transforming agricultural education colleges' programs to offer education and specialized skills for university graduates to become capable leaders in satisfying the needs for agricultural development, food security goals, and industry entrepreneurship endeavors.
- Nour et al. (2011) indicated that a major problem the education system in the GCC countries face is stagnant enrollment.
- Lack of support for pre-college agricultural education to increase enrollment at the higher education level.
- Lacking an emphasis on relevant courses on water resource management, land ownership, environment, advanced controlled agricultural production system (e.g., greenhouse, hydroponic, and aquaculture systems), sustainable cropping, and mixed farming (plants and animals) practices that are suitable to the arid climate of the GCC countries.
- Few qualified educators develop modern curricula highlighting contemporary subjects such as climate change, food security, agriculture sustainability, food supply chain management, and agricultural policy.
- Few agricultural education programs with curricula modifications include advanced technology-driven fieldwork opportunities and training activities.
- Few agricultural education programs' learning outcomes are being measured and assessed to address students' self-learning, problem-solving, and communication skills (e.g., the need for both English and Arabic speaking skills).
- The GCC countries share typical constraints of effective teaching and learning for reading literacy. For example, males in the GCC countries significantly underperform females in reading literacy (El-Saharty et al., 2020).
- Lack of educational programs and training in agricultural extension and community development.
- Few means for assessing post-graduate (MSc. and Ph.D.) programs in the GCC countries and evaluating progress towards the fulfillment of specialization in arid-land agricultural practices, food supply chain management, and sustainable agricultural and food systems.

- GCC countries rarely offer scholarships for nationals to study abroad at western universities such as in the U.S., Europe, and Australia. The expansion and support of such scholarship programs are highly needed to encourage graduates to return to serve as faculty members, researchers, and leaders in the field of agricultural and food sciences in the GCC countries.
- Lack of overall research funding (less than 0.5% of the universities budget) – Daghir, 2018)
- Underdeveloped agricultural extension services and community development programs.

HISTORICAL OVERVIEW OF AGRICULTURAL EDUCATION IN THE GULF COOPERATIVE COUNCIL (GCC) COUNTRIES

Agricultural education and research are vital in supporting food security and sustainability in developing and growing global economies. Worldwide, colleges of agriculture are estimated to be approximately one thousand colleges, graduating over one hundred students per year for each college. In the Arab countries (22 countries), there are more than seventy-five colleges of food and agriculture. However, in the GCC countries, there are five publicly funded colleges of food and agriculture, three colleges in Saudi Arabia, one in the United Arab Emirates, and one in the Sultanate of Oman (Table 2). The first college of food and agricultural science in this region is the King Saud University in Riyadh, Saudi Arabia, established in 1957. The latest started college is also in Saudi Arabia at Qassim University, established in 2004. There are no colleges of agriculture in Kuwait, Qatar, and Bahrain (Daghir, 2018).

Table 2. Colleges of agriculture in the Gulf Cooperation Council (GCC) countries

Country	City	University	College	Year Established
Kingdom of Saudi Arabia	Riyadh	King Saud University	College of Food and Agricultural Sciences	1957
	Dammam	King Faisal University	College of Agricultural and Food Sciences	1975
	Qassim	Qassim University	Faculty of Agriculture and Veterinary Medicine	2004
United Arab Emirates	Al Ain	United Arab Emirates University	College of Agriculture and Veterinary Medicine	1977
Sultanate of Oman	Muscat	Sultan Qaboos University	College of Agricultural and Marine Sciences	1986

Source: Daghir, N. J. (2018). Higher Agricultural Education in the Arab World: Past, present, and future. Universities in Arab Countries: An Urgent Need for Change, 209–224. https://doi.org/10.1007/978-3-319-73111-7_12

Higher institutions' agricultural education contribution to growth and development cannot be analyzed in isolation from agricultural research and extension services. Colleges of agriculture must have solid and effective research project outreach and extension services programs. Daghir (2018) indicated that most of the research and nearly all agricultural extension activities in the Arab World, including the GCC countries, are carried out by government agencies. It is crucial to design research programs and projects to address specific agricultural needs strategically and further include agricultural extension services and community development. In general, faculty members at the colleges of agriculture sciences devote 50% of their time to education and 50% to research and services. However, faculty members often complain that this is not the case for them in the Arab world, with GCC countries included, as the teaching load is often very high, leaving little room for high-quality research. This varies from one university to another (Daghir, 2018).

HIGHER AGRICULTURAL EDUCATION ENROLLMENT AND GRADUATION DURING ACADEMIC YEARS (AY) 2012/13 TO 2018/19

This section discusses trends in admission, enrollment, and graduation across the six colleges of agriculture in the GCC countries. Furthermore, this section explores the macro and micro challenges of agriculture higher education in the GCC countries. Macro challenges are relevant to capital investment in higher education, Education Development Index (EDI), and the employability of graduates. Meanwhile, micro-challenges include the need for improvement in the teaching-learning environment, students' prior knowledge and skills, curricula reform and update, and learning quality assessment. On the macro level, Shah (2017) showed that government expenditure on education for the GCC countries had been compared to many developed countries when measured as a percentage of the Gross Domestic Product (GDP). However, such government spending has not provided the expected dividends. Daghir (2018) showed that one problem facing agricultural colleges in the Arab region, including the GCC countries, is low student enrollment. The author suggested such a problem can be resolved by introducing new emerging fields of applied science such as natural resources, environmental sciences, earth sciences, and atmospheric sciences to be developed in colleges of agriculture and become part of curricula to attract students to such colleges.

Undergraduate Academic Programs Enrollment and Graduation

Table 3 and Table 4 show the most recent data for undergraduate and graduate students' admittance, enrollment, and graduation in the GCC countries' Colleges of agricultural and food sciences. The total number of newly admitted undergraduate students was 1,556 students. The enrolled number of students was reported to be 5,126, and graduated students were reported to be 523 in the GCC countries in Academic Year 2018/19. The graduation rate was reported to be 10% of total enrolled students on average, ranging from 9% in Saudi Arabia colleges to 14% in Oman.

Table 3. Gulf Cooperation Council (GCC) Countries' Undergraduate Students Admittance, Enrollment, and Graduation in College of Agricultural and Food Sciences during 2018-2019

Country	University	Newly Admitted Students	Enrolled Students	Graduated Students	Percentage of Graduated to Enrolled
Kingdom of Saudi Arabia	King Saud University King Faisal University Qassim University	1,079	3,009	264	9%
United Arab Emirates	United Arab Emirates University	203	798	93	12%
Sultanate of Oman	Sultan Qaboos University	264	1,237	173	14%
Total		**1,556**	**5,126**	**523**	10%

Source: Ministry of Education, Saudi Arabia (2021), Sultan Qaboos University Oman (2021) – Al-Hashmi, 2019. and the United Arab Emirates University (UAEU).

Trends in Enrollment and Graduation at Undergraduate Programs

Despite the effort to establish undergraduate programs in the GCC countries, the overall trends of admission and enrollment in the colleges of agriculture in the tables showed either stagnant or declining trends over the academic years 2013-14 to 2018-19 in Saudi Arabia and Oman except for an overall increasing enrollment trend in the United Arab Emirates (UAE), as the College of Food and Agriculture offers a new department of Veterinary Medicine. The name of the College changed to the College of Agriculture and Veterinary Medicine.

For Arab countries, including GCC countries, enrollment and graduation ratios in medical sciences, natural sciences, engineering, and agriculture accounted for

35% and 39% compared to 63% and 60% for art, the humanities, law, and social sciences, respectively. This imbalance is particularly significant in Saudi Arabia, Qatar, Oman, and the UAE (Nour, 2011). Overall, enrollment in the colleges of agricultural sciences in the GCC countries is relatively low compared to other disciplines and has been notably stagnant over the last two decades.

Figure 1. College of Agriculture and Veterinary Medicine, United Arab Emirates University (UAEU), United Arab Emirates

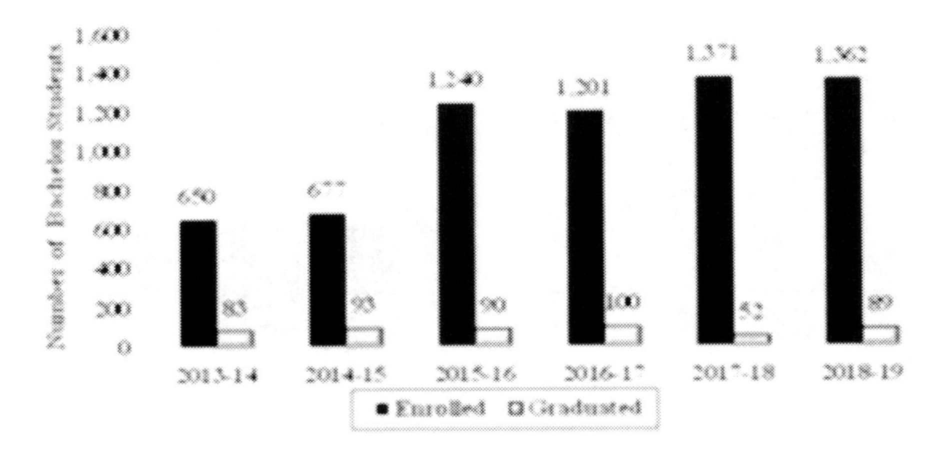

Figure 2. Enrollment and Graduation at the College of Food and Agricultural Sciences, King Saud University, Saudi Arabia - Undergraduate (Bachelor of Science B.Sc.)

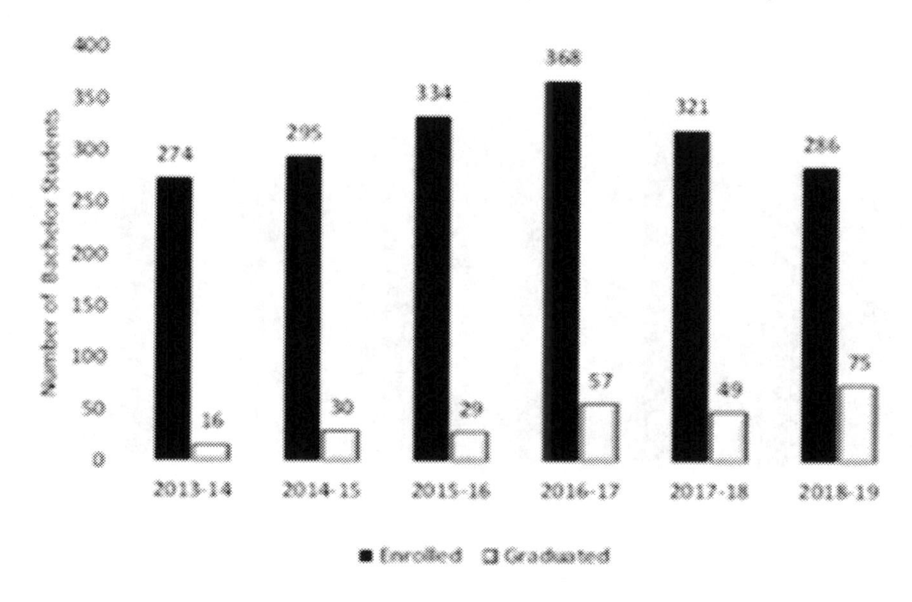

Figure 3. Enrollment and Graduation at the College of Agricultural and Food Sciences, King Faisal University, Saudi Arabia - Undergraduate (Bachelor of Science B.Sc.)

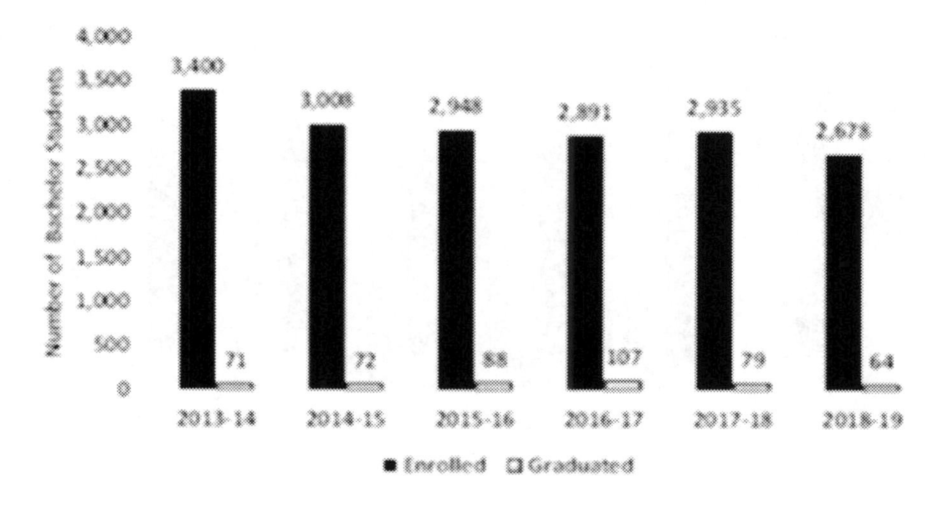

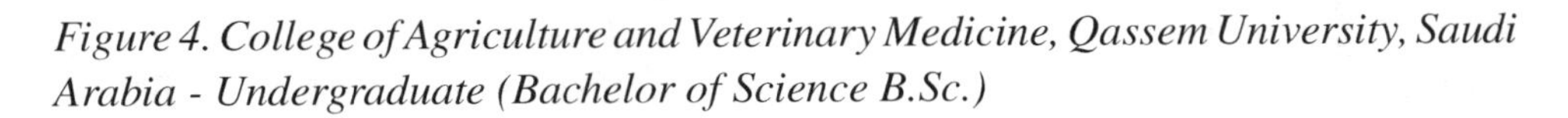

Figure 4. College of Agriculture and Veterinary Medicine, Qassem University, Saudi Arabia - Undergraduate (Bachelor of Science B.Sc.)

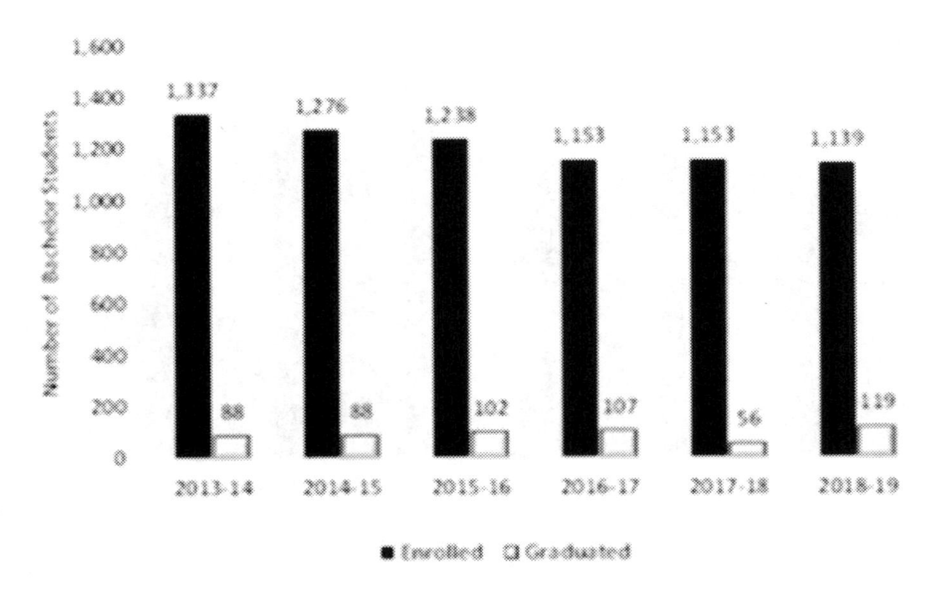

Figure 5. College of Agriculture and Marine Sciences, Sultan Qaboos University, Sultanate of Oman - Undergraduate (Bachelor of Science B.Sc.)

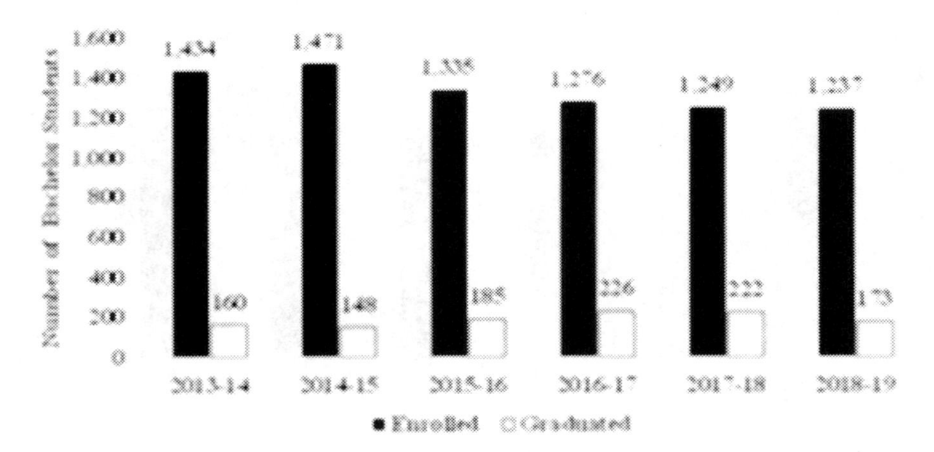

Current and Trends in Enrollment and Graduation at Graduate Programs

Graduate programs in the GCC countries are minimal and primarily include Master of Science programs without the Ph.D. program. These programs often graduate the professionals who mostly end up as government employees to support governments' institution missions such as Ministries of Agriculture and Environment. Table 4 shows that in the GCC countries, the number of graduate students admitted in 2018/19 is 113 in all five colleges of agriculture, 80 students (or 70%) are master of science students. The number of students who graduated in the Academic Year 2018/19 is 64; 43 are master of science students (or 71%). This is due to two factors; all the Ph.D. programs in the GCC colleges are relatively new programs developed within the last decade, and the GCC countries support scholarship programs to study in the United States, Europe, and Australia (Graduate students prefer studying abroad as opposed to being enrolled at the GCC countries higher education institutions).

Table 4. Gulf Cooperation Council (GCC) countries' graduate students (Master and Ph.D.) admittance, enrollment, and graduation rates in agricultural and food sciences in the academic year 2018-2019.

Country	University		Newly Admitted Students	Enrolled Students	Graduated Students
Kingdom of Saudi Arabia	· King Saud University · King Faisal University · Qassim University	Master	41	212	21
		Ph. D.	14	50	10
		Total	55	262	31
United Arab Emirates	United Arab Emirates University	Master	5	52	4
		Ph. D.	2	29	3
		Total	7	81	7
Sultanate of Oman	Sultan Qaboos University	Master	34	76	18
		Ph. D.	17	29	8
		Total	51	105	26
Total number of Ph.D. Students			33	108	21
Total number of Master of Sciences Students			80	340	43
Total number of Graduate Students			113	448	64

Source: Ministry of Education, Saudi Arabia (2021), Sultan Qaboos University, Oman (2021), United Arab Emirates University, UAEU

Figure 6. below shows, different than undergraduate enrollment, Ph. D. agricultural sciences programs have had an increasing number of graduate students in the past few years. The total number of Ph. D. candidates in the three countries, Saudi Arabia, UAE, and Oman, increased from 47 students in 2014 to 90 students in 2019 (more the double). However, all the Ph.D. or master's programs are going through curriculum changes, updates, and internal accreditations to meet high-quality education standards, as these programs were developed less than two decades ago.

Figure 6. The Number of Enrolled Graduate Students at the GCC Countries at the Colleges of Agriculture.

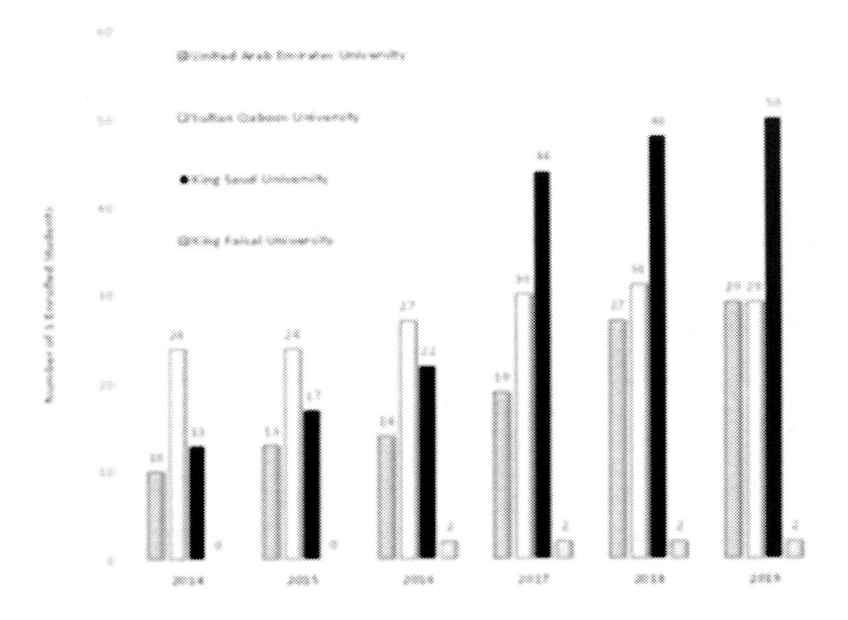

Mobility of Graduate Students in the GCC Countries

The growth of inward learning mobility seems different from the growth of foreign universities in the region. Scholarships predominantly drive the growth of outward learning mobility, and the potential for growth in inward and outward learning mobility seems likely to increase in the future. The GCC countries may enhance learning mobility through multilateral learning and research cooperation through scholarships and regulated commercial activities. To emerge as an international education hub, the GCC countries must support international scholarship and become an attractive destination for students from outside of the region. For example, Saudi Arabia ranks fourth among the 25 leading countries of origin of students who choose the United States for their tertiary education, after China, India, and South Korea. In the 2016/17 academic year, 52 611 Saudi students were studying in the U.S. The govern-funded scholarship program for students to study abroad in Saudi Arabia is seen primarily to enhance the performance of the Saudi economy, especially within the framework of Saudi Vision 2030. There is a Presidential Scholarship Program, started in 1998, in the UAE that supports undergraduate and graduate students studying abroad in the United States, United Kingdom, New Zealand, and Canada. However, there is no UAE information about the number of graduates and their specializations (Karolak & Allam, 2020).

AGRICULTURAL EDUCATION QUALITY

In the GCC countries and all other Arab countries, there is a debate about the quality of the human capital development agenda (Maroun et al., 2008). The debates question whether agricultural education improvement progress in these countries is sufficient to address future challenges, including food security and sustainability. In its most common use, education quality refers to the extent that an education system can achieve the accepted educational goals central to cognitive knowledge and skills development (Randall, 2006). For the most part, education systems are deemed to be of higher quality when students demonstrate higher levels of learning (Chapman & Miric, 2009).

A study examining the teaching-learning challenges of higher education in the GCC countries (Shah, 2017) found that despite the significant capital investments to expand educational opportunities in the GCC countries, the quality of education and learning remains a major concern. The author concluded that the quality of higher education (learning achievements) could not be improved unless there is a focus on students' needs, students' characteristics, instructors' needs, instructors' qualifications and experience development, curricula, teaching pedagogies, and the ability to offer more diverse programs and specialization – See appendix for the GCC countries colleges of agriculture ad specializations offered.

Education quality is assessed through the measurement of education outcomes. Higher education assessment includes assessment of achieving strategic and Institutional Learning Outcomes (ILOs) and Course Learning Outcomes (CLOs) following the well-known Qualification Frameworks (QFs). There are sectoral, national, and regional QFs offering coordinative opportunities in all regions of the world (Caspersen, Frølich, and Muller, 2017). GCC countries' agricultural education institutions follow the QF system approach to assess agricultural learning outcomes.

Davidson and Mackenzie (2012) showed that most of the higher education programs at the GCC universities are being designed to reflect the typical design of colleges and universities in the United States so they would qualify for accreditation by regional and professional accrediting associations in the U.S. That would facilitate transfers to US institutions and entrance to U.S. graduate programs for students pursuing graduate studies.

AGRICULTURAL RESEARCH PUBLIC-PRIVATE SECTOR PARTNERSHIPS, AGRICULTURE EXTENSION, AND COMMUNITY OUTREACH IN THE GULF COOPERATIVE COUNCIL (GCC) COUNTRIES

Daghir (2018) described the major challenges Colleges of Agriculture in Arab Countries, including the GCC countries, face. Those colleges were almost unsuccessful not only in fully contributing to food production but have also been unable to assist with the production of safe food and help with protecting the environment.

The barriers to addressing Agricultural Research, Extension, and Community Development in the GCC countries are listed below:

- Access to research policies, priority-setting systems, and research resources such as facilities and funding.
- Limited funding for agricultural research and extension services.
- Lack of linkages between education, research, extension institutions, the private agricultural and food sector, and farmers.
- Faculty members in higher education agricultural colleges with high teaching loads have limited time to research and extension projects.
- Weak linkages between GCC countries' research and education institutes and external, regional, and international research institutes.

Keskin (2014) showed that due to the sharp rise in demand for processed food commodities and the fact that GCC countries are heavily dependent on food imports which is highly elastic in food prices as food prices witnessed several prices shocks in the past posting a significant threat to food security in the region. However, Keskin (2014) suggested that:

GCC countries should bring professional managers, scientists, and researchers. Yet, an integral part of this endeavor should be in local capacity, training and producing young scientists and researchers in the food and agriculture sector.

It is also recommended to develop general strategies to improve participation in extension programs considering the specific circumstances and resources-availability in the GCC countries (Hassan and Aenis, 2016).

A study about extension agents' perception and the role of agricultural extension services in organic agriculture by Alotaibi et al. (2021) showed that future programs in agriculture extension might include and enable planners, policymakers, and related ministries to devise viable and workable policies and plans that genuinely reflect concerns and challenges of extension services that need to be improved.

HUMAN CAPITAL DEVELOPMENT IN THE FOOD AND AGRICULTURAL SECTORS IN THE GULF COOPERATIVE COUNCIL (GCC) COUNTRIES

What is "human capital"? The term refers to people's knowledge, skills, and health over their lifetimes. By investing in human capital, countries in the GCC can enable people to maximize their potential as productive members of society (El-Saharty et al., 2020). Issa and Siddiek (2012) highlighted that the destiny of any nation begins in its classroom, where young people are equipped with the knowledge and skills to lead the country. Higher education is always the cornerstone in development, where the workforce is trained to lead social, economic, political, and cultural change. In a competitive global economy, human capital is the most valuable asset, as it is the training units where the needed labor force is made. More specifically, higher education institutions contribute toward industry and investment development, deployment, and diffusion of new knowledge among the largest number of individuals at local and global levels. Higher education institutions can also contribute towards raining positive intellectual orientations of the students and society. In the past few decades, the Gulf Cooperative Council (GCC) countries heavily invested in infrastructure and human capital development in all areas of science and humanities disciplines based on the belief that higher education institutions' development is a cornerstone for economic growth. Figure 5 shows that all GCC countries are ranked among the world's very high human development groups using the Human Development Index (HDI), scoring higher than the average of all Arab States in 2019. Expected years of schooling and mean years of education showed higher indices than the Arab States' average. The United Arab Emirates is ranked 31 globally and first among the GCC countries scoring 0.890, followed by Saudi Arabia, which ranked 40 globally at 0.854 HDI (Table 5).

Table 5. Human Development Index (HDI) and its components in the Gulf Cooperation Council (GCC) countries 2019.

Country/Region	HDI Rank Within GCCC	HDI Global Rank	HDI Value	Life expectancy at birth	Expected years of schooling	Mean years of schooling
United Arab Emirates	1	31	0.890	78	14.3	12.1
Saudi Arabia	2	40	0.854	75	16.1	10.2
Bahrain	3	42	0.852	77	16.3	9.5
Qatar	4	45	0.848	80	12.0	9.7
Oman	5	60	0.813	80	14.2	9.7
Kuwait	6	64	0.806	76	14.2	7.3
Arab States	-	-	0.705	72	12.1	7.3
World Average	-	-	0.737	73	12.7	8.5

Source: United Nations Development Program UNDP, 2021, Human Development Report 2020.

Issa and Siddiek (2012) indicated that domestic and regional job markets need graduates with specific skills. Academic institutions' role is to prepare students for education and specialized scientific agricultural research, provide settings in highly specialized areas in line with economic and social life needs, and offer agricultural extension and outreach to all relevant stakeholders and farmers. Furthermore, it is crucial to prepare graduates to fulfill such needs. There is significant progress in human capital development, infrastructural development, higher institutional structure, the introduction of several new food and agricultural education programs, and curricula assessment in the GCC countries. However, challenges exist due to the lack of harmony between the institutions of higher education admission and graduation on the one hand and the need to satisfy the labor market needs on the other. Although public spending on higher education in many universities in the GCC countries continues to grow, there is a need to support public-private partnerships to expand self-financing future alternative education tailored toward fulfilling the private sector's job market needs only credentials but professional skills. There is tension between credentials and graduates' skills needed in the private sector. Gulf Cooperative Council (GCC) governments have placed little pressure on educational institutions to ensure that graduates possess skills relevant to the private sector labor market. Attaining credentials such as a diploma, degree, or certificate is emphasized more than acquiring skills and is reinforced by the solid historical preference for public sector employment. In many GCC countries, such employment is guaranteed for anyone who obtains enough credentials, resulting in a credentialism preference in which the public sector employers communicate their strong demand for credentials. In contrast, the private sector is unable to transmit signals for the relevant skills needed (El-Saharty et al., 2020).

Davidson and Mackenzie (2012) identified two interrelated policies concerning human capital development in the GCC countries. These two policies need a holistic approach that considers higher education in relation to political, economic, and sociocultural dynamics and a policy that focuses on the intersection of higher education and citizenship.

SOLUTIONS AND RECOMMENDATIONS

In the last two decades, significant progress has been made in Saudi Arabia, Oman, and United Arab Emirates in agricultural education and human capital development. The government's financial support to the colleges of agriculture leads to promising results. However, the challenges to maintaining food security and sustainability necessitate the transformation of agricultural education programs, research, and agricultural extension services. Agricultural higher education colleges in the GCC countries should transform education programs to satisfy the need for development professionals, not technicians (Daghir, 2018).

Agricultural higher education institutions in the GCC countries may offer advanced competencies in curricula for undergraduate agricultural and food sciences in fundamental subjects that are relevant to water scarcity, food security, and sustainability. Agricultural higher education institutions in the GCC countries may also offer agricultural dual-track specializations, address quality education (teaching-learning), and offer these specializations to meet job-market needs. Such dual-track education prepares graduates to meet job market demands for skilled educators, scientists, and technicians, all of which are highly needed in the agricultural field and industry.

Due to stagnant enrollment, it is imperative for the GCC colleges of agriculture to link increases in enrollment by establishing a link to pre-college agricultural education for students in the GCC countries. Pre-college education plays an essential role as a feeder source for students to be enrolled at the colleges of agriculture level. Incentives and magnet facilities should attract academic institutions in GCC countries' higher agricultural education. Governments must develop innovative strategies that encourage the private sector from the national and international market to invest in agricultural education (e.g., offering scholarships and fellowships) and research (e.g., co-funding research) with more substantial intellectual property rights for improved crops varieties and other innovation (David et al. 2017).

For sustainable progress, the GCC countries need new technology and tailor-made solutions resulting from more indigenous research and development activities in the food and agriculture sector. The government could fund support for public-private partnerships in research and extension services and national colleges of agriculture in the GCC countries. Still, it should also encourage the pursuit of the commercialization of agricultural and food technology.

At agricultural higher education institutions, faculty should be encouraged to update course material and introduce advanced pedagogy methods such as digital manuals, blended courses, online exercises, self-learning, field practices, and developing internship programs in partnership with the private sector. Agricultural colleges should transform their education system towards interdisciplinary sciences, combining biological, chemical, physical, environmental, and socio-economic sciences (Daghir, 2018).

Higher agricultural education institutions may expand agricultural university campuses in remote rural areas of the countries for more equitable distribution of knowledge, technology diffusion, and rural community development services to all relevant stakeholders. Furthermore, there is a need for more collaboration, communication, and exchange of ideas. Such cooperation should be ongoing among colleges of agriculture at regional and international universities and agencies.

FUTURE RESEARCH DIRECTIONS

Future research may consider the assessment of learning following the assessment of agricultural education learning outcomes using the indicators of the Higher-Education Learning Outcomes (HELOs) approach. Applying HELOs is a relatively new approach that attracts policymakers, quality assurance agencies, university administrators, and educators. The primary objective of the HELOs approach is to analyze how learning outcomes are understood, interpreted, and practiced in varying higher education contexts, most notably across different fields of science at higher education institutions. Another primary objective is to clarify to what extent and under what conditions learning outcomes result in changes in administrative arrangements or teaching and learning activities. The secondary objective of the HELOs approach is to use these insights to develop more sophisticated, contextualized, and valid performance indicators that convincingly reflect higher education learning outcomes (Caspersen, Frølich, and Muller, 2017). Future research on education quality assessment would adopt an approach to link education outcomes to effective teaching-learning processes, targeting enhanced education quality to address strategic goals such as food security and agricultural sustainability.

CONCLUSION

Significant progress has been made in agricultural education infrastructure in Saudi Arabia, Oman, and United Arab Emirates. These countries invested considerable capital in expanding educational opportunities in the GCC countries and human capital development in the last two decades. However, higher agricultural education faces several challenges, including a lack of clear strategies for developing education programs that specify measurable, targeted agricultural development, food security, and economic growth. As well, few agricultural education programs' learning outcomes have been measured and assessed to address students' self-learning, problem-solving, and communication skills (i.e., enhanced education quality) and stagnant admission and enrollment within the last two decades.

Agricultural research in the GCC countries lacks sufficient access to strategic research policies. There is a lack of strong linkages between education and research and extension services at government institutions, the private food sector, and among farmers. There are also weak linkages between GCC countries' research and education institutes and external regional and international research institutes.

To address agricultural education and research challenges, the agricultural higher education institutions in the GCC countries may offer more diverse programs, expand graduate programs, and enhance competencies in curricula for undergraduate and graduate agricultural and food sciences programs in fundamental subjects that are relevant to water scarcity, food security, and agricultural sustainability.

Higher agricultural education institutions may expand agricultural university campuses beyond large cities to remote rural areas of the countries for more equitable distribution of knowledge and technology diffusion and rural community development services to all relevant stakeholders.

REFERENCES

Al-Hashmi, S. Z., Al-Bahrani, A. A., Al-Harthi, S. M., Al-Bahlani, F. S., & Al-Shukaili, A. M. A. (2019). *Annual Statistics Book*. Sultan Khaboos Department of Planning and Statistics. https://www.squ.edu.om/Portals/0/DNNGalleryPro/uploads/2020/9/3/AnnualStatisticsBOOK_2019-2020_compressed.pdf

Alotaibi, B. A., Yoder, E., & Kassem, H. S. (2021). Extension agents' perceptions of the role of extension services in organic agriculture: A case study from Saudi Arabia. *Sustainability*, *13*(9), 4880. doi:10.3390u13094880

Alston, J. M., Norton, G. W., & Pardey, P. G. (1995). *Science under scarcity: Principles and practice for Agricultural Research Evaluation and priority setting*. Cornell University Press.

Caspersen, J., Frølich, N., & Muller, J. (2017). Higher education learning outcomes–Ambiguity and change in higher education. *European Journal of Education*, *52*(1), 8–19. doi:10.1111/ejed.12208

Chapman, D. W., & Miric, S. L. (2009). Education quality in the Middle East. *International Review of Education*, *55*(4), 311–344. doi:10.100711159-009-9132-5

Daghir, N. J. (2018). Higher Agricultural Education in the Arab World: Past, present, and future. Universities in Arab Countries: An Urgent Need for Change, 209–224. doi:10.1007/978-3-319-73111-7_12

David, S. A., Taleb, H., Scatolini, S. S., Al-Qallaf, A., Al-Shammari, H. S., & George, M. A. (2017). An exploration into student learning mobility in higher education among the Arabian Gulf Cooperation Council countries. *International Journal of Educational Development*, *55*, 41–48. doi:10.1016/j.ijedudev.2017.05.001

Davidson, C., & Mackenzie, P. (Eds.). (2012). *Higher Education in the Gulf States: Shaping Economies, Politi and Culture* (Vol. 7). Saqi.

El-Saharty, S., Kheyfets, I., Herbst, C., & Ajwad, M. I. (2020). *Fostering human capital in the Gulf Cooperation Council countries*. World Bank Publications. doi:10.1596/978-1-4648-1582-9

Faurès, J. M., Hoogeveen, J., Winpenny, J., Steduto, P., & Burke, J. (2012). *Coping with water scarcity: An action framework for agriculture and food security. FAO water reports, 38*. Food and Agriculture Organization of the United Nations.

Hassan, N. A., & Aenis, T. (2016). Participation in extension program planning for an improvement of smallholders' livelihoods in the MENA region. In *12th European International Farming Systems Association (IFSA) Symposium, Social and technological transformation of farming systems: Diverging and converging pathways, 12-15 July 2016, Harper Adams University, Newport, Shropshire, UK* (pp. 1-11). International Farming Systems Association (IFSA) Europe.

Hussain, S., & Al Barwani, T. (2015). *Sustainable development at universities in the Sultanate of Oman: The interesting case of Sultan Qaboos University (SQU). In Transformative Approaches to Sustainable Development at Universities*. Springer. doi:10.1007/978-3-319-08837-2_1

Issa, A. T., & Siddiek, A. G. (2012). Higher education in the Arab world & challenges of labor market. *International Journal of Business and Social Science, 3*(9).

Karolak, M., & Allam, N. (Eds.). (2020). *Gulf Cooperation Council Culture and Identities in the New Millennium: Resilience, Transformation, (Re) Creation and Diffusion*. Springer Nature. doi:10.1007/978-981-15-1529-3

Keskin, H. (2014). Agricultural research and development and food security in the Gulf Region. In *Environmental Cost and Face of Agriculture in the Gulf Cooperation Council Countries* (pp. 163–179). Springer. doi:10.1007/978-3-319-05768-2_10

Maroun, N., Samman, H., Moujaes, C. N., Abouchakra, R., & Insight, I. C. (2008). How to succeed at education reform: The case for Saudi Arabia and the broader GCC region. Ideation Center, Booz & Company.

Medeiros, C. O. (2017). Sustainability challenges and educating people involved in the agrofood sector. Sustainability Challenges in the Agrofood Sector, 660-674. doi:10.1002/9781119072737.CH28

Nour, S. S. O. M. (2011). National, regional and global perspectives of higher education and science policies in the Arab region. *Minerva, 49*(4), 387–423. doi:10.100711024-011-9183-1

Randall, J. (2006). *Strategy for the development of education in the Sultanate of Oman, Theme Paper 1*. The Strategy for Education in the Sultanate of Oman, 2020.

Romani, V. (2009). *The politics of higher education in the Middle East: Problems and prospects*. Middle East Brief, Brandeis University Crown Center for Middle East Studies.

Sabry, M. (2009). Funding policy and higher education in Arab Countries. *Journal of Comparative & International Higher Education, 1*(Fall), 11–12.

Shah, I. A. (2017). Teaching-learning challenges of higher education in the Gulf Cooperation Council countries. *Afro Asian Journal of Social Sciences, 8(1)*, 1–18.

Shahid, S. A., & Ahmed, M. (2014). Changing face of agriculture in the Gulf Cooperation Council countries. In *Environmental cost and face of agriculture in the Gulf Cooperation Council Countries* (pp. 1–25). Springer. doi:10.1007/978-3-319-05768-2_1

The Economist Group. (n.d.). *Global Food Security Index (GFSI)*. https://foodsecurityindex.com/

The Ministry of Education. (n.d.). *Premium education to build universally competitive knowledge society*. https://www.moe.gov.sa/en/Pages/default.aspx

The World Bank. (n.d.). *Agriculture & Rural Development Indicators*. https://data.worldbank.org/indicator

United Arab Emirates University. (n.d.). *The Institutional Research Unit. Statistical Yearbook 2018-2019*. https://uaeu.ac.ae/en/vc/oie/iru/statistics-yearbook-2018-19.pdf

United Nations Development Program (UNDP). (2021). *Human Development Report 2020. The Next frontier human development and the Anthropocene*. https://hdr.undp.org/en/composite/HDI

Varma, S., Hussain, S. S., Ollero, A. M., Shaukat Khan, T., Maseeh, A. N., Alhmoud, K. B. R., … Herbst, C. H. (2019). Building the foundations for economic sustainability: Human capital and growth in the GCC (No. 136435). The World Bank.

Weber, A. S. (2011). The role of education in knowledge economies in developing countries. *Procedia: Social and Behavioral Sciences*, *15*, 2589–2594. doi:10.1016/j.sbspro.2011.04.151

ADDITIONAL READING

Asanuma, S. (2015). Are current agriculture education models suitable to meet global challenges? Case study: Japan. In I. Romagosa, M. Navarro, S. Heath, & A. Lopez-Francos (Eds.), *Series A, Number 113 - Options Mediterraneenes, agricultural education in the 21st Century* (pp. 59–66). CIHEAM.

Bashshur, M. (2004). Higher education in the Arab states. UNESCO Regional Bureau for Education in the Arab States.

Daghir, N. J. (2012). *Agriculture at AUB: A century of progress*. AUB Press. Accessed from https://www.aub.edu.lb/aubpress/Pages/agricultureataub.aspx

Knauft, D. A. (2015). Ethics in the agricultural curriculum. In I. Romagosa, M. Navarro, S. Heath, & A. Lopez-Francos (Eds.), *Options Mediterraneenes, agricultural education in the 21st century. Series A, Number 113* (pp. 35–42). CIHEAM.

Meulendijks, L. (2015). Agriculture higher education in the 21st century, student views: attractiveness and employability. In I. Romagosa, M. Navarro, S. Heath, & A. Lopez-Francos (Eds.), *Options Mediterraneenes, agricultural education in the 21st century, Series A, Number 113* (p. 43). CIHEAM.

Zaglul, J. (2015). EARTH University educational model: A case study for agricultural educational models for the 21st century. *Options Méditerranéennes. Série A: Séminaires Méditerranéens*, (113), 81–85.

KEY TERMS AND DEFINITIONS

Biocapacity (BC): The calculation of a country's biocapacity begins with the total amount of bio-productive land and sea and refers to areas of land and water that support significant photosynthetic activity and biomass accumulation.

Course Learning Outcome (CLO): Course learning outcomes (CLOs) are central to your course's curriculum. They articulate to students, faculty, and other stakeholders what students will achieve in each course and how their learning will be measured.

Ecological Footprint (FT): The most straightforward way to define ecological footprint would be to call it the impact of human activities measured in the area of biologically productive land and water required to produce the goods consumed and to assimilate the wastes generated.

Food Security: Food security exists when all people, at all times, have physical and economic access to sufficient, safe, and nutritious food that meets their dietary needs and food preferences for an active and healthy life.

Institutional Learning Outcome (ILO): Institutional learning outcomes are the knowledge, skills, abilities, and attitudes that students are expected to develop due to their overall experiences with any aspect of the college, including courses, programs, and student services.

Learning Outcome Assessment: Learning outcome assessment is the process of systematically collecting evidence that indicates the extent to which students learning has occurred and matches predefined expectations.

Compilation of References

Abdel, G. A. (1997). *Comparative Education "Curriculum and its Application"*. Dar al-Thakr al-Arabi.

Abdel-Ghany, M., & Diab, M. (2013). Reforming agricultural extension in Egypt from the viewpoint of central level extension employees. *Arab Univ. J. Agric. Sci.*, *21*(2), 143–154. doi:10.21608/ajs.2013.14816

Abdel-Hay, A. A. I. (2008). *Some requirements for applying the system of community colleges in Egypt in the light of the experiences of some Arab and foreign countries*, [Unpublished master's thesis, College of Education, Mansoura University].

Abu-Hola, I. R., & Tareef, A. B. (2009). Teaching for sustainable development in higher education institutions: University of Jordan as a case study. *College Student Journal*, *43*(4), 1287–1306.

Abu-Ismail, K. (2018). *Extreme poverty in Arab states: a growing cause for concern.* The Forum ERF Policy Portal. https://theforum.erf.org.eg/2018/10/16/extreme-poverty-arab-states-growing-cause-concern/

Action plan for herbal medicines 2010–2011 . (2010). European Medicines Agency. Retrieved from https://www.ema.europa.eu/en/documents/other/action-plan-herbal-medicines-2010-2011_en.pdf

Adam, A. H. M. (2019). The Future Perspectives of Agricultural Graduates and Sustainable Agriculture in Sudan. *Journal of Agronomy Research*, *1*(4), 36–43. doi:10.14302/issn.2639-3166.jar-19-2732

Agirreazkuenaga, L. (2019). Embedding sustainable development goals in education. Teachers' perspective about education for sustainability in the Basque Autonomous Community. *Sustainability*, *11*(5), 1496. doi:10.3390u11051496

Ahlam, R. A. (1983) *Agricultural Education in Egypt in Light of Development Demands and Trends, Calendar Study,* [Doctoral Thesis, Faculty of Fayoum Education, Cairo University].

Ahmad, Z. H. A. (1993). *Technical Education and Development Requirements in Saudi Society "Applied Study on Samples of Students, Graduates and Educators of the Taif Industrial Institute"* [Master's Thesis, Institute of Arab Studies and Research].

Akça, F. (2019). Sustainable development in teacher education in terms of being solution oriented and self-efficacy. *Sustainability, 11*(23), 6878. doi:10.3390u11236878

Akiba, M., & Guodong, L. (2016). Effects of teacher professional learning activities on student achievement growth. *The Journal of Educational Research, 109*(1), 99–110. doi:10.1080/00220671.2014.924470

Al Balghiti, A., & Mouaaid, A. (2010). Country profile report on agricultural research, extension and information services: the case of Morocco. Sub-Regional Workshop on Knowledge Exchange Management System for Strengthening Rural Community Development, Cairo, Egypt.

Al Jazirah Newspaper. (2008). *G Issue,* 12981.

Al-Absi, M. (2017). *The First Conference, The Future of Engineering and Technical Education.* MDPI.

Alani, T. A., & Saadallah, G. (1986). Vocational Education in the Arab World. Arab Educational, Cultural and Scientific Organization, Tunisia.

Alaoui, H., & Springborg, R. (2021). *Education in the Arab World: A Legacy of Coming Up Short.* The Wilson Center. https://www.wilsoncenter.org/article/education-arab-world-legacy-coming-short

Al-Aqeel, A. (2005). Education Policy and System in the Kingdom of Saudi Arabia. Al-Rushd Library, Riyadh.

Al-Bahnasawy, F. A. (2006). *The Higher Education System in the United States of America,* 228. World of Books.

Alexandria University. (2018). *Alexandria University receives a delegation from the American Ocean Society.* International Relations at Alexandria University. https://bit.ly/3J7EDH3

Alexandria University. (2018). *Meeting with a delegation from Alexandria University with the President of the American Ocean County University to establish a community college.* Alexandria University. https://bit.ly/34EJNLv

Al-Ghamdi, A. A. (2018). Problems facing students of transitional programs in the College of Applied Studies and Community Service at King Saud University. *International Journal of Educational and Psychological Studies, 4* (3), 416. www.refaad.com

Al-Hashmi, S. Z., Al-Bahrani, A. A., Al-Harthi, S. M., Al-Bahlani, F. S., & Al-Shukaili, A. M. A. (2019). *Annual Statistics Book.* Sultan Khaboos Department of Planning and Statistics. https://www.squ.edu.om/Portals/0/DNNGalleryPro/uploads/2020/9/3/AnnualStatisticsBOOK_2019-2020_compressed.pdf

Ali, F. (1996). Some problems of vocational preparatory education in Egypt and ways to overcome them in the light of Japanese experience. *Journal of Contemporary Education, Egypt*, 41.

Aliyah, B. M. O. (2007). *Problems Faced in Agricultural Education in Areas of the Palestinian Authority from the Perspective of Agricultural School Teachers and Remedies,* [Master's Thesis, Islamic University, Gaza].

Al-Mahdy, Y. F. H., & Sywelem, M. M. G. (2016, October). Teachers' perspectives on professional learning communities in some Arab countries. *International Journal of Research Studies in Education, 5*(4), 45–57. doi:10.5861/ijrse.2016.1349

Almannie, M. (2015). Proposed Model for Innovation of Community colleges to Meet Labor Market Needs in Saudi Arabia. *Journal of Education and Practice, 6*(20).

Al-Maqtari, Y. A. (2000). The needs of faculty members: Ibb University training on the use of educational aids and their attitudes towards them. *The Seventh Scientific Conference Learning Technology System in Schools and Universities: Reality 0/17-1000 and the Aspiration.* The Egyptian Association for Educational Technology.

Alotaibi, B. A., Yoder, E., & Kassem, H. S. (2021). Extension agents' perceptions of the role of extension services in organic agriculture: A case study from Saudi Arabia. *Sustainability, 13*(9), 4880. doi:10.3390u13094880

Al-Shayaa, M., Baig, B., & Straquadine, G. (2012). Agricultural extension in the Kingdom of Saudi Arabia: Difficult present and demanding future. *Journal of Animal Plant Sci., 22*(1), 239–246.

Alsolaimi, K. S. (1996). *Quality Effectiveness of Technical Colleges as Perceived by Students, unpublished master theses, college of education.* Umulgra University.

Alston, J. M., Norton, G. W., & Pardey, P. G. (1995). *Science under scarcity: Principles and practice for Agricultural Research Evaluation and priority setting.* Cornell University Press.

Al-Zahrani, K., Aldosari, F., Baig, M., Shalaby, M., & Straquadine, G. (2016). Role of agricultural extension service in creating decision-making environment for the farmers to realize sustainable agriculture in Al-Qassim and Al-Kharj regions. *J. Anim. Plant Sci., 26*(4), 1063–1071.

Al-Zahrani, K., Aldosari, F., Baig, M., Shalaby, M., & Straquadine, G. (2017). Assessing the competencies and training needs of agricultural extension workers in Saudi Arabia. *Journal of Agricultural Science and Technology, 19*, 33–46.

Al-Zahrani, K., Baig, M., Russell, M., Hasan, A., Abduaziz, H., Dabiah, M., & Al-Zahrani, K. A. (2021, May). Biological yields through agricultural extension activities and services: A case study from Al-Baha region – Kingdom of Saudi Arabia. *Saudi Journal of Biological Sciences, 28*(5), 2789–2794. doi:10.1016/j.sjbs.2021.02.009 PMID:34012320

Al-Zaidi, A. (1982). *Adequacies of Curriculum and Training in Agriculture Provided at Three Saudi Institutions as Assessed by Administrators, Instructors, Senior Students, and Regional Directors.* [Unpublished doctoral dissertation. Oklahoma State University, Stillwater, Oklahoma, United States].

Amadi, E. C. (2008). *Introduction to Educational Administration: A Module.* Harey Publications.

Amadi, N. S., & Solomon, U. E. (2020). *Issue 3 Ser* (Vol. 10). Assessment of Quality Instruction Indicators in Vocational Agricultural Education in South –South Universities Nigeria. Journal of Research & Method in Education (IOSR-JRME). May - June.

Amanah, S., Seprehatin, S., Iskandar, E., Eugenia, L., & Chaidirsyah. *(2021). Investing in farmers through public – private – producer partnerships – Rural Empowerment and Agricultural Development Scaling-up Initiative in Indonesia.* FAO Investment Center Country Highlights, No. 7. FAO and IFPRI.

Ambusaidi, A., & Al Washahi, M. (2016). Prospective teachers' perceptions about the concept of sustainable development and related issues in Oman. *Journal of Education for Sustainable Development, 10*(1), 3–19. doi:10.1177/0973408215625528

Ambusaidi, A., & Al-Rabaani, A. (2009). Environmental education in the Sultanate of Oman: Taking sustainable development into account. In *Environmental education in context* (pp. 37–50). Brill.

Annan-Diab, F., & Molinari, C. (2017). Interdisciplinarity: Practical approach to advancing education for sustainability and for the Sustainable Development Goals. *International Journal of Management Education, 15*(2), 73–83. doi:10.1016/j.ijme.2017.03.006

Anne, M.H. (2020). List of medicines made from plants. *Science, Tech, Math.*

AOAD. (n.d.). *Arab Organization and Development.* https://www.aoad.org/Eabout.htm

APEFEL. (2022). *APEFEL: Moroccan Association of Producers and Exporters of Fruits and Vegetables.* https://www.agroligne.com/news-entreprises/14674-apefel-association-marocaine-des-producteurs-et-producteurs-exportateurs-des-fruits-et-legumes.html

Aquastat, F. (2016). Country Profile-Egypt. Rome, Italy: Food and Agriculture Organization of the United Nations (FAO).

Arab Encyclopedia. (2023). *Community college – Category, education, and psychology, 16.* Arab Encyclopedia. arab-ency.com.sy

Arab News. (2022). *Saudi Arabia takes important steps in securing farming sector.* https://www.arabnews.com/node/1597711

Armstrong, M. (2002). *Employee reward.* CIPD Publishing.

Asharq Al-Awsat. (2022). *IMF: 141 million people across Arab world are exposed.* https://english.aawsat.com/home/article/3910051/imf-141-million-people-across-arab-world-are-exposed-food-insecurity

Austin, M. J., & Rust, D. Z. (2015). Developing an Experiential Learning Program: Milestones and Challenges. *International Journal on Teaching and Learning in Higher Education, 27*(1), 143–153.

Babou, A. I., Jiyed, O., El Batri, B., Maskour, L., El Mostapha Aouine, A. A., & Zaki, M. (2020). Education for sustainable development and teaching biodiversity in the classroom of the sciences of the Moroccan school system: A case study based on the ministry's grades and school curricula from primary to secondary school and qualifying. *Brock Journal of Education, 8*(2), 13–21. doi:10.37745/bje/vol8.no2.pp13-21.2020

Baig, M. B., & Straquadine, G. (2014). Sustainable agriculture and rural development in the Kingdom of Saudi Arabia: Implications for agricultural extension and education. In M. Behnassi (Eds.), *Vulnerability of agriculture, water and fisheries to climate change: Toward sustainable adaptation strategies*. Springer. DOI doi:10.1007/978-94-017-8962-2_7

Bandura, A. (1997). *Self-efficacy: The exercise of control*. Freeman.

Bannerman, R. H. (1982). Traditional medicine in modern health care. *World Health Forum, 3*, 8–13.

Barrick, R. K., Ladewig, H. W., & Hedges, L. E. (1983). Development of a systematic approach to identifying technical inservice needs of teachers. *The Journal of the American Association of Teacher Educators in Agriculture, 24*(1), 13–19. doi:10.5030/jaatea.1983.01013

Barrick, R. K., Samy, M. M., Gunderson, M. A., & Thoron, A. C. (2009). A model for developing a well-prepared agricultural workforce in an international setting. *Journal of International Agricultural and Extension Education, 16*(3), 25–32. doi:10.5191/jiaee.2009.16303

Barrick, R. K., Samy, M. M., Roberts, T. G., Thoron, A. C., & Easterly, R. G. (2011). Assessment of Egyptian agricultural technical school instructor's ability to implement experiential learning activities. *Journal of Agricultural Education, 52*(3), 6–15. doi:10.5032/jae.2011.03006

Batty, M. (2022). The conundrum of 'form follows function.' *Environment and Planning B: Urban Analytics and City Science, 49*(7), 1815-1819. Doi:10.1177/23998083221120313

Bautista-Baños, S., Romanazzi, G., & Jiménez-Aparicio, A. (Eds.). (2016). *Chitosan in the preservation of agricultural commodities*. Academic Press.

Beijaard, D., Meijer, P. C., & Verloop, N. (2004). Reconsidering research on teachers' professional identity. *Teaching and Teacher Education, 20*(2), 107–128. doi:10.1016/j.tate.2003.07.001

Ben Haman, O. (2020). The Moroccan education system, dilemma of language and think-tanks: the challenges of social development for the North African country. *The Journal of North African Studies, 26*(4), 709-732.

Berl, D. A. (2001). *Field study of some of the problems of school administration at the Agricultural High School in the Arab Republic of Egypt,* [Master's thesis, Bakfar el-Sheikh University of Tanta].

Bertrand, S., Roberts, A.S., & Walker, E. (2022). *Cover Crops for Climate Change Adaptation and Mitigation.* Agriculture and Climate Series, Environmental and Energy Study Institute (EESI).

Bloom, B. S. (1956). *Taxonomy of Education Objectives: The Classification of Education Goals.* Edwards Bros.

Board Source. (2021). *Advisory councils nine keys to success.* Author.

Boeve-de Pauw, J., Olsson, D., Berglund, T., & Gericke, N. (2022). Teachers' ESD self-efficacy and practices: A longitudinal study on the impact of teacher professional development. *Environmental Education Research, 28*(6), 867–885. doi:10.1080/13504622.2022.2042206

Boston Consulting Group. (2021). *Addressing the Middle East's growing skills mismatch.* Author.

Bryan, A., & Bracken, M. (2011). *Learning to read the world? Teaching and Learning about Global Citizenship and International Development in Post-primary Schools.* Irish Aid.

Bubshait, G. I. (1997). *Establishing Community Colleges for Girls in the Kingdom of Saudi Arabia: Justifications, Objectives and Suggested Programs,* [Unpublished Ph.D. thesis, College of Education, Umm Al-Qura University, Makkah Al-Mukarramah].

Buong, A. (2020). Achieving SDGs through higher educational institutions: A case study of the University of Bahrain. In *Sustainable Development and Social Responsibility—Volume 2* (pp. 19–23). Springer. doi:10.1007/978-3-030-32902-0_3

Burnette, B. (2002). How we formed our community. *National Staff Development Council, 23*(1), 51–54.

CAAEE. (2018). *The Annual Report.* The Central Administration for Agricultural Extension and Environment. http://www.caaes.org/

Cabrera, R. I., & Solis-Perez, A. R. (2017). *Mineral nutrition and fertilization management.* Academic Press.

Caena, F. (2011). Literature review Quality in Teachers' continuing professional development. *European Commission, 2,* 20.

Cameron, R. W. F., & Emmett, M. R. (2003). Production Systems and Agronomy| Nursery Stock and Houseplant Production. Encyclopedia of Applied Plant Sciences, 949-956.

Campbell, J., Martin, R., & Diab, A. (2008). *Impact Assessment Report: The AERI Linkage Project in Retrospect and Prospect.* MUCIA.

Canrinus, E. T., Helms-Lorenz, M., Beijaard, D., Buitink, J., & Hofman, A. (2012). Self-efficacy, Job Satisfaction, Motivation and Commitment: Exploring the Relationships between Indicators of Teachers' Professional Identity. *European Journal of Psychology of Education, 27*(1), 115–132. doi:10.100710212-011-0069-2

Cantor, J. A. (1995). ASHEERIC Higher Education Report: (Vol. 7): *Experiential Learning in Higher Education. Washington, D.C.*

Carnevale, A. P., Jayasundera, T., & Hanson, A. R. (2012). *Five ways that pay.* Georgetown Public Policy Institute.

Caspersen, J., Frølich, N., & Muller, J. (2017). Higher education learning outcomes–Ambiguity and change in higher education. *European Journal of Education, 52*(1), 8–19. doi:10.1111/ejed.12208

Center for Innovative Teaching and Learning, Northern Illinois University. (2022). *Experiential Learning.* https://www.niu.edu/citl/resources/guides/instructional-guide/experiential-learning.shtml

Center for Teaching Innovation. Cornell University. (2022). *Active Learning.* https://teaching.cornell.edu/teaching-resources/active-collaborative-learning/active-learning

Central Agency for Public Mobilization and Statistics (CAPMAS). (2022). *Egypt in figures.* Retrieved from https://www.capmas.gov.eg/Pages/Publications.aspx?page_id=5104&YearID=23602

Chaaban, J., Chalak, A., Ismail, T., & Khedr, S. (2018). *Agriculture, Water and Rural Development in Egypt: A Bottom-Up Approach in Evaluating European Trade and Assistance Policies.* Academic Press.

Chaanin, A. (2003). Breedingl Selection Strategies for Cut Roses. Encyclopedia of Rose Science, 33-41.

Chalekian, P. (2013, December). POSDCORB: Core patterns of administration. In *Proceedings of the 20th Conference on Pattern Languages of Programs. The Hillside Group* (Vol. 17). Academic Press.

Chapman, D. W., & Miric, S. L. (2009). Education quality in the Middle East. *International Review of Education, 55*(4), 311–344. doi:10.100711159-009-9132-5

Chickering, A. W., & Gamson, Z. F. (1987, March). *Seven principles for good practice in undergraduate education.* Washington, DC: AAHE Bulletin.

Chuang, W. H. (2002). An Innovation Teacher Training Approach: Combine live instruction with a web-based Reflection system. *British Journal of Educational Technology, 33*(2), 229–232. doi:10.1111/1467-8535.00256

Chuma, L. L. (2020). The Role of Information Systems in Business Firms Competitiveness: Integrated Review Paper from Business Perspective. *International Research Journal of Nature Science and Technology, 2*(4). www.scienceresearchjournals.org

Connecticut's Official State Website. (2022). *A Guide to Curriculum Development: Purposes, Practices,* Procedures. https://portal.ct.gov/-/media/SDE/Health-Education/curguide_generic.pdf

Conway, P. (2001). Anticipatory reflection while learning to teach: From a temporally truncated to a temporally distributed model of reflection in teacher education. *Teaching and Teacher Education, 17*(1), 89–106. doi:10.1016/S0742-051X(00)00040-8

Craft, A. (2000). *Continuing professional development: A practical guide for teachers and schools.* London: Routledge/Falmer.

Creemers, B., Kyriakides, L., & Antoniou, P. (2013). *Teacher Professional Development for Improving Quality of Teaching*. Springer. doi:10.1007/978-94-007-5207-8

Czyżewski, B., Sapa, A., & Kułyk, P. (2021). Human Capital and Eco-Contractual Governance in Small Farms in Poland: Simultaneous Confirmatory Factor Analysis with Ordinal Variables. *Agriculture, MDPI, 11*(1), 1–16. doi:10.3390/agriculture11010046

Daghir, N. (2018). Higher Agricultural Education in the Arab World: Past, Present, and Future. In A. Badran, A. Baydoun, & R. Hillman (Eds.), *Universities in Arab Countries: An Urgent Need for Change* (pp. 209–224). American University of Beirut. doi:10.1007/978-3-319-73111-7_12

Dale, E. (1969). *Audiovisual methods in teaching* (3rd ed.). Dryden Press.

Darling-Hammond, L. (2010). Teacher education and the American future. *Journal of Teacher Education, 61*(1/2), 35–47. doi:10.1177/0022487109348024

David, S. A., Taleb, H., Scatolini, S. S., Al-Qallaf, A., Al-Shammari, H. S., & George, M. A. (2017). An exploration into student learning mobility in higher education among the Arabian Gulf Cooperation Council countries. *International Journal of Educational Development, 55*, 41–48. doi:10.1016/j.ijedudev.2017.05.001

Davidson, C., & Mackenzie, P. (Eds.). (2012). *Higher Education in the Gulf States: Shaping Economies, Politi and Culture* (Vol. 7). Saqi.

Davis, K., Gammelgaard, J., Preissing, J., Gilbert, R., & Ngwenya, H. (2021). *Investing in farmers agriculture human capital investment strategies*. Food and Agriculture Organization of the United Nations. https://www.fao.org/family-farming/detail/en/c/ 1470656/

Deanship of Community Service and Continuing Education. (2000). *The annual report.* King Saud University in Riyadh, Saudi Arabia.

Dembicki, M. (2006). Community colleges fill local economic needs. *Community college Times*. http://www.communitycollegetimes.com/article.cfm?ArticleId=7 5

Demir, N. K. (2020). The Need of Adult Education and Training Administration in Lifelong Learning. *Mediterranean Journal of Social & Behavioral Research, 4*(3), 41–45. doi:10.30935/ mjosbr/9600

Desimone, L. M. (2009). Improving impact studies of teachers' professional development: Toward better conceptualizations and measures. *Educational Researcher, 38*(3), 181–199. doi:10.3102/0013189X08331140

DiBenedetto, C. A., Willis, V. C., & Barrick, R. K. (2018). Needs assessments for school-based agricultural education teachers: A review of literature. *Journal of Agricultural Education, 59*(4), 52–71. doi:10.5032/jae.2018.04052

Dinesen, I. G., & van Zaayen, A. (1996). Potential of Pathogen Detection Technology for Management of Diseases in glasshouse Ornamental Crops. *Advances in Botanical Research*, *23*, 137–170. doi:10.1016/S0065-2296(08)60105-6

Djomo, N., & Sikod, F. (2012). The Effects of human capital on agricultural productivity and farmer's income in Cameroon. *Journal of International Business Research.*, *5*(4). Advance online publication. doi:10.5539/ibr.v5n4p149

Dougherty, K. J. & Townsend, B. K. (2006): Community College Missions: A Theoretical and Historical Perspective. *New Directions For Community Colleges*, (136), p. 8.

Ebner, P., Ghimire, R., Joshi, N., & Saleh, W. D. (2020). Employability of Egyptian agriculture university graduates: Skills gaps. *Journal of International Agricultural and Extension Education*, *27*(4), 128–143. doi:10.5191//jiaee.2020.274128

Economic discussion. (2022). *Placement: Meaning, Definition, Importance, Principles, Benefits, Problems*. Economic Discussion. https://www.economicsdiscussion.net/human-resource-managemen t/placement/placement/ 32361.

Ed100. (2022). *Measures of Success*. Ed100. https://ed100.org/lessons/measures

Egypt's Human Development Report. (2021). *Development, a right for all: Egypt's pathways and prospects*. Retrieved from https://www.undp.org/egypt/publications/egypt-human-developm ent-report-2021

El Ashmawi, A. (2015). *The skills mismatch in the Arab world: A critical view. British Council Cairo Symposium.*

El Bilali, H., Driouech, N., Berjan, S., Capone, R., Abouabdillah, L., Azim, K., & Najid, K. (2013). *Agricultural extension system in Morocco: problems, resources, governance and reform*. https://www.researchgate.net/publication/260890282

Elhami, M.G. (2000). Developing vocational education in Egypt in light of contemporary global trends. *Educational Sciences*, (1).

El-Hanafy, H. S. (2007). Alternative additives to vase solution that can prolong vase life of carnation (Dianthus caryophyllus) flowers. *J. Product Dev*, *12*(1), 263–276. doi:10.21608/jpd.2007.44957

Elmenofi, G., El Bilali, H., & Berjan, S. (2014). *Contribution of extension and advisory services to agriculture development In Egypt. Sixth International Scientific Agricultural Symposium Agrosym*, Egypt. doi:10.7251/Agsy15051874e

El-Saharty, S., Kheyfets, I., Herbst, C., & Ajwad, M. I. (2020). *Fostering human capital in the Gulf Cooperation Council countries*. World Bank Publications. doi:10.1596/978-1-4648-1582-9

El-Shafie, E. (2009). *Improving agricultural extension in Egypt, a need for new institutional arrangements*. 19th European Seminar on Extension Education, Theory and practice of advisory work in a time of turbulences, Assisi, Italy.

El-Shafie, E. M., Azam, A., & Ibrahim, R. H. (2022). *Participatory Extension and Advisory Services (PEAS) as an extension approach to achieve food security in Egypt: Lessons learned from 5 Governorates.* Retrieved from https://esciencepress,net/Journals/IJAE

Epstein, J. L. (2008, February). *Improving family and community involvement in secondary schools.* Condensed from *Principal Leadership, 2007.* National Association of Secondary School Principals.

Erikson, E. H. (1979). *Dimensions of a new identity.* WW Norton & Company.

FAO. (2012). *Greening the economy with agriculture.* FAO.

Fasih, T. (2008). *Linking education policy to labor market outcomes.* The World Bank. doi:10.1596/978-0-8213-7509-9

Faurès, J. M., Hoogeveen, J., Winpenny, J., Steduto, P., & Burke, J. (2012). *Coping with water scarcity: An action framework for agriculture and food security. FAO water reports, 38.* Food and Agriculture Organization of the United Nations.

Fifi, N. (2013). *External Efficiency of the Community Colleges in Saudi Arabia,* [Unpublished PhD thesis, Faculty of Social Sciences, the Islamic University of Imam Muhammad bin Saud].

Food and Agriculture Organization of the United Nations (FAO). (2018). Agricultural Development Economics [Rome, Italy.]. *Policy Brief*, 7.

Food and Agriculture Organization of the United Nations (FAO). (2021a). *Workforce development.* https://www.ruralhealthinfo.org/toolkits/sdoh/2/economic-stability/workforce-development

Food and Agriculture Organization of the United Nations (FAO). (2021b). *Regional Overview of Food Security and Nutrition; Statistics and Trends.* Food And Agriculture Organization of the United Nations.

Food and Agriculture Organization of the United Nations (FAO). (2022). *The State of Food Security and Nutrition in the World.* Food And Agriculture Organization of the United Nations.

Frala, J. (2022). *How important is an advisory committee to your vocational program success?* Academic Senate for California Community Colleges.

Freeman, S., Eddy, S. L., McDonough, M., Smith, M. K., Okoroafor, N., Jordt, H., & Wenderoth, M. P. (2014). Active learning increases student performance in science, engineering, and mathematics. [NATL ACAD SCIENCES. United States.]. *Proceedings of the National Academy of Sciences of the United States of America*, *111*(23), 8410–8415. doi:10.1073/pnas.1319030111 PMID:24821756

Garmston, R. J. (2007). Results-oriented agendas transform meetings into valuable collaborative events. *The Learning Professional*, *28*(2), 55.

Garton, B. L., Miller, G., & Torres, R. M. (1992). Enhancing student learning through teacher behaviors. *The Agricultural Education Magazine*, *65*(3), 10–11.

Gehem, M., van Duijme, F., Ilko, I., Mukena, J., & Castelion, N. (2015). *Maooing agribusiness opportunities in the MENA: Exploring favorable conditions and challenges for agribusiness in the Middle East and North Africa*. The Hague Centre for Strategic Studies.

General Authority of Statistics. (2018). *Bureau of Labor Market Statistics, Third Quarter, 2018*. Saudi Arabia.

Gerry, J., & Quirke-Bolt, N. (2019). Teachers' professional identities and development education, *Policy & Practice: A Development. Education Review*, *29*, 110–120.

Gibb, N. (2016). knowledge-rich curriculum to the Social Market Foundation [Paper presentation]. The Association of School and College Leaders (ASCL) Event 'Taking ownership of your curriculum: a national summit' in 2016. London, UK.

Gitsaki, C., Donaghue, H., & Wang, P. (2012). *Using a community of practice for teacher professional development in the United Arab Emirates. In The international handbook of cultures of professional development for teachers: Comparative international issues in collaboration, reflection, management and policy*. Analytrics.

Global Forum for Rural Advisory Services (GFRAS). (2013). *Morocco*. https://www.g-fras.org/en/world-wide-extension-study/africa/northern-africa/morocco.html#extension-providers

Grabe, M., & Grabe, C. (2004). *Integrating technology for meaningful learning* (4th ed.). Houghton Mifflin.

Greene, M. G. (2022). *Cover Crops Play a Starring role in Climate Change mitigation*. U.S. Department of Agriculture. Retrieved from https://www.farmers.gov/blog/cover-crops-play-starring-role-in-climate-change-mitigation#:~:text=Cover%20crops%20offer%2
0agricultural%20producers,resilient%20to%20a%20changing%20cl
imate

Grund, J., & Brock, A. (2020). Education for sustainable development in Germany: Not just desired but also effective for transformative action. *Sustainability*, *12*(7), 2838. doi:10.3390u12072838

Gudrun, L., & Gerhard, B. (2011). A Review on Recent Research Results (2008-2010) on Essential Oils as antimicrobials and Antifungals. A review. *Flavour and Fragrance Journal*, *27*, 13–39.

Guskey, T. R., & Yoon, K. S. (2009). What works in professional development? *Phi Delta Kappan*, *90*(7), 495–500. doi:10.1177/003172170909000709

Habib, A. S. (2014). Technical Education in Egypt: Problems and Solutions, Administration (Consortium of Administrative Development Associations). Egypt, 51(1).

Hajloo, N. (2014). Relationships between self-efficacy, self-esteem and procrastination in undergraduate psychology students. *Iranian Journal of Psychiatry and Behavioral Sciences, 8*(3), 42–49. PMID:25780374

Hamad, T. A. (1984). *Graduates of Libraries and Documentation of Community Colleges, 1*(19), 21-26. Risalah-Jordan Library.

Hammoud, J., & Tarabay, M. (2019). Higher Education for Sustainability in the Developing World: A Case Study of Rafik Hariri University1 in Lebanon. *European Journal of Sustainable Development, 8*(2), 379–379. doi:10.14207/ejsd.2019.v8n2p379

Hanafi, M. M. M. (2010). The role of American community colleges in meeting the requirements of the labor market and how to benefit from them in Egypt. *Journal of the College of Education in Port Said,* (7), 235 - 274 http://web.ebscohost.com/ehost/pdf?vid=5&hid=102&sid=52bdfe16-6a3a4799-922b-b43e71782327%40sessionmgr107

Hanafi, M. T. (1996). *A calendar study of Egypt's agricultural secondary school in light of contemporary trends.* Fourth Annual Conference (Future of Education in the Arab World between Regional and Global), Helwan University. h://portal.moe.gov.eg

Harby, M. K. (1965). Technical education in the Arab States. UNESCO.

Hartl, M. (2009). Technical and vocational education and training (TVET) and skills development for poverty reduction – Do rural women benefit? Presented at the *FAO-IFAD-ILO Workshop on Gaps, Trends and Current Research in Gender Dimensions of Agricultural and Rural Employment: Differentiated Pathways Out of Poverty*, Rome, Italy.

Hassan, N. A., & Aenis, T. (2016). Participation in extension program planning for an improvement of smallholders' livelihoods in the MENA region. In *12th European International Farming Systems Association (IFSA) Symposium, Social and technological transformation of farming systems: Diverging and converging pathways, 12-15 July 2016, Harper Adams University, Newport, Shropshire, UK* (pp. 1-11). International Farming Systems Association (IFSA) Europe.

Hassanin, E. H., & Abdelaziz, R. Z. (1994). *Study on the reality of agricultural secondary education and ways of developing it in the Arab Republic of Egypt.* General Directorate of Agricultural Education.

Hatch, C., Burkhart-Kriesel, C., & Sherin, K. (2018). Ramping up rural workforce development: An extension-centered model. *Journal of Extension, 56*(2).

Huffman, W. E., & Orazem, P. (2007). The role of agriculture and human capital in economic growth: Farmers, schooling, and health. Staff General Research Papers Archive 12003, Iowa State University, Department of Economics.

Hussain, S., & Al Barwani, T. (2015). *Sustainable development at universities in the Sultanate of Oman: The interesting case of Sultan Qaboos University (SQU). In Transformative Approaches to Sustainable Development at Universities.* Springer. doi:10.1007/978-3-319-08837-2_1

IFAD. (2021). *Country document.* https://www.ifad.org/en/web/ operations/w/country/morocco

IFPRI. (2012). *Agricultural extension and advisory services worldwide: Morocco*. International Food Policy Research Institute (IFPRI). http://www.worldwide-extension.org/africa/morocco

Iiso.org. (2015). The International Organization for Standardization.

ILO & FAO. (2020). *Skills development for inclusive growth in the Lebanese agriculture sector – Policy Brief*. FAO.

International Trade Administration. (2022). *Morocco-Country Commercial Guide*. https://www.trade.gov/country-commercial-guides/morocco-agricultural-sector#:~:text=Agriculture%20contributes%20almost%2013%25%20to,about%2031%25%20of%20Morocco's%20workforce

Introduction to Technology Park. (2019). *Vision 2030*. Kingdom of Saudi Arabia. https://www.vision2030.gov.sa/media/rc0b5oy1/saudi_vision203.pdf

Irby, D. M. (2018). *Improving Environments for Learning in the Health Professions. Proceedings of a conference on Improving Environments for Learning in the Health Professions*. Atlanta, Georgia. United States.

Ismail, Y. M. (1991). *The contribution of the second cycle of education mainly in preparing pupils for technical secondary education*, [Master's thesis, Faculty of Education, Ain Shams University, Egypt].

Issa, A. T., & Siddiek, A. G. (2012). Higher education in the Arab world & challenges of labor market. *International Journal of Business and Social Science*, *3*(9).

Johnson, E. C., & Khalidi, R. (2005). Communities of practice for development in the Middle East and North. *Africa KM4D Journal, 1*(1), 96–110.

Karolak, M., & Allam, N. (Eds.). (2020). *Gulf Cooperation Council Culture and Identities in the New Millennium: Resilience, Transformation, (Re) Creation and Diffusion*. Springer Nature. doi:10.1007/978-981-15-1529-3

Kelchtermans, G. (2006). Teacher collaboration and collegiality as workplace conditions. A review. *Zeitschrift fur Padagogik*, *52*(2), 220–237.

Kelchtermans, G. (2017). 'Should I stay or should I go?': Unpacking teacher attrition/retention as an educational issue. *Teachers and Teaching*, *23*(8), 961–977. doi:10.1080/13540602.2017.1379793

Kerby, A. (1991). *Narrative and the self*. Indiana University Press.

Keskin, H. (2014). Agricultural research and development and food security in the Gulf Region. In *Environmental Cost and Face of Agriculture in the Gulf Cooperation Council Countries* (pp. 163–179). Springer. doi:10.1007/978-3-319-05768-2_10

King, E. (2011, Jan. 28). Education is fundamental to Development and Growth. *Education for Global Development.*

Kingdom of Saudi Arabia. (2022). *Vision 2030.* https://www.vision2030.gov.sa/v2030/ overview/

Kolb, D. A. (1984). *Experiential learning: Experience as the source of learning and development* (Vol. 1). Prentice-Hall. United States.

Krueger, D., & Mundt, J. (1991). Change: Agricultural education in the 21st century. *The Agricultural Education Magazine.*, *64*(1), 7–9.

Kwasi, S., Jakkie C., & Yeboua, K. (2022). *Race to Sustainability? Egypt's Challenges and Opportunities to 2050.* Academic Press.

Lambert, J., Srivastava, J., & Vietmayer, N. (1997). *Technical Paper no. 355. Medicinal plants – rescuing a global heritage.* Washington, DC: World Bank.

LeBaron, H., McFarland, J., & Burnside, O. (2008). The Triazine Herbicides. 50 years Revolutionizing Agriculture. Academic Press.

Leicht, A., Heiss, J., & Byun, W. J. (2018). *Issues and trends in education for sustainable development* (Vol. 5). UNESCO Publishing.

Mahmoud M, D., & SP, B. (2022). Science and Quality Education for Sustainability Development in Libya. *INTI Journal, 2022*(18).

Mahmoud, M. E. K. K. (1999). *The Linguistic Concept for keeping up with civilization.* Conference of Cairo University, Egypt, for Development of University Education. Vision to the Future University. Republished in the site: www.alhiwartoday.net

Mahmoud, M. E. K. K. (2005). The Language of a Nation is Their Tool for science. Conference of the Society of the Tongue of Arabs, held in The League of Arabic States, Cairo, Egypt.

Mahmoud, M. E. K. K. (2022). *A Call towards Amendment of Scientific Research and Education in the Modern Republic.* A Letter for state stakeholders in Egypt.

Mahmoud, M. E.K.K. (2016). *The Way to Get out of The Crisis of Deterioration of Education in Egypt, with a Guarantee of The Occurrence of a Distinct Scientific Renaissance within a Few Years.* A letter for state stakeholders in Egypt.

Mahmoud, M. E. K. K. (1998). *Arabization and Westernization between what is Common and what is Required (translated from Arabic).* Published in the Policy Newspaper.

Maiga, W. H. E., Porgo, M., Zahonogo, P., Amengnalo, C. J., Coulibaly, D. A., Flynn, J., Seogo, W., Traore, S., Kelly, J. A., & Chimwaza, G. (2020, October). A systematic review of employment outcomes from youth skills training programmes in agriculture in low- and middle-income countries. *Nature Food, 1*(1), 605–619. doi:10.103843016-020-00172-x

Malaysian Qualifications Register. (2008). *List of Qualification – Community Colleges.* Wayback Machine.

Mansour, H. M. M. (2012). *Developing Technical Education in Egypt in the Light of Malaysian Experience,* [Master's Thesis, University of Tanta]

Maroun, N., Samman, H., Moujaes, C. N., Abouchakra, R., & Insight, I. C. (2008). How to succeed at education reform: The case for Saudi Arabia and the broader GCC region. Ideation Center, Booz & Company.

Mbowane, C. K., De Villiers, J. J., & Braun, M. W. (2017). Teacher participation in science fairs as professional development in South Africa. *South African Journal of Science, 113*(7-8), 1–7. doi:10.17159ajs.2017/20160364

McHenry County College. (n.d.). *Career and technical advisory committee manual.* Author.

McKinley, B. G., Birkenholz, R. J., & Stewart, B. R. (1993). Characteristics and experiences related to the leadership skills of agriculture students in college. *Journal of Agricultural Education, 34*(3), 76–83. doi:10.5032/jae.1993.03076

McKinsey & Co. (2011). *Linking jobs and education in the Arab world.* Author.

Medeiros, C. O. (2017). Sustainability challenges and educating people involved in the agrofood sector. Sustainability Challenges in the Agrofood Sector, 660-674. doi:10.1002/9781119072737.CH28

Meyers, C., & Jones, T. B. (1993). *Promoting Active Learning. Strategies for the College Classroom.* Jossey-Bass Inc., Publishers.

Miller, M. T., & Kissinger, D. B. (2007, Spring). Connecting Rural Community Colleges to Their Communities. *New Directions for Community Colleges, 2007*(137), 27–34. doi:10.1002/cc.267

Ministerial Resolution No. 228. (2012). *Concerning the Establishment of an Integrated Educational Complex in Fayoum Governorate.*

Ministerial Resolution No. 483. (2014). *Admission Regulations for Young Education Three and Five Years.*

Ministry of Education - General Directorate of Agricultural Education - Report on Agricultural Education. (1890 -1970)

Ministry of Education Report. (2016). *Acceptance of students with technical certificates, the three-year system.* Ministry of Education.

Ministry of Education. (1963). Agricultural Education. *Ministry's Press.*

Ministry of Education. (1965). Agricultural Secondary School Book. *Ministry Press.*

Ministry of Education: Act No. 233 of 1988 amending certain provisions of Act No. 139 of 1981.

Ministry of Higher Education and Scientific Research. (2019). National Strategy for Science, *Technology, and Innovation.* Arab Republic of Egypt. Retrieved from http://mohesr.gov.eg/en-us/Documents/sr_strategy.pdf

Ministry of Higher Education. (2003). *The comprehensive national report on higher education in the Kingdom of Saudi Arabia.*

Miqdad, M. (2004). Educational Preparation for the University Professor, Seminar on Faculty Development in Higher Education Institutions, Challenges. King Saud University.

Mordor Intelligence. (2022). *Agriculture in Saudi Arabia - growth, trends, covid-19 impact, and forecasts (2023 – 2028).* https://www.mordorintelligence.com/industry-reports/agriculture-in-the-kingdom-of-saudi-arabia-industry

Morriso, G. R. (2002). *Integration Computer technology into the classroom.* Person Education.

Morse, L. (2017-2020). *UCCDs develop students' and graduates' job-seeking skills to strengthen lifelong career prospects.* USAID. Retrieved from https://2017-2020.usaid.gov/egypt/higher-education/university-centers-career-development

Moshashai, D., & Bazoobandi, S. (2020). The Complexities of Education Reform in Saudi Arabia. *The Gulf Monitor.* https://castlereagh.net/the-complexities-of-education-reform-in-saudi-arabia/

Mostafa, Y. (2021). Educational Reform Movement in Egypt towards 2030 Vision: Learning from History to Incorporate New Education. *Journal of School Improvement and Leadership, 3*, 115–125.

MUCIA. (2006). *Request for Second Amendment to the AERI Linkage Project.* Unpublished Manuscript.

MUCIA. (2008). *Final Technical Report: AERI Institutional Linkage Project.* Unpublished Manuscript.

MUCIA. (2012). *The Second Quarter Report, Year 4:* The Value Chain Project. Unpublished Manuscript.

Mulà, I., & Tilbury, D. (2011). National journeys towards education for sustainable development: reviewing national experiences from Chile, Indonesia, Kenya, the Netherlands, Oman. In United Nations Decade of Education for Sustainable Development 2005-2014. UNESCO.

Muneer, S. (2014). Agricultural extension and the continuous progressive farmers' bias and laggards blame: The Case of date palm producers in Saudi Arabia. *International Journal of Agricultural Extension, 2*(3).

National Association of Colleges and Employers (NACE). (2011, July). *Positions Statement on United States Internships.* Retrieved from https://www.naceweb.org/about-us/advocacy/position-statements/position-statement-us-internships/

National Association of Colleges and Employers (NACE). (2022). *15 Best Practices for Internship Programs.* Retrieved from https://www.naceweb.org/talent-acquisition/internships/15-best-practices-for-internship-programs/

National Education Association. (2022). *The Teacher Professional Growth.* NEA. https://www.nea.org/professional-excellence/professional-learning/teachers

Norton, P., & Sprague, D. (2002). Timber Lane technology tales: A design experiment in alternative field experiences for preservice candidates. *Journal of Computing in Teacher Education*, 40–60.

Nour, S. S. O. M. (2011). National, regional and global perspectives of higher education and science policies in the Arab region. *Minerva*, *49*(4), 387–423. doi:10.100711024-011-9183-1

O'Brien, A. (2012). *The importance of community involvement in schools*. Edutopia.

Ohio Department of Education. (n.d.). *Developing a local advisory committee*. Author.

Othayman, M. B., Mulyata, J., Meshari, A., & Debrah, Y. (2022). The challenges confronting the training needs assessment in Saudi Arabian higher education. *International Journal of Engineering Business Management*, *14*, 1–13. doi:10.1177/18479790211049706

Oxford Business Group. (2012). *Economic update. Morocco: Boosting agriculture.* http://www.oxfordbusinessgroup.com/economic_updates/maroc-dynamiser-lagriculture#englishMorocco: Boosting agriculture

Oxford Business Group. (2022). *How Egypt is boosting food production and exports.* Retrieved from https://oxfordbusinessgroup.com/overview/seeds-growth-crop-exports-rise-research-and-training-programmes-seek-support-next-generation-farmers

Popova, A., Evans, D. K., Breeding, M. E., & Arancibia, V. (2022). Teacher professional development around the world: The gap between evidence and practice. *The World Bank Research Observer*, *37*(1), 107–136. doi:10.1093/wbro/lkab006

Posti-Ahokas, H., Idriss, K., Hassan, M., & Isotalo, S. (2022). Collaborative professional practice for strengthening teacher educator identities in Eritrea. *Journal of Education for Teaching*, *48*(3), 300–315. doi:10.1080/02607476.2021.1994838

Power School. (2022). *The Three Most Common Types of Teacher Professional Development and How to Make Them Better*. Power School. https://www.powerschool.com/blog/the-three-most-common-types-of-teacher-professional-development-and-how-to-make-them-better

Promethean. (2022). *12 active learning strategies in the classroom.* Promethean World. https://www.prometheanworld.com/gb/resource-centre/blogs/12-active-learning-strategies-in-the-classroom/)

Puckett, J., Hoteit, L., Perapechka, S., Loshkareva, E., & Bikkulova, G. (2022, January). *Fixing the global skills mismatch.* Boston Consulting Group.

Puckett, J., Hoteit, L., Perapechka, S., Loshkareva, E., & Bikkulova, G. (2022, January). *Fixing the global skills mismatch.* Boston, MA: Boston Consulting Group.

Purdue University. (2022). *Active Learning Strategies*. Purdue University. Error! Hyperlink reference not valid.

Qamar, K. (2005). Modernizing national agricultural extension systems: a practical guide for policy-makers of developing countries. Food and Agriculture Organization (FAO).

Qi, W., Sorokina, N., & Liu, Y. (2021). The Construction of Teacher Identity in Education for Sustainable Development: The Case of Chinese ESP Teachers. *International Journal of Higher Education, 10*(2), 284–298. doi:10.5430/ijhe.v10n2p284

Rabah N. (2017). Introducing the profession of social work. *Social Science Magazine Forum – Palestine*. Wayback Machine website.

Randall, J. (2006). *Strategy for the development of education in the Sultanate of Oman, Theme Paper 1*. The Strategy for Education in the Sultanate of Oman, 2020.

Rasmussen, C., Pardello, R. M., Vreyens, J. R., Chazdon, S., Teng, S., & Liepold, M. (2017). Building social capital and leadership skills for sustainable farmer associations in Morocco. *Journal of International Agricultural and Extension Education, 24*(2), 35–49. doi:10.5191/jiaee.2017.24203

Rates, S. M. K. (2001). Plants as source of drugs. *Toxicon, 39*(5), 603–613. doi:10.1016/S0041-0101(00)00154-9 PMID:11072038

Reliefweb. (2023). *Regional Overview of Food Security and Nutrition; Statistics and trends*. Food And Agriculture Organization of the United Nations. https://reliefweb.int/report/algeria/near-east-and-north-africa-regional-overview-food-security-and-nutrition-2021

Richer, R. A. (2014). Sustainable development in Qatar: Challenges and opportunities. *QScience Connect, 2014*(1), 22. doi:10.5339/connect.2014.22

Rivera, W., Qamar, K., & Mwandemere, H. (2005). *Enhancing coordination among AKIS/RD actors: an analytical and comparative review of country studies on agricultural knowledge and information systems for rural development*. FAO.

Roberts, T. G., Thoron, A. C., Barrick, R. K., & Samy, M. M. (2008). Lessons learned from conducting workshops with university agricultural faculty and secondary school agricultural teachers in Egypt. *Journal of International Agricultural and Extension Education, 15*(1), 85–87. doi:10.5191/jiaee.2008.15108

Rogers, A., & Taylor, P. (1998). *Participatory Curriculum Development in Agricultural Education: A training guide*. Food and Agriculture Organization of the United Nations.

Romani, V. (2009). *The politics of higher education in the Middle East: Problems and prospects*. Middle East Brief, Brandeis University Crown Center for Middle East Studies.

Rosenshine, B., & Furst, N. (1971). Research on Teacher Performance Criteria. In B. Othanel Smith (Ed.), *Symposium on Research In Teacher Education* (pp. 37-72). Englewood Cliffs, New Jersey: Prentice-Hall, Inc.

Saab, N., Badran, A., & Sadik, A. K. (2019). *Environmental Education for Sustainable Development in Arab Countries*. Arab Forum for Environment and Development.

Sabry, M. (2009). Funding policy and higher education in Arab Countries. *Journal of Comparative & International Higher Education, 1*(Fall), 11–12.

Sadiku, M. N., Ashaolu, T. J., & Musa, S. M. (2020). Emerging technologies in agriculture. *International Journal of Scientific Advances, 1*(1), 31–34. doi:10.51542/ijscia.v1i1.6

Salim, R., & Hassan, J. (2005) The experience of technical higher education in Egypt. General Directorate of Cultural Research of Egypt.

Saraya, A. A. M. (2003). Designing a training program in the field of employing technology in education for faculty members in teacher colleges in the Kingdom of Saudi Arabia. *The Fourteenth Scientific Conference - Education Curricula in the Light of the Concept of Performance,* (*Volume 1*, pp. 265 – 306). Ain Shams University - The Egyptian Association for Curricula and Teaching Methods.

Saric, M., & Steh, B. (2017). Critical reflection in the professional development of teachers: Challenges and possibilities. *CEPS Journal, 7*(3), 67-85.

Sewilam, H., McCormack, O., Mader, M., & Abdel Raouf, M. (2015). Introducing education for sustainable development into Egyptian schools. *Environment, Development and Sustainability, 17*(2), 221–238. doi:10.100710668-014-9597-7

Shah, I. A. (2017). Teaching-learning challenges of higher education in the Gulf Cooperation Council countries. *Afro Asian Journal of Social Sciences, 8(1)*, 1–18.

Shahid, S. A., & Ahmed, M. (2014). Changing face of agriculture in the Gulf Cooperation Council countries. In *Environmental cost and face of agriculture in the Gulf Cooperation Council Countries* (pp. 1–25). Springer. doi:10.1007/978-3-319-05768-2_1

Sheldon, S. B., & Epstein, J. L. (2004, October). Getting students to school: Using family and community involvement to reduce chronic absenteeism. *School Community Journal, 14*, 39–56.

Shoulders, C. W., Barrick, R. K., & Myers, B. E. (2011). An assessment of the impact of internship programs in the Agricultural Technical Schools of Egypt as perceived by participant groups. *Journal of International Agricultural and Extension Education, 18*(2), 18–29. doi:10.5191/jiaee.2011.18202

Shubenkova, E. V., Badmaeva, S. V., Gagiev, N. N., & Pirozhenko, E. (2017). Adult education and lifelong learning as the basis of social and employment path of the modern man. *Espacios, 38*(25), 25–30.

Sims, S., & Fletcher-Wood, H. (2021). Identifying the characteristics of effective teacher professional development: A critical review. *School Effectiveness and School Improvement*, *32*(1), 47–63. doi:10.1080/09243453.2020.1772841

Sinakou, E., Boeve-de Pauw, J., & Van Petegem, P. (2019). Exploring the concept of sustainable development within education for sustainable development: Implications for ESD research and practice. *Environment, Development and Sustainability*, *21*(1), 1–10. doi:10.100710668-017-0032-8

Skyepack. (2021). *Curriculum Development: Complete Overview & 6 Steps*. Skyepack. Error! Hyperlink reference not valid. curriculum-development.

Skyline College. (n.d.). *CTE program advisory committees*. Author.

Sprott, R. A. (2019). Factors that foster and deter advanced teachers' professional development. *Teaching and Teacher Education*, *77*, 321–331. doi:10.1016/j.tate.2018.11.001

Statista.com. (2022). *Population growth*. statista.com.

Statistics Times. (2022). *Demographics of Morocco*. https://statisticstimes.com/ demographics/ country /morocco-population.php

Steinhoff, B. (2005). Laws and regulation on medicinal and aromatic plants in Europe. *Acta Horticulturae*, (678), 13–22. doi:10.17660/ActaHortic.2005.678.1

Strategic Media. (2009). *The process of Improving Farming in Saudi Arabia*. https://borgenproject.org/farming-in-saudi-arabia/

Svendsen, B. (2020). Inquiries into teacher professional development – What matters? *Education*, *140*(3), 111–131.

Swanson, B. E., Barrick, R. K., & Samy, M. M. (2007, May). Transforming higher agricultural education in Egypt: Strategy, approach and results. In *Proceedings of the 23th annual conference of Association for International Agricultural Extension Education (AIAEE), Polson* (pp. 20-24). Academic Press.

Swanson, B. E., Cano, J., Samy, M. M., Hynes, J. W., & Swan, B. (2007). *Introducing active Teaching–Learning methods and materials into Egyptian agricultural technical secondary.*

Swanson, B., & Davis, K. (2012). *Status of agricultural extension and rural advisory services worldwide: Summary Report.* https://www.g-fras.org/en/knowledge/gfras-publications.html?download=391:status-of-agricultural-extension-and-rural-advi
sory-services-worldwide

Swanson, B. E., Barrick, R. K., & Samy, M. M. (2007). *Transforming higher education in Egypt: Strategy, approach and results. Proceedings of the 23rd Annual AIAEE Conference*, Polson, MN.

Swanson, B. E., Cano, J., Samy, M. M., Hynes, J. W., & Swan, B. (2007). Introducing active teaching-learning methods and materials into Egyptian agricultural technical secondary schools. *Proceedings of the Association for International Agricultural Extension and Education Research Conference*. Polson, MT.

Swanson, B. E., Cano, J., Samy, M. M., Hynes, J. W., & Swan, B. (2007). Introducing active teaching-learning methods into Egyptian agricultural technical secondary schools. *Proceedings of the 23rd Annual AIAEE Conference*, Polson, MN.

Sykes, G. (1996). Reform of and as professional development. *Phi Delta Kappan*, *77*(7), 464.

Symoneaux, R., Segond, N., & Maignant, A. (2022). Sensory and consumer sciences applicated on ornamental plants. In *Nonfood Sensory Practices* (pp. 291–311). Woodhead Publishing. doi:10.1016/B978-0-12-821939-3.00007-5

The Economist Group. (n.d.). *Global Food Security Index (GFSI)*. https://foodsecurityindex.com/

The Economist. (2022). *Arab governments are worried about food security*. https://www.economist.com/middle-east-and-africa/2021/04/15/arab-governments-are-worried-about-food-security

The International Fund for Agricultural Development (IFAD). (2022). *Country documents*. https://www.ifad.org/en/web/operations/w/country/egypt

The Ministry of Agriculture and Land Reclamation Egypt (MALR). (2009). A strategy for sustainable agricultural development up to 2030. Ministry of Agriculture and Land Reclamation, Agricultural Research and Development Council.

The Ministry of Agriculture and Land Reclamation Egypt (MALR). (2021). *Projects and Initiatives*. https://moa.gov.eg/en/projects-and-initiatives/

The Ministry of Agriculture, Fisheries, Rural Development, Water and Forests, Morocco. (2022). https://panorama.solutions/en/organisation/moroccan-ministry-agriculture-fisheries-rural-development-water-and-forests

The Ministry of Education. (n.d.). *Premium education to build universally competitive knowledge society*. https://www.moe.gov.sa/en/Pages/default.aspx

The Ministry of Environment, Water, and Agriculture (MEWA). (2012). *A Glance on Agricultural Development in the Kingdom of Saudi Arabia*. https://www.moa.gov.sa/files /Lm_eng.pdf

The Poultry Site. (2019). *King Salman unveils major Sustainable Agricultural Rural Development Program*. https://www.thepoultrysite.com/news/2019/01/king-salman-unveils-major-sustainable-agricultural-rural-development-program
me

The Saudi Arabian Market Information Resource (SAMIRAD). (2005). *Agricultural Developments in Saudi Arabia*. http://www.saudinf.com/main/f41.htm.http: //www.saudinf.com/main/f41.htm

The Wing Institute. (2203). *Evidence-Based Education.* The Wing institute at the Morningside Academy. https://www.winginstitute.org/evidence-based-education

The World Bank Data (2022). https://data.worldbank.org/indicator/SP. RUR.TOTL. ZS?locations=1A.

The World Bank Data. (2015). https://data.worldbank.org/indicator/AG.SRF. TOTL. K2?locations=SA

The World Bank Group. (2015). *Egypt PROMOTING POVERTY REDUCTION AND SHARED PROSPERITY: A Systematic Country Diagnostic.* (P151429). World Bank. https:// documents1.worldbank.org/curated/en/853671468190130279/pdf/99722-CAS-P151429-SecM2015-0287-IFC-SecM2015-0142-MIGA-SecM2015-0093-Box393212B-OUO-9.pdf

The World Bank Group. (2020). *Morocco: A case for building a stronger education system in the post Covid-19 era*. World Bank. https://www.worldbank.org/en/news/feature/ 2020/10/27/a-case-for-building-a-stronger-education-system-in-the-post-covid-19-era

The World Bank Group. (2022). *Overview*. https://www.worldbank.org/en/country/egypt

The World Bank. (2008). *Sector Brief: Agriculture & Rural Development in MENA*. The World Bank. http://go.worldbank.org/WMLZXRV380

The World Bank. (2018). *Growing Morocco's Agricultural Potential.* https://www.worldbank.org/en/news/feature/2016/02/18/growing-morocco-s-agricultural-potential1

The World Bank. (2023). *Data*. World Bank. https://data.worldbank.org/indicator/SP

The World Bank. (n.d.). *Agriculture & Rural Development Indicators*. https://data.worldbank.org/indicator

The World Economic Forum. (2013). *The Human Capital Report*. Prepared in collaboration with Mercer. Error! Hyperlink reference not valid.

Theobald, E. J., Hill, M. J., Tran, E., Agrawal, S., Arroyo, E. N., Behling, S., Chambwe, N., Cintrón, D. L., Cooper, J. D., Dunster, G., Grummer, J. A., Hennessey, K., Hsiao, J., Iranon, N., Jones, L. II, Jordt, H., Keller, M., Lacey, M. E., Littlefield, C. E., & Freeman, S. (2020). Active learning narrows achievement gaps for underrepresented students in undergraduate science, technology, engineering, and math. *Proceedings of the National Academy of Sciences of the United States of America*, *117*(12), 6476–6483. doi:10.1073/pnas.1916903117 PMID:32152114

Thoron, A. C., Barrick, R. K., Roberts, T. G., & Samy, M. M. (2008). *Establishing technical internship programs for Agricultural Technical School students in Egypt*. Paper presented at the Association for International Agricultural and Extension Education Conference, Costa Rica. Retrieved from https://www.academia.edu/51021584/Establishing_technical_internship_programs_for_agricultural_technical_school_students_in_Egypt

Thoron, A. C., Barrick, R. K., Roberts, T. G., Gunderson, M. A., & Samy, M. M. (2010). Preparing for, conducting and evaluating workshops for agricultural technical school instructors in Egypt. *Journal of Agricultural Education*, *51*(1), 75–87. doi:10.5032/jae.2010.01075

Timm, J. M., & Barth, M. (2021). Making education for sustainable development happen in elementary schools: The role of teachers. *Environmental Education Research*, *27*(1), 50–66. doi:10.1080/13504622.2020.1813256

TRT World. (2022). *Nearly one-third of the Arab world is experiencing food insecurity*. https://www.trtworld.com/magazine/nearly-one-third-of-the-arab-world-is-experiencing-food-insecurity-52711

Turner, F. J. (2005). Encyclopedia of Canadian Social Work. Wilfrid Laurier Univ. Press.

U. S. Department of Health and Human Services. (2006). *Linking education and employment for brighter futures.* Author.

UN. (2002). *The millennium development goals and the United Nations role*. Fact Sheet. United Nations Department of Public Information. https://www.un.org/millenniumgoals/MDGs-FACTSHEET1.pdf

United Arab Emirates University. (n.d.). *The Institutional Research Unit. Statistical Yearbook 2018-2019*. https://uaeu.ac.ae/en/vc/oie/iru/statistics-yearbook-2018-19.pdf

United Nations Development Program (UNDP). (2010). *Saudi Arabia progress towards environmental sustainability*. United Nations Development Program. http: //www.undp.org/energyandenvironment/sustainabledifference/PDFs/ ArabKingdomKingdoms/Sa udiArabia.pdf

United Nations Development Program (UNDP). (2021). *Human Development Report 2020. The Next frontier human development and the Anthropocene*. https://hdr.undp.org/en/composite/HDI

United Nations Educational, Scientific and Cultural Organization. (2020). *Education for sustainable development: A roadmap.* Retrieved from: https://unesdoc.unesco.org/ark:/48223/pf0000374802.locale=en

United Nations. (2015). *Sustainable Development Goals*. Retrieved from: https://sdgs.un.org/goals

United Nations. (2022), *Sustainable Development*. https://www.un.org/sustainable development/. https://www.un.org/sustainabledevelopment/development-agenda/

United Nations. (2022). *Sustainable Development*: UN. https://www.un.org/sustainable development/

United Nations. (2023). *Quality Education*. Department of Economic and Social Affairs. Sustainable Development. https://sdgs.un.org/goals/goal4

United States Department of Education (2008). *Building an effective advisory committee. Mentoring Resource Center Fact Sheet No. 21*. Building an Effective Advisory Committee (Fact Sheet) (educationnorthwest.org)

United States Department of Education (2008). *Building an effective advisory committee. Mentoring Resource Center Fact Sheet. No. 21*. educationnorthwest.org

University of California Policy. (2012). *Ownership of Course Materials*. Ucal. (https://policy.ucop.edu/ doc/2100004/CourseMaterials)

University of California. Davis. (2022). *Experiential Learning*. UCal. https://www.experientiallearning.ucdavis.edu/ module1/el1_40-5step-definitions.pdf

USAID/Egypt. (2022). *Agriculture and Food Security*. https://www.usaid.gov/ egypt/agriculture-and-food-security

USDA Foreign Agricultural Service (FAS). (2022). *Data and Analysis*. https://www.fas.usda.gov/data

Vangrieken, K., Dochy, F., Raes, E., & Kyndt, E. (2015). Teacher collaboration: A systematic review. *Educational Research Review*, *15*, 17–40. doi:10.1016/j.edurev.2015.04.002

Varma, S., Hussain, S. S., Ollero, A. M., Shaukat Khan, T., Maseeh, A. N., Alhmoud, K. B. R., . . . Herbst, C. H. (2019). Building the foundations for economic sustainability: Human capital and growth in the GCC (No. 136435). The World Bank.

Vescio, V., Ross, D., & Adams, A. (2008). A review of research on the impact of professional learning communities on teaching practice and student learning. *Teaching and Teacher Education*, *24*(1), 80–91. doi:10.1016/j.tate.2007.01.004

Vreyens, J. R., & Shaker, M. H. (2005). Preparing market-ready graduates: Adapting curriculum to meet the agriculture employment market in Egypt. *Proceedings of the 21st Annual AIAEE Conference*, San Antonio, TX.

Vreyens, J., & Shaker, M. H. (2005). *Preparing Market-Ready Graduates: Adapting curriculum to meet the agriculture employment market in Egypt. Proceedings of the 21st Annual Conference of the Association for International Agricultural and Extension Education*, Egypt.

Vygotsky, L. S. (1978). *Mind in society: The development of higher psychological processes* (M. Cole, V. John-Steiner, S. Scribner, & E. Souberman, Eds. & Trans.). Harvard University Press.

Weber, A. S. (2011). The role of education in knowledge economies in developing countries. *Procedia: Social and Behavioral Sciences*, *15*, 2589–2594. doi:10.1016/j.sbspro.2011.04.151

Webster-Wright, A. (2009). Reframing professional development through understanding authentic professional learning. *Review of Educational Research*, *79*(2), 702–739. doi:10.3102/0034654308330970

Wenger, E., McDermott, R. A., & Snyder, W. (2002). *Cultivating communities of practice: A guide to managing knowledge*. Harvard Business Press.

Wiggins, G., & McTighe, J. (1998). *Understanding by Design*. Association for Supervision and Curriculum Development.

Wopertz, E. (2017). *Agriculture and development in the wake of the Arab Spring*. International Development Policy, Revue international de politique de developpement. https://journals.openedition.org/poldev/2274

Working Group for International Cooperation in Skills Development. (2001). *Linking work, skills and knowledge*. Author.

World Bank. (2010). *Study/Reviews of National Higher Education Policies - Higher Education in Egypt*. OECD.

World Bank. (2022). *Egypt - Country Climate and Development Report*. World Bank Group. Retrieved from https://documents.worldbank.org/curated/en/099510011012235419/P17729200725ff0170ba05031a8d4ac26d7

World Bank. (2022). *Human capital plan for the Middle East and North Africa*. Retrieved from: https://www.worldbank.org/en/region/mena/publication/2022-human-capital-plan-for-the-middle-east-and-north-africa

World Commission on Environment and Development. (1987). *Our Common Future: A Report from the United Nations World Commission on Environment and Development*. WCED.

World Economic Forum. (2017, May). *The future of jobs and skills in the Middle East and North Africa*. World Economic Forum, Geneva, Switzerland.

World Economic Forum. (2017, May). *The future of jobs and skills in the Middle East and North Africa*. World Economic Forum.

World Food Program. (2023). *Global Food Crisis*. WFP. https://www.wfp.org/ emergencies/global-food-crisis

Ya'qub, S., & al-Tourist, O. (1988). Secondary and higher agricultural education and its role in meeting the needs and requirements of rural development in selected Arab countries. UNESCO Regional Bureau of Education in Arab States, Yundbas.

Yoon, K. S., Duncan, T., Lee, S. W. Y., Scarloss, B., & Shapley, K. L. (2007). Reviewing the evidence on how teacher professional development affects student achievement. *Regional Educational Laboratory Southwest (NJ1)*.

Zguir, M. F., Dubis, S., & Koç, M. (2021). Embedding Education for Sustainable Development (ESD) and SDGs values in curriculum: A comparative review on Qatar, Singapore and New Zealand. *Journal of Cleaner Production*, *319*, 128534. doi:10.1016/j.jclepro.2021.128534

Zguir, M. F., Dubis, S., & Koç, M. (2022). Integrating sustainability into curricula: Teachers' perceptions, preparation and practice in Qatar. *Journal of Cleaner Production*, *371*, 133167. doi:10.1016/j.jclepro.2022.133167

About the Contributors

Mohamed M. Samy earned his doctorate in International Agricultural Development at the University of Illinois, Urbana-Champaign. Dr. Samy is an academic and administrator with over 30 years of experience designing and managing educational and institutional capacity-building projects and higher education and technical agricultural training programs in the United States, United Arab Emirates (UAE), and Egypt. Dr. Samy served as the Chief of Party for two leading educational and agricultural development projects for the United States Agency for International Development in Egypt with a total budget of US$ 20 million. He directed the University of Illinois and the Midwest Universities Consortium for International Activities' efforts to support six Egyptian universities and the Ministries of Education and Agriculture in modernizing educational programs and implementing effective technical training that prepared thousands of students for careers in global agriculture. Dr. Samy previously served as coordinator for the University of Illinois project "Increasing Farm Income through the Production of Value-Added Products. He worked with Illinois producers to increase their income.

R. Kirby Barrick earned the Ph.D. in Agricultural Education from Ohio State University and joined the faculty where he taught courses in teaching and learning. He was awarded the OSU Distinguished Teaching Award, the Gamma Sigma Delta Teaching Award, and the NACTA Distinguished Educator Award. His research interests include teaching and learning theory and practice in domestic and international settings. He has authored/presented more than 275 articles, papers, posters, and workshops, and served as principal investigator for more than $8 million in grants. He is a Fellow in the AAAE and is a Fulbright Senior Specialist. Barrick was a department chair Ohio State, Associate Dean for Academic Programs at the University of Illinois, and Dean of the College of Agricultural and Life Sciences at the University of Florida. He has taught graduate-level courses in teacher education, instructional design, and university governance, and conducted professional development activities for practicing educators. He has worked extensively in Egypt, conducting teaching and learning and curriculum development workshops for university and technical school faculty. .

* * *

Mohamed Abdelgawad is a professor, College of Agriculture, Cairo University, Egypt and associate professor, deanship of skills development (DSD) & deanship of development and quality (DDQ), King Saud University (KSU). Published 40 articles in animal nutrition career and other articles, chapters and translation in HE development and capacity building. Awarded (PGCAP) from Kings college, UK in 2010 and Fellowship of Higher Education Academy (HEA) in UK. Organized overseas training programs for academic leaders and faculty in HE development and capacity building with more than 30 global university centers in USA, Canada, UK, Australia, Singapore and Newzeland. Launched YouTube channel containing up to 60 video clips for university development, capacity building and accreditation. Professional international certified trainer by different interties (ACI, PCT, IAC, ASTD and IACET), and trained more than 20,000 participants of universities faulty. Learning quality and accreditation consultant, trainer and peer consultation for KSU programs (Under & Post) for national and global accreditation.

Ghaleb Ali Al Hadrami Al Breiki is a scientist researcher specializing in the field of food science and nutrition. He is currently a member of the United Arab Emirates Scientists Council and the acting Vice-Chancellor of the United Arab Emirates University (UAEU). Previously he held the position of Deputy Vice-Chancellor for Research and Graduate Studies at the UAEU and Deputy Vice-Chancellor for Admissions and Student Affairs. Professor Al Breiki is the former Dean for the College of Food and Agriculture at the same university. Professor Al Breiki obtained his Ph.D. in Nutrition Sciences from Arizona State University, USA, in 1991. He has published over sixty research papers at various scientific journals and international conferences.

Safia Hamdy Mahmoud Shamardal El-Hanafy is Professor of Ornamental and Medicinal Plants in the Department of Floriculture, Faculty of Agriculture, Cairo University, Egypt. She earned a Bachelor of Agricultural Sciences, a Master's Degree in Floriculture, and the PhD also in Floriculture, from the College of Agriculture, Cairo University, Egypt. She holds two patents in plant production and preservation and has received certificates of completion for several professional development programs. Professor El-Hanafy is the author of more than 30 scientific agricultural publications in the field of Ornamental, Medicinal and Aromatic Plants. She teaches a number of courses in her field of expertise and has supervised more than 20 thesis works for fulfilment of the Master's Degree or Philosophy Degree of postgraduate students in Cairo University and other universities in Egypt.

Sherine Fathy Mansour, Associate professor of Economic Agricultural in Desert Research Center, Cairo, Egypt since August 2000, has a PhD in agriculture economic from Faculty of agriculture Cairo university in 2012, M.Sc. in agriculture economic from Faculty of Agriculture, Tanta university, in 2004 Her expertise is focusing on has been participating in a member of many projects in partnership with another divisions of DRC. she has contributed and supervised all administration and coordination socioeconomic activities within all implemented projects funded by USAID, IFAD and ICBA, her research is focus on rural sustainable development in the various governorates of the Egyptian deserts and development of some agricultural techniques for improving farmers livelihood. She has published many papers in crop production, finance economics and visibility studies. She is a membership of the jury for scientific research of the institution of academic journals, she worked as director of Climate Smart Agriculture Entrepreneurship - Development of Quinoa Value Chain in Egypt project funded by HSBC and International Center for Biosaline Agriculture ICBA (2019 – 2021), this project aimed to cultivated quinoa in marginal lands in Egypt like quinoa and Salicornia to improve the likelihood of people in Egypt. she obtains an Arab Women Leaders in Agriculture (AWLA) fellowship program 2019-2021, I got the title of ambassador for sustainable development in Egypt through training at the Ministry of Planning 2021.

Eihab Fathelrahman is an associate professor and his research focuses on economics of animals' biosecurity, food security, precision agriculture in arid land conditions and climate change adaptations on the UAE, animals fisheries sector biosecurity and return to investment in education, R&D and innovation. Before joining UAEU, Dr. Fathelrahman worked for the U.S. Department of Agriculture (USDA) at both the Agricultural Research Services (ARS) and the Animal and Plant Health Inspection Services (APHIS), Colorado State University (CSU), Washington State University (WSU)and the U.S. Department of Energy Pacific Northwest National Laboratory (PNNL) in Richland, Washington operated by Battelle. He also actively participated on the World Bank initiatives on the economics of climate change adaptations in West Africa and the World Bank Initiative on climate Change Global Track of cost and benefits of climate change adaptations. He published over fifty articles and books chapters as well applied risk management textbooks and software for agriculture, agricultural sustainability, and agriculture and food sustainability and trade.

Paula E. Faulkner earned a Bachelor's of Science in Agricultural Technology (Animal Husbandry) and a Master's of Science in Agricultural Education from North Carolina Agricultural and Technical State University. She earned a doctorate in Agricultural and Extension Education from Pennsylvania State University. As a Professor in the Department of Agribusiness, Applied Economics and Agriscience Education, she instructs undergraduates and graduate students. She presents research in various professional settings as well as publishes research in numerous journals. She serves as a board member for the North Carolina Alumni Fulbright Association. Faulkner is a member of the European Scientific Journal Editorial Board.

Daniel Foster earned undergraduate degrees in Agricultural Technology Management and Agricultural Education at the University of Arizona in Tucson, Arizona. He earned a Masters in Human Community Resource Development and a Ph.D. in Agricultural and Extension Education at The Ohio State University in Columbus, Ohio. He taught secondary agriculture in Willcox, Arizona where he facilitated the exponential growth of the program enrollment and the building of a new agriscience education facility. Dr. Foster is currently an Associate Professor of Agricultural Education a The Pennsylvania State University in University Park, PA where he contributes to the agricultural teacher education programming. He is co-founder of the Global Teach Ag Network and his research agenda focuses on exploring educator professional development around global learning in agriculture. He is also a professional parliamentarian and is active in many professional organizations related to teaching and learning, agriculture and parliamentary procedure.

Melanie Miller Foster currently serves as Associate Teaching Professor of International Agriculture in the College of Agricultural Sciences at The Pennsylvania State University. She is the co-founder of the Global Teach Ag Network with the mission of empowering educators to address the world's most pressing issues in agriculture and food security. Her focus is on identifying ways for all disciplines to play a role in finding new and innovative solutions to food security, and for helping students, educators, and programs tell their story of global impact. At the heart of Dr. Miller Foster's research program are themes of global learning, agriculture and food security.

Mohammad Fawzi Saeed Shahine is Emeritus Professor at Desert Research Center Researcher in ministry of planning/good production sector (1975 – 1998). Participates in the preparation of annual, five-year and rolling plans Researcher at the Desert Research Center (1999-2001), Department of Agricultural Economics for Development in Desert Regions Emeritus Professor, Department of Agricultural

Economics (2012/2022) in the field of studying projects and development in desert areas.

Andrew Thoron is Professor and Department Head for the Agricultural Education and Communication Department at Abraham Baldwin Agricultural College in Tifton, Georgia. Andrew began teaching school-based agricultural courses at Mt. Pulaski High School in Illinois. He then took on a state leadership role with the Illinois State Board of Education for agricultural education where his primary responsibility was to aid agriscience teachers, administrators, and community members in developing and maintaining agricultural education programs. Dr. Thoron served on the faculty at the University of Illinois at Urbana-Champaign and University of Florida where he served as the teacher education coordinator and provided state-wide leadership for teacher professional development. His research expertise is in the area of teaching and learning that focuses on effective teaching and argumentation skill development. He has authored/co-authored over 50 referred journal articles, over 125 paper/poster presentations, and has been awarded over $3 million in grants. Dr. Thoron has been recognized as the Journal of Agricultural Education author of the year four times and been recognized nationally and internationally for his teaching, advising, and research. Dr. Thoron has done international work in the Arab Republic of Egypt among other international settings. Dr. Thoron holds degrees from Lincoln Land Community College (AS), Illinois State University (BS) – Agricultural Education, and the University of Florida (MS & Ph.D.) in Agricultural Education & Communication – Teacher Preparation.

Index

A

B

C

D

E

F

G

H

I

L

M

O

P

R

S

T

U

V